BIOMEDICAL
INSTRUMENTATION
AND MEASUREMENTS

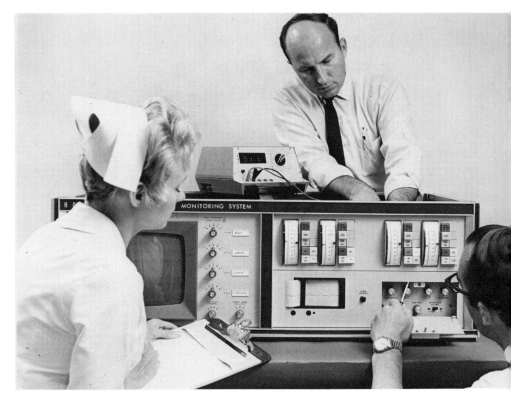

Servicing a patient monitoring system. (Courtesy
of Honeywell Biomedical Electronic Products, Denver, Colo.)

BIOMEDICAL
INSTRUMENTATION
AND MEASUREMENTS

Leslie Cromwell

California State University, Los Angeles

Fred J. Weibell
Erich A. Pfeiffer
Leo B. Usselman

Veterans Administration
Biomedical Engineering and Computing Center
Sepulveda, California

Prentice-Hall, Inc., Englewood Cliffs, New Jersey

Library of Congress Cataloging in Publication Data
CROMWELL, LESLIE.
    Biomedical Instrumentation and measurements.

    Bibliography:
      1. Biomedical engineering. 2. Physiological
apparatus. 3. Physiology. I. Title. [DNLM:
1. Biomedical engineering—Instrumentation. QT 34
B6122 1973]
QP55. C76   610'.28   73-6
ISBN 0-13-077131-7

10 9 8 7 6 5 4 3 2

Printed in the United States of America

Prentice-Hall International, Inc., *London*
Prentice-Hall of Australia, Pty. Ltd., *Sydney*
Prentice-Hall of Canada, Ltd., *Toronto*
Prentice-Hall of India Private Limited, *New Delhi*
Prentice-Hall of Japan, Inc., *Tokyo*

CONTENTS

PREFACE

As the world's population grows, the need for health care increases. In recent years progress in medical care has been rapid, especially in such fields as neurology and cardiology. A major reason for this progress has been the marriage of two important disciplines: medicine and engineering.

There are similarities between these two disciplines and there are differences, but there is no doubt that cooperation between them has produced excellent results. This fact can be well attested to by the man or woman who has received many more years of useful life because of the help of a prosthetic device, or from careful and meaningful monitoring during a critical illness.

The disciplines of medicine and engineering are both broad. They encompass people engaged in a wide spectrum of activities from the basic maintenance of either the body, or a piece of equipment, to research on the frontiers of knowledge in each field. There is one obvious common denominator: the need for instrumentation to make proper and accurate measurements of the parameters involved.

Personnel involved in the design, use, and maintenance of biomedical instrumentation come from either the life sciences or from engineering and technology, although most probably from the latter areas. Training in the life sciences includes physiology and anatomy, with little circuitry, electronics, or instrumentation. For the engineer or electronics technician the reverse is true, and anything but a meager knowledge of physiology is usually lacking on the biomedical side.

Unfortunately for those entering this new field, it is still very young and few reference books are available. This book has been written to help fill the gap. It has grown out of notes prepared by the various authors as reference material for educational courses. These courses have been presented at many levels in both colleges and hospitals. The participants have included engineers, technicians, doctors, dentists, nurses, psychologists, and many others covering a multitude of professions.

This book is primarily intended for the reader with a technical background in electronics or engineering, but with not much more than a casual familiarity with physiology. It is broad in its scope, however, covering a major portion of what is known as the field of biomedical instrumentation. There is depth where needed, but, in general, it is not intended to be too sophisticated. The authors believe in a down-to-earth approach. There are ample illustrations and references to easily accessible literature where more specialization is required. The presentation is such that persons in the life sciences with some knowledge of instrumentation should have little difficulty using it.

The introductory material is concerned with giving the reader a perspective of the field and a feeling for the subject matter. It also introduces the concept of the man-instrument system and the problems encountered in attempting to obtain measurements from a living body. An overall view of the physiological systems of the body is presented and then is later reinforced by more detailed explanations in appropriate parts of the text. The physiological material is presented in a language that should be readily understood by the technically trained person, even to the extent of using an engineering-type analysis. Medical terminology is introduced early, for one of the problems encountered in the field of biomedical instrumentation is communication between the doctor and the engineer or technician. Variables that are meaningful in describing the body system are discussed, together with the type of difficulties that may be anticipated.

It should also be noted that although reference works on physiology are included for those needing further study, enough fundamentals are presented within the context of this book to make it reasonably self-sufficient.

All measurements depend essentially on the detection, acquisition, display, and quantification of signals. The body itself provides a source of many types of signals. Some of these types—the bioelectric potentials responsible for the electrocardiogram, the electroencephalogram, and the electromyogram—are discussed in Chapter 3. In later chapters the measurement of each of these forms of biopotentials is discussed. One chapter is devoted to electrodes—the transducers for the biopotential signals.

With regard to the major physiological systems of the body, each segment is considered as a unit but often relies on material presented in the earlier chapters. The physiology of each system is first discussed in general, followed by an analysis of those parameters that have clinical importance. The fundamental principles and methods of measurement are discussed, with descriptions of equipment actually in use today. This is done in turn for the cardiovascular, respiratory, and nervous systems. There are certain physical measurements that do not belong to any specific system but could relate to any or all of them. These physical variables, including temperature, displacement, force, velocity and acceleration, are covered in Chapter 9.

One of the novel ideas developed in this book is the fact that, together with a discussion of the nervous system, behavioral measurements are covered as well as the interaction between psychology and physiology.

The latter chapters are devoted to special topics to give the reader a true overall view of the field. Such topics include the use of remote monitoring by radio techniques commonly known as biotelemetry; radiation techniques, including X rays and radioisotopes; the clinical laboratory; and the digital computer as it applies to the medical profession, since this is becoming a widely used tool.

The final chapter is one of the most important. Electrical safety in the hospital and clinic is of vital concern. The whole field becomes of no avail unless this topic is understood.

For quick reference, a group of appendices are devoted to medical terminology, an alphabetical glossary, a summary of physiological measurements, and some typical values.

The book has been prepared for a multiplicity of needs. The level is such that it could be used by those taking bioinstrumentation in a community college program, but the scope is such that it can also serve as a text for an introductory course for biomedical engineering students. It should also prove useful as a reference for medical and paramedical personnel with some knowledge of instruments who need to know more.

The background material was developed by the authors in courses presented at California State University, Los Angeles, and at various centers and hospitals of the United States Veterans Administration.

In a work of this nature, it is essential to illustrate commercial systems in common use. In many examples there are many manufacturers who produce similar equipment, and it is difficult to decide which to use as illustrative material. All companies have been most cooperative, and we apologize for the fact that it is not possible always to illustrate alternate examples. The authors wish to thank all the companies that were willing to supply illustrative material, as well as the authors of other textbooks for some borrowed descriptions and drawings. All of these are acknowledged in the text at appropriate places.

The authors wish to thank Mrs. Irina Cromwell, Mrs. Elissa J. Schrader, and Mrs. Erna Wellenstein for their assistance in typing the manuscript; Mr. Edward Francis, Miss Penelope Linskey, and the Prentice-Hall Company for their help, encouragement, and cooperation; and Mr. Joseph A. Labok, Jr., for his efforts in encouraging us to write this book.

Los Angeles, California

LESLIE CROMWELL
FRED J. WEIBELL
ERICH A. PFEIFFER
LEO B. USSELMAN

BIOMEDICAL
INSTRUMENTATION
AND MEASUREMENTS

• 1 •

INTRODUCTION TO BIOMEDICAL INSTRUMENTATION

Science has progressed through many gradual states. It is a long time since Archimedes and his Greek contemporaries started down the path of scientific discoveries, but a technological historian could easily trace the trends through the centuries. Engineering has emerged out of the roots of science, and since the Industrial Revolution the profession has grown rapidly. Again, there are definite stages that can be traced.

1.1. THE AGE OF BIOMEDICAL ENGINEERING

It is common practice to refer to developmental time eras as "ages." The age of the steam engine, of the automobile, and of radio communication each spanned a decade or so of rapid development. Since World War II there have been a number of overlapping technological ages. Nuclear engineering and aerospace engineering are good examples. Each of these fields reached a peak of activity and then settled down to a routine, orderly progression. The age of computer engineering, with all its ramifications, has been developing rapidly and still has much momentum. The time for the age of biomedical engineering has now arrived.

1

The probability is great that the 1970s will be known as the decade in which the most rapid progress was made in this highly important field. There is one vital advantage that biomedical engineering has over many of the other fields that preceded it: the fact that it is aimed at keeping people healthy and helping to cure them when they are ill. Thus it may escape many of the criticisms aimed at progress and technology. Many purists have stated that technology is an evil. Admittedly, although the industrial age introduced many new comforts, conveniences, and methods of transportation, it also generated many problems. These problems include air and water pollution, death by transportation accidents, and the production of such weapons of destruction as guided missiles and nuclear bombs. However, even though biomedical engineering is not apt to be criticized as much for producing evils, some new problems have been created, such as shock hazards in the use of electrical instruments in the hospital. Yet these side effects are minor compared to the benefits that mankind can derive from it.

One of the problems of "biomedical engineering" is defining it. The prefix *bio-,* of course, denotes something connected with life. Biophysics and biochemistry are relatively old "interdisciplines," in which basic sciences have been applied to living things. One school of thought subdivides bioengineering into different engineering areas—for example, biomechanics and bioelectronics. These categories usually indicate the use of that area of engineering applied to living rather than to physical components. *Bioinstrumentation* infers measurement of biological variables, and this field of measurement is often referred to as *biometrics,* although the latter term is also used for mathematical and statistical methods applied to biology.

Naturally committees have been formed to define these terms, and professional societies have become involved. The latter include the IEEE Engineering in Medicine and Biology Group, the ASME Biomechanical and Human Factors Division, the Instrument Society of America, and the American Institute of Aeronautics and Astronautics. Many new "cross-disciplinary" societies have also been formed.

A few years ago an engineering committee was formed to define bioengineering. This was Subcommittee B (Instrumentation) of the Engineers Joint Council Committee on Engineering Interactions with Biology and Medicine. Their recommendations are as follows:

Bioengineering is the application of the knowledge gained by a cross fertilization of engineering and the biological sciences so that both will be more fully utilized for the benefit of man.

Bioengineering has at least six areas of application, which are defined below.

Medical Engineering is the application of engineering to medicine to provide replacement for damaged structures.

Environmental Health Engineering is the application of engineering principles to control the environment so that it will be healthful and safe.

Agricultural Engineering is the application of engineering principles to problems of biological production and to the external operations and environment that influence it.

Bionics is the study of the function and principles of operation of living systems with application of the knowledge gained to the design of physical systems.

Fermentation Engineering is engineering related to microscopic biological systems that are used to create new products by synthesis.

Human Factors Engineering is the application of engineering, physiology, and psychology to the optimization of the man-machine relationship.

These definitions are all noteworthy, but whatever the name, this age of the marriage of engineering to medicine and biology is destined to benefit all concerned. Improved communication among engineers, technicians, and doctors, better and more accurate instrumentation to measure vital physiological parameters, and the development of interdisciplinary tools to help fight the effects of body malfunctions and diseases, are all a part of this new field. Remembering that Shakespeare once wrote "A rose by any other name . . . ," it must be realized that the name is actually not too important; however, what the field can accomplish *is* important. With this point in mind, the authors of this book use the term *biomedical engineering* for the field in general and the term *biomedical instrumentation* for the methods of measurement within the field.

Another major problem of biomedical engineering involves communication between the engineer and the medical profession. The language and "jargon" of the physician are quite different from those of the engineer. In some cases, the same word is used by both disciplines, but with entirely different meanings. Although it is important for the physician to understand enough engineering terminology to allow him to discuss problems with the engineer, the burden of bridging the communication gap usually falls on the latter. The result is that the engineer, or technician, must learn the doctor's language, as well as some anatomy and physiology, in order that the two disciplines can work effectively together.

To help acquaint the reader with this special aspect of biomedical engineering, a basic introduction to medical terminology is presented in Appendix A. This appendix is in two parts: Appendix A.1 is a list of the more common roots, prefixes, and suffixes used in the language of medicine, and Appendix A.2 is a glossary of some of the medical terms frequently encountered in biomedical instrumentation.

In addition to the language problem, other differences may affect communication between the engineer or technician and the doctor. Since the physician is often self-employed, whereas the engineer is usually salaried, a different concept of the fiscal approach exists. Thus, some physicians are reluctant to consider engineers as professionals and would tend to place them in a subservient position rather than as equals. Also, engineers, who are accustomed to precise quantitative measurements, based on theoretical principles, may find it difficult to accept the often imprecise, empirical, and qualitative methods employed by their counterparts.

Since the development and use of biomedical instrumentation must be a joint effort of the engineer or technician and the physician (or nurse), every effort must be exerted to avoid or overcome these "communication" problems. By being aware of their possible existence, the engineer or technician can take steps to avert these pitfalls by adequate preparation and care in establishing his relationship with the medical profession.

1.2. DEVELOPMENT OF BIOMEDICAL INSTRUMENTATION

The field of medical instrumentation is by no means new. Many instruments were developed as early as the nineteenth century—for example, the electrocardiograph, first used by Einthoven at the end of that century. Progress was rather slow until after World War II, when a surplus of electronic equipment, such as amplifiers and recorders, became available. At that time many technicians and engineers, both within industry and on their own, started to experiment with and modify existing equipment for medical use. This process occurred primarily during the 1950s and the results were disappointing, for the experimenters soon learned that physiological parameters are not measured in the same way as physical parameters. They also encountered a severe communication problem with the medical profession.

During the next decade many instrument manufacturers entered the field of medical instrumentation, but development costs were high and the medical profession and hospital staffs were suspicious of the new equipment and often uncooperative. Many developments with excellent potential seemed to have become lost causes. It was during this period that some progressive companies decided that rather than modify existing hardware, they would design instrumentation specifically for medical use. Although it is true that many of the same components were used, the philosophy was changed; equipment analysis and design were applied directly to medical problems.

A large measure of help was provided by the U.S. government, in particular by NASA (National Aeronautics and Space Administration). The Mercury, Gemini, and Apollo programs needed accurate physiological monitoring for the astronauts; consequently, much research and development money went into this area. The aerospace medicine programs were expanded considerably, both within NASA facilities, and through grants to universities and hospital research units. Some of the concepts and features of patient-monitoring systems presently used in hospitals throughout the world evolved from the base of astronaut monitoring. The use of adjunct fields, such as biotelemetry, also finds some basis in the NASA programs.

Also, in the 1960s, an awareness of the need for engineers and technicians to work with the medical profession developed. All the major engineering technical societies recognized this need by forming "Engineering in Medicine and Biology" subgroups, and new societies were organized. Along with the medical research programs at the universities, a need developed for courses and curricula in biomedical engineering, and today almost every major university or college has some type of biomedical engineering program. However, much of this effort is not concerned with biomedical instrumentation per se.

1.3. BIOMETRICS

The branch of science that includes the measurement of physiological variables and parameters is known as *biometrics*. Biomedical instrumentation provides the tools by which these measurements can be achieved.

In later chapters each of the major forms of biomedical instrumentation will be covered in detail, along with the physiological basis for the measurements involved. The physiological measurements themselves are summarized in Appendix B, which also includes such information as amplitude and frequency range where applicable.

Some physiological measurements can be obtained passively in that no external energy is required to produce electrical signals representing the information desired. Such measurements include the bioelectric potentials that form the electrocardiogram, the electroencephalogram, and the electromyogram. Similarly, microphone devices that measure heart sounds and vibrations, and thermocouples used for body temperature measurements require no external excitation. Many physiological measuring devices, however, require the application of energy in order to obtain data. For example, a physiological variable may first be converted into variations of some circuit element, such as resistance, which in turn requires the application of an external voltage or current for measurement. In other instances, the measurements may involve the reflection or refraction of light which must be furnished by the measuring system.

Some forms of biomedical instrumentation are unique to the field of medicine but many are adaptions of widely used physical measurements. A thermistor, for example, changes its electrical resistance with temperature, regardless of whether the temperature is that of an engine or the human body. The principles are the same. Only the shape and size of the device might be different. Another example is the strain gage, which is commonly used to measure the stress in structural components. It operates on the principle that electrical resistance is changed by the stretching of a wire or a piece of semiconductor material. When suitably excited by a source of constant voltage, an electrical output can be obtained that is proportional to the amount of the strain. Since pressure can be translated into strain by various means, blood pressure can be measured by an adaption of this device. When the transducer is connected into a typical circuit, such as a bridge configuration, and this circuit is excited from a source of constant input voltage, the changes in resistance are reflected in the output as voltage changes. For a thermistor, the temperature is indicated on a voltmeter calibrated in degrees Centrigrade or Fahrenheit.

In the design or specification of medical instrumentation systems, each of the following factors should be considered:

RANGE. The range of an instrument is generally considered to include all the levels of input amplitude and frequency over which the device is expected to operate. The objective should be to provide an instrument that will give a usable reading from the smallest expected value of the variable or parameter being measured to the largest.

SENSITIVITY. The sensitivity of an instrument determines how small a variation of a variable or parameter can be reliably measured. This factor differs from the instrument's range in that sensitivity is not concerned with the absolute levels of the parameter but rather with the minute changes that can be detected. The sensitivity directly determines the *resolution* of the device, which is the minimum variation that can accurately be read. Too high a sensitivity often results in nonlinearities or instability. Thus the optimum sensitivity must be determined for any given type of measurement. Indications of sensitivity are frequently expressed in terms of scale length per quantity to be measured—for example, inches per microampere in a galvanometer coil or inches per millimeter of mercury. These units are sometimes expressed reciprocally. A sensitivity of 0.025 inch per millimeter of mercury (in./mm Hg) could be expressed as 40 millimeters of mercury per inch.

LINEARITY. The degree to which variations in the output of an instrument follow input variations is referred to as the linearity of the device. In a *linear* system the sensitivity would be the same for all absolute levels of input, whether in the high, middle, or low portion of the range. In some instruments a certain form of nonlinearity is purposely introduced to create a desired effect, whereas in others it is desirable to have linear scales as much as possible over the entire range of measurements. Linearity should be obtained over the most important segments, even if it is impossible to achieve it over the entire range.

HYSTERESIS. Hysteresis (from the Greek, *hysterein,* meaning "to be behind" or "to lag") is a characteristic of some instruments whereby a given value of the measured variable results in a different reading when reached in an ascending direction from that obtained when it is reached in a descending direction. Mechanical friction in a meter, for example, can cause the movement of the indicating needle to lag behind corresponding changes in the measured variable, thus resulting in a hysteresis error in the reading.

FREQUENCY RESPONSE. The frequency response of an instrument is its variation in sensitivity over the frequency range of the measurement. It is important to display a waveshape that is a faithful reproduction of the original physiological signal. An instrument system should be able to respond rapidly enough to reproduce all frequency components of the waveform with equal sensitivity. This condition is referred to as a "flat response" over a given range of frequencies.

ACCURACY. Accuracy is a measure of systemic error. Errors can occur in a multitude of ways. Although not always present simultaneously, the following errors should be considered:

1. Errors due to tolerances of electronic components
2. Mechanical errors in meter movements
3. Component errors due to drift or temperature variation
4. Errors due to poor frequency response
5. In certain types of instruments, errors due to change in atmospheric pressure or temperature
6. Reading errors due to parallax, inadequate illumination, or excessively wide ink traces on a pen recording.

Two additional sources of error should not be overlooked. The first concerns correct instrument zeroing. In most measurements, a zero, or a baseline, is necessary. It is often achieved by balancing the Wheatstone Bridge or a similar device. It is very important that, where needed, balancing or zeroing is done prior to each set of measurements. Another source of error is the effect of the instrument on the parameter to be measured, and vice versa. This is especially true in measurements in living organisms and is further discussed in Chapter 2.

SIGNAL-TO-NOISE RATIO. It is important that the signal-to-noise ratio be as high as possible. In the hospital environment, power-line frequency noise or interference is common and is usually picked up in long leads. Also, interference due to electromagnetic, electrostatic, or diathermy equipment is possible. Poor grounding is often a cause of this kind of noise problem.

Such "interference noise," however, which is due to coupling from other energy sources, should be differentiated from thermal and shot noise, which originate within the elements of the circuit itself because of the discontinuous nature of matter and electrical current. Although thermal noise is often the limiting factor in the detection of signals in other fields of electronics, interference noise is usually more of a problem in biomedical systems.

It is also important to know and control the signal-to-noise ratio in the actual environment in which the measurements are to be made.

STABILITY. In control engineering, stability is the ability of a system to resume a steady-state condition following a disturbance at the input rather than be driven into uncontrollable oscillation. This is a factor that varies with the amount of amplification, feedback, and other features of the system. The overall system must be sufficiently stable over the useful range. *Baseline stability* is the maintenance of a constant baseline value without drift.

ISOLATION. Often measurements must be made on patients or experimental animals in such a way that the instrument does not produce a direct electrical connection between the subject and ground. This requirement is often necessary for reasons of electrical safety (see Chapter 16) or to avoid interference between different instruments used simultaneously. Electrical isolation can be achieved by using magnetic or optical coupling techniques, or radio telemetry. Telemetry is also used where movement of the person or animal to be measured is essential, and thus the encumbrance of connecting leads should be avoided.

SIMPLICITY. All systems and instruments should be as simple as possible to eliminate the chance of component or human error.

Most instrumentation systems require calibration before they are actually used. Each component of a measurement system is usually calibrated individually at the factory against a standard. When a medical system is assembled, it should be calibrated as a whole. This step can be done external to the living organism or in situ (connected to or within the body). This point is discussed in later chapters. Calibration should always be done by using error-free devices of the simplest kind for references. An example would be that of a complicated, remote blood pressure monitoring system, which is calibrated against a simple mercury manometer.

1.4. THE OBJECTIVES OF THE BOOK

The purpose of this book is to relate specific engineering and instrumentation principles to the task of obtaining physiological data.

Each of the major body systems is discussed by presenting physiological background information. Then the variables to be measured are considered, followed by the principles of the instrumentation that could be used. Finally, applications to typical medical, behavioral, and biological use are given.

The subject matter is presented in such a way that it could be extended to classes of instruments that will be used in the future. Thus the material can be used as building blocks for the health-care instrumentation systems of tomorrow.

· 2 ·

THE
MAN-INSTRUMENT
SYSTEM

A classic exercise in engineering analysis involves the measurement of outputs from an unknown system as they are affected by various combinations of inputs. The object is to learn the nature and characteristics of the system. This unknown system, often referred to as a *black box,* may have a variety of configurations for a given combination of inputs and outputs. The end product of such an exercise is usually a set of input–output equations intended to define the internal functions of the box. These functions may be relatively simple or extremely complex.

One of the most complex "black boxes" conceivable is a living organism, especially the living human being. Within this box can be found electrical, mechanical, acoustical, thermal, chemical, optical, hydraulic, pneumatic, and many other types of systems, all interacting with each other. It also contains a powerful computer, several types of communication systems, and a great variety of control systems. To further complicate the situation, upon attempting to measure the inputs and outputs, an engineer would soon learn that none of the input–output relationships is deterministic. That is, repeated application of a given set of input values will not always produce the same output values. In fact, many of the outputs seem to show a wide range of responses to a given set of inputs, depending on some seemingly irrelevant conditions, whereas others appear to be completely random and totally unrelated to any of the inputs.

The living black box presents other problems, too. Many of the important variables to be measured are not readily accessible to measuring devices. The result is that some key relationships cannot be determined or that less accurate substitute measures must be used. Furthermore, there is a high degree of interaction among the variables in this box. Thus it is often impossible to hold one variable constant while measuring the relationship between two others. In fact, it is sometimes difficult to determine which are the inputs and which are the outputs, for they are never labeled and almost inevitably include one or more feedback paths. The situation is made even worse by the application of the measuring device itself, which often affects the measurements to the extent that they may not represent normal conditions reliably.

At first glance an assignment to measure and analyze the variables in a living black box would probably be labeled "impossible" by most engineers; yet this is the very problem facing those in the medical field who attempt to measure and understand the internal relationships of the human body. The function of medical instrumentation is to aid the medical clinician and researcher in devising ways of obtaining reliable and meaningful measurements from a living human being.

Still other problems are associated with such measurements: the process of measuring must not in any way endanger the life of the person on whom the measurements are being made, and it should not require the subject to endure undue pain, discomfort, or any other undesirable conditions. This means that many of the measurement techniques normally employed in the instrumentation of nonliving systems cannot be applied in the instrumentation of humans.

Additional factors that add to the difficulty of obtaining valid measurements are (a) safety considerations, (b) the environment of the hospital in which these measurements are often performed, (c) the medical personnel usually involved in the measurements, and (d) occasionally even ethical and legal considerations.

Because special problems are encountered in obtaining data from living organisms, especially humans, and because of the large amount of interaction between the instrumentation system and the subject being measured, it is essential that the person on whom measurements are made be considered an integral part of the instrumentation system. In other words, in order to make sense out of the data to be obtained from the black box (the human), the internal characteristics of the black box must be considered in the design and application of any measuring instruments. Consequently, the overall system, which includes both the human and the instrumentation required for measurement of the human, is called *the man-instrument system*. This chapter deals with the properties of this system and the problems of measuring within the living human.

2.1. INTRODUCTION TO THE MAN-INSTRUMENT SYSTEM

An instrumentation system is defined as the set of instruments and equipment utilized in the measurement of one or more characteristics or phenomena, plus the presentation of information obtained from those measurements in a form that can be read and interpreted by man. In some cases, the instrumentation system includes components that provide a stimulus or drive to one or more of the inputs to the device being measured. There may also be some mechanism for automatic control of certain processes within the system, or of the entire system. As indicated earlier, the complete man-instrument system must also include the human subject on whom the measurements are being made.

The basic objectives of any instrumentation system generally fall into one of the following major categories:

1. *Information gathering.* In an information-gathering system, instrumentation is used to measure natural phenomena and other variables to aid man in his quest for knowledge about himself and the universe in which he lives. In this setting, the characteristics of the measurements may not be known in advance.
2. *Diagnosis.* Measurements are made to help in the detection and, hopefully, the correction of some malfunction of the system being measured. In some applications, this type of instrumentation may be classed as "troubleshooting equipment."
3. *Evaluation.* Measurements are used to determine the ability of a system to meet its functional requirements. These could be classified as "proof-of-performance" or "quality control" tests.
4. *Monitoring.* Instrumentation is used to monitor some process or operation in order to obtain continuous or periodic information about the state of the system being measured.
5. *Control.* Instrumentation is sometimes used to automatically control the operation of a system based on changes in one or more of the internal parameters or in the output of the system.

The general field of biomedical instrumentation involves, to some extent, all the preceding objectives of the general instrumentation system. Instrumentation for biomedical research can generally be viewed as information-gathering instrumentation, although it sometimes includes some monitoring and control devices. Instrumentation to aid the physician in the diagnosis of disease and other disorders also has widespread use. Similar instrumentation is used in evaluation of the physical condition of patients in routine physical examinations. Also, special instrumentation systems are used for monitoring of patients undergoing surgery or under intensive care.

Biomedical instrumentation can generally be classified into two major types: clinical and research. *Clinical instrumentation* is basically devoted to the diagnosis, care, and treatment of patients, whereas *research instrumentation* is used primarily in the search for new knowledge pertaining to the various systems that compose the human organism. Although some instruments can be used in both areas, clinical instruments are generally designed to be more rugged and easier to use. Emphasis is placed on obtaining a limited set of reliable measurements from a large group of patients and on providing the physician with enough information to permit him to make clinical decisions. On the other hand, research instrumentation is normally more complex, more specialized, and often designed to provide a much higher degree of accuracy, resolution, and so on. Clinical instruments are used by the physician or nurse, whereas research instruments are generally operated by skilled technologists whose primary training is in the operation of such instruments. The concept of the man-instrument system applies to both clinical and research instrumentation.

Measurements in which biomedical instrumentation is employed can also be divided into two categories: in vivo and in vitro. An *in vivo* measurement is one that is made on or within the living organism itself. An example would be a device inserted into the bloodstream to measure the pH of the blood directly. An *in vitro* measurement is one performed outside the body, even though it relates to the functions of the body. An example of an in vitro measurement would be the measurement of the pH of a sample of blood that had been drawn from a patient. Literally, the term in vitro means "in glass," thus implying that in vitro measurements are usually performed in test tubes. Although the man-instrument system described here applies mainly to in vivo measurements, problems are often encountered in obtaining appropriate samples for in vitro measurements and in relating these measurements to the living human.

2.2. COMPONENTS OF THE MAN-INSTRUMENT SYSTEM

A block diagram of the man-instrument system is shown in Figure 2.1. The basic components of this system are essentially the same as in any instrumentation system. The only real difference is in having a living human being as the subject. The system components are given below.

2.2.1. THE SUBJECT. The subject is the human on whom the measurements are made. Since it is the subject who makes this system different from other instrumentation systems, the major physiological systems that constitute the human body are treated in much greater detail in Section 2.3.

2.2.2. STIMULUS. In many measurements, the response to some form of external stimulus is required. The instrumentation used to generate and present this stimulus to the subject is a vital part of the man-instrument system whenever responses are measured. The stimulus may be visual (e.g. a flash of light), auditory (e.g. a tone), tactile (e.g. a blow to the Achilles tendon), or direct electrical stimulation of some part of the nervous system.

2.2.3. THE TRANSDUCER. In general, a transducer is defined as a device capable of converting one form of energy or signal to another. In the man-instrument system, each transducer is used to produce an electric signal that is an analog of the phenomenon being measured. The transducer may measure temperature, pressure, flow, or any of the other variables that can be found in the body, but its output is always an electric signal. As indicated in Figure 2.1, two or more transducers may be used simultaneously to obtain relative variations between phenomena.

2.2.4. SIGNAL-CONDITIONING EQUIPMENT. The part of the instrumentation system that amplifies, modifies, or in any other way changes the electric output of the transducer is called signal-conditioning (or sometimes signal-processing) equipment. Signal-conditioning equipment is also used to combine or relate the outputs of two or more transducers. Thus, for each item of signal-conditioning equipment, both the input and the output are electric signals, although the output signal is often greatly modified with respect to the input. In essence, then, the purpose of the signal-conditioning equipment is to process the signals from the transducers in order to satisfy the functions of the system and to prepare signals suitable for operating the display or recording equipment that follows.

2.2.5. DISPLAY EQUIPMENT. In order to be meaningful, the electrical output of the signal-conditioning equipment must be converted into a form that can be perceived by one of man's senses and that can convey the information obtained by the measurement in a meaningful way. The input to the display device is the modified electric signal from the

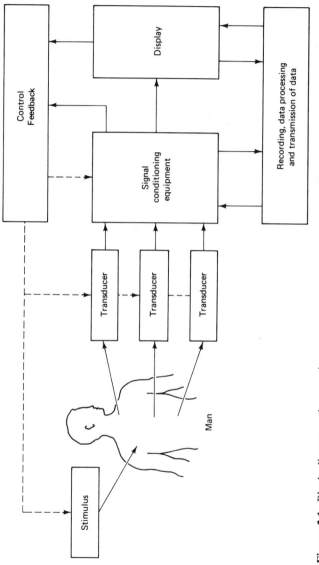

Figure 2.1. Block diagram—the man-instrument system.

signal-conditioning equipment. Its output is some form of visual, audible, or possibly tactile information. In the man–instrumentation system, the display equipment may include a graphic pen recorder that produces a permanent record of the data.

2.2.6. RECORDING, DATA PROCESSING AND TRANSMISSION EQUIPMENT. It is often necessary, or at least desirable, to record the measured information for possible later use or to transmit it from one location to another, whether across the hall of the hospital or halfway around the world. Equipment for these functions is often a vital part of the man-instrument system. Also, where automatic storage or processing of data is required, or where computer control is employed, an on-line analog or digital computer may be part of the instrumentation system. It should be noted that the term *recorder* is used in two different contexts in biomedical instrumentation. A graphic pen recorder is actually a display device used to produce a paper record of analog waveforms, whereas the recording equipment referred to in this paragraph includes devices by which data can be recorded for future playback, as in a magnetic tape recorder.

2.2.7. CONTROL DEVICES. Where it is necessary or desirable to have automatic control of the stimulus, transducers, or any other part of the man-instrument system, a control system is incorporated. This system usually consists of a feedback loop in which part of the output from the signal-conditioning or display equipment is used to control the operation of the system in some way.

2.3. PHYSIOLOGICAL SYSTEMS OF THE BODY

From the previous sections it should be evident that, in order to obtain valid measurements from a living human being, it is necessary to have some understanding of the subject on which the measurements are being made. Within the human body can be found electrical, mechanical, thermal, hydraulic, pneumatic, chemical, and various other types of systems, each of which communicates with an external environment, and internally with the other systems of the body. By means of a multilevel control system and communications network, these individual systems are organized to perform many complex functions. Through the integrated operation of all these systems, and their various subsystems, man is able to sustain life, learn to perform useful tasks, acquire personality and behavioral traits, and even reproduce himself.

Measurements can be made at various levels of man's hierarchy of organization. For example, the human being as a whole (the highest level of organization) communicates with his environment in many ways. These methods of communicating could be regarded as the inputs and outputs of the black box and are illustrated in Figure 2.2. In addition, these various

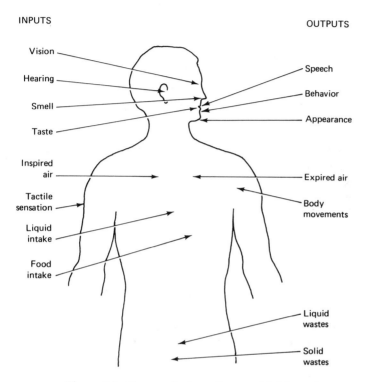

INPUTS

OUTPUTS

Vision

Hearing

Smell

Taste

Inspired air

Tactile sensation

Liquid intake

Food intake

Speech

Behavior

Appearance

Expired air

Body movements

Liquid wastes

Solid wastes

Figure 2.2. Communication of man with his environment.

inputs and outputs can be measured and analyzed in a variety of ways. Most are readily accessible for measurement, but some, such as speech, behavior, and appearance, are difficult to analyze and interpret.

Next to the whole being in the hierarchy of organization are the major functional systems of the body, including the nervous system, the cardiovascular system, the pulmonary system, and so on. Each major system is discussed later in this chapter, and most are covered in greater detail in later chapters. Just as the whole person communicates with his environment, these major systems communicate with each other as well as with the external environment.

These functional systems can be broken down into subsystems and organs, which can be further subdivided into smaller and smaller units. The process can continue down to the cellular level and perhaps even to the molecular level. The major goal of biomedical instrumentation is to make possible the measurement of information communicated by these various elements. If all the variables at all levels of the organization hierarchy could be measured, and all their interrelationships determined, the functions of the mind and body of man would be much more clearly understood and could probably be completely defined by presently known laws of physics, chemistry, and other sciences. The problem is, of course, that many of the inputs at the various organizational levels are not accessible for measurement. The interrelationships among elements are sometimes so complex and involve so many systems that the "laws" and relationships thus far derived are inadequate to define them completely. Thus the models in use today contain so many assumptions and constraints that their application is often severely limited.

Although each of the systems is treated in much more detail in later chapters, a brief engineering-oriented description of the major physiological systems of the body is given below in order to illustrate some of the problems to be expected in dealing with a living organism.

2.3.1. THE BIOCHEMICAL SYSTEM. The human body has within it an integrated conglomerate of chemical systems that produce energy for the activity of the body, messenger agents for communication, materials for body repair and growth, and substances required to carry out the various body functions. All operations of this highly diversified and very efficient chemical factory are self-contained in that from a single point of intake for fuel (food), water, and air, all the source materials for numerous chemical reactions are produced within the body. Moreover, the chemical factory contains all the monitoring equipment needed to provide the degree of control necessary for each chemical operation, and it incorporates an efficient waste disposal system.

2.3.2. THE CARDIOVASCULAR SYSTEM. To an engineer, the cardiovascular system can be viewed as a complex, closed hydraulic system with a four-chamber pump (the heart), connected to flexible and sometimes elastic tubing (blood vessels). In some parts of the system (arteries, arterioles), the tubing changes its diameter to control pressure. Reservoirs in the system (veins) change their volume and characteristics to satisfy certain control requirements, and a system of gates and variable hydraulic resistances (vasoconstrictors, vasodilators) continually alters the pattern of fluid flow. The four-chamber pump acts as two synchronized but functionally isolated two-stage pumps. The first stage of each pump (the atrium) collects fluid (blood) from the system and pumps it into the second stage (the ventricle). The action of the second stage is so timed that the fluid is pumped into the system immediately after it has been received from the first stage. One of the two-stage pumps (right side of the heart) collects fluid from the main hydraulic system (systematic circulation) and pumps it through an oxygenation system (the lungs). The other pump (left side of the heart) receives fluid (blood) from the oxygenation system and pumps it into the main hydraulic system. The speed of the pump (heart rate) and its efficiency (stroke volume) are constantly changed to meet the overall requirements of the system. The fluid (blood), which flows in a laminar fashion, acts as a communication and supply network for all parts of the system. Carriers (red blood cells) of fuel supplies and waste materials are transported to predetermined destinations by the fluid. The fluid also contains mechanisms for repairing small system punctures and for rejecting foreign elements from the system (platelets and white blood cells, respectively). Sensors provided to detect changes in the need for supplies, the buildup of waste materials, and out-of-tolerance pressures in the system are known as *chemoreceptors, P_{CO_2} sensors,* and *baroreceptors,* respectively. These and other mechanisms control the pump's speed and efficiency, the fluid flow pattern through the system, tubing diameters, and other factors. Because part of the system is required to work against gravity at times, special one-way valves are provided to prevent gravity from pulling fluid against the direction of flow between pump cycles. The variables of prime importance in this system are pump (cardiac) output and pressure, flow rate, and volume of the fluid (blood) at various locations through the system.

2.3.3. THE RESPIRATORY SYSTEM. Whereas the cardiovascular system is the major hydraulic system in the body, the respiratory system is the pneumatic system. An air pump (diaphragm), which alternately creates negative and positive pressures in a sealed chamber (thoracic cavity), causes air to be sucked into and forced out of a pair of elastic bags (lungs) located within the compartment. The bags are connected to the outside environment through a passageway (nasal cavities, pharynx, larynx, trachea, bronchi, and bronchioles), which at one point is in common with the tubing that carries liquids and solids to the stomach. A special valving arrangement interrupts the pneumatic system whenever liquid or solid matter passes through the common region. The passageway divides to carry air into each of the bags, wherein it again subdivides many times to carry air into and out of each of many tiny air spaces (pulmonary alveoli) within the bags. The dual air input to the system (nasal cavities) has an alternate vent (the mouth) for use in the event of nasal blockage and for other special purposes. In the tiny air spaces of the bags is a membrane interface with the body's hydraulic system through which certain gases can diffuse. Oxygen is taken into the fluid (blood) from the incoming air, and carbon dioxide is transferred from the fluid to the air, which is exhausted by the force of the pneumatic pump. The pump operates with a two-way override. An automatic control center (respiratory center of the brain) maintains pump operation at a speed that is adequate to supply oxygen and carry off carbon dioxide as required by the system. Manual control can take over at any time either to accelerate or to inhibit the operation of the pump. Automatic control will return, however, if a condition is created that might endanger the system. System variables of primary importance are respiratory rate, respiratory airflow, respiratory volume, and concentration of CO_2 in the expired air. This system also has a number of relatively fixed volumes and capacities, such as tidal volume (the volume inspired or expired during each normal breath), inspiratory reserve volume (the additional volume that can be inspired after a normal inspiration), expiratory reserve volume (the additional amount of air that can be forced out of the lungs after normal expiration), residual volume (amount of air remaining in the lungs after all possible air has been forced out), and vital capacity (tidal volume, plus inspiratory reserve volume, plus expiratory reserve volume).

2.3.4. THE NERVOUS SYSTEM. The nervous system is the communication network for the body. Its center is a self–adapting central information processor, or computer (the brain) with memory, computational power, decision–making capability, and a myriad of input-output channels. The computer is self adapting in that if a certain section is damaged, other sections can adapt and eventually take over (at least in part) the function of the damaged section. By use of this computer, a person is able to make decisions, solve complex problems, create art, poetry, and music, "feel" emotions, integrate input information from all parts of the body, and coordinate output signals to produce meaningful behavior. Almost as fascinating as the central computer are the millions of communication lines (afferent and efferent nerves) that bring sensory information into, and transmit control information out of, the brain. In general, these lines are not single long lines but often complicated networks with many interconnections that are continually changing to meet the needs of the system. By means of the interconnection patterns, signals from a large number of sensory devices, which detect light, sound, pressure, heat, cold, and certain chemicals, are channeled to the appropriate parts of the computer where they can be acted upon. Similarly, output control signals are channeled to specific motor devices (motor units of the muscles), which respond to the signals with some type of motion or force. Feedback regarding every action controlled by the system is provided to the computer through appropriate sensors. Information is usually coded in the system by means of electrochemical pulses (nerve action potentials) that travel along the signal lines (nerves). The pulses can be transferred from one element of a network to another in one direction only, and frequently the transfer takes place only when there is the proper combination of elements acting on the next element in the chain. Action by some elements tends to inhibit transfer by making the next element less sensitive to other elements that are attempting to actuate it. Both serial and parallel coding are used, sometimes together in the same system. In addition to the central computer, a large number of simple decision-making devices (spinal reflexes) are present to control directly certain motor devices from certain sensory inputs. A number of feedback loops are accomplished by this method. In many cases, only situations where important decision making is involved require that the central computer be utilized.

2.4. PROBLEMS ENCOUNTERED IN MEASURING A LIVING SYSTEM

The previous discussions of the man-instrument system and the physiological systems of the body imply measurements on a human subject. In some cases, however, animal subjects are substituted for humans in order to permit measurements or manipulations that cannot be performed without some risk. Although ethical restrictions sometimes are not as severe with animal subjects, the same basic problems can be expected in attempting measurements from any living system. Most of these problems were introduced in earlier sections of the chapter. However, they can be summarized as follows:

2.4.1. INACCESSIBILITY OF VARIABLES TO MEASUREMENT. One of the greatest problems in attempting measurements from a living system is the difficulty in gaining access to the variable being measured. In some cases, such as in the measurement of dynamic neurochemical activity in the brain, it is impossible to place a suitable transducer in a position to make the measurement. Sometimes the problem stems from the required physical size of the transducer as compared to the space available for the measurement. In other situations the medical operation required to place a transducer in a position from which the variable can be measured makes the measurement impractical on human subjects, and sometimes even on animals. Where a variable is inaccessible for measurement, an attempt is often made to perform an indirect measurement. This process involves the measurement of some other related variable that makes possible a usable estimate of the inaccessible variable under certain conditions. In using indirect measurements, however, one must be constantly aware of the limitations of the substitute variable and must be able to determine when the relationship is not valid.

2.4.2. VARIABILITY OF THE DATA. Few of the variables that can be measured in the human body are truly deterministic variables.

In fact, such variables should be considered as stochastic processes. A *stochastic process* is a time variable related to other variables in a non-deterministic way. Physiological variables can never be viewed as strictly deterministic values but must be represented by some kind of statistical or probabilistic distribution. In other words, measurements taken under a fixed set of conditions at one time will not necessarily be the same as similar measurements made under the same conditions at another time. The variability from one subject to another is even greater. Here, again, statistical methods must be employed in order to estimate relationships among variables.

2.4.3. LACK OF KNOWLEDGE ABOUT INTERRELATIONSHIPS. The foregoing variability in measured values could be better explained if more were known and understood about the interrelationships within the body. Physiological measurements with large tolerances are often accepted by the physician because of a lack of this knowledge and the resultant inability to control variations. Better understanding of physiological relationships would also permit more effective use of indirect measurements as substitutes for inaccessible measures and would aid the engineer or technician in his job of coupling the instrumentation to the physiological system.

2.4.4. INTERACTION AMONG PHYSIOLOGICAL SYSTEMS. Because of the large number of feedback loops involved in the major physiological systems, a severe degree of interaction exists both within a given system and among the major systems. The result is that stimulation of one part of a given system generally affects all other parts of that system in some way (sometimes in an unpredictable fashion) and often affects other systems as well. For this reason, "cause-and-effect" relationships become extremely unclear and difficult to define. Even when attempts are made to open feedback loops, collateral loops appear and some aspects of the original feedback loop are still present. Also, when one organ or element is rendered inactive, another organ or element sometimes takes over the function. This situation is especially true in the brain and other parts of the nervous system.

2.4.5. EFFECT OF THE TRANSDUCER ON THE MEASUREMENT. Almost any kind of measurement is affected in some way by the presence of the measuring transducer. The problem is greatly compounded in the measurement of living systems. In many situations the physical presence of the transducer changes the reading significantly. For example, a large flow transducer placed in a bloodstream partially blocks the vessel and changes the pressure-flow characteristics of the system. Similarly, an attempt to measure the electrochemical potentials generated within an individual cell requires penetration of the cell by a transducer. This penetration can easily kill the cell or damage it so that it can no longer function normally. Another problem arises from the interaction discussed earlier. Often the presence of a transducer in one system can affect responses in other systems. For example, local cooling of the skin, to estimate the circulation in the area, causes feedback that changes the circulation pattern as a reaction to the cooling. The psychological effect of the measurement can also affect the results. Long-term recording techniques for measuring blood pressure have shown that some individuals who would otherwise have normal pressures show an elevated pressure reading whenever they are in the physician's office. This is a fear response on the part of the patient, involving the autonomic nervous system. In designing a measurement system, the biomedical instrumentation engineer or technician must exert extreme care to ensure that the effect of the presence of the measuring device is minimal. Because of the limited amount of energy available in the body for many physiological variables, care must also be taken to prevent the measuring system from "loading" the source of the measured variable.

2.4.6. ARTIFACTS. In medicine and biology, the term *artifact* refers to any component of a signal that is extraneous to the variable represented by the signal. Thus random noise generated within the measuring instrument, electrical interference (including 60-Hz pickup), cross-talk, and all other unwanted variations in the signal are considered artifacts. A major source of artifact in the measuring of a living system is the movement of the subject, which in turn results in movement of the measuring device. Since many transducers are sensitive to movement, the movement of the subject often produces variations in the output signal. Sometimes these variations are indistinguishable from the measured variable; at other times they may be sufficient to obscure the desired information completely. Application of anesthesia to reduce movement may itself cause unwanted changes in the system.

2.4.7. ENERGY LIMITATIONS. Many physiological measurement techniques require that a certain amount of energy be applied to the living system in order to obtain a measurement. For example, resistance measurements require the flow of electric current through the tissues or blood being measured. Some transducers generate a small amount of heat due to the current flow. In most cases, this energy level is so low that its effect is insignificant. However, in dealing with living cells, care must continually be taken to avoid the possibility of energy concentrations that might damage cells or affect the measurements.

2.4.8. Safety considerations. As previously mentioned, the methods employed in measuring variables in a living human subject must in no way endanger the life or normal functioning of the subject. Recent emphasis on hospital safety requires that extra caution must be taken in the design of any measurement system to protect the patient. Similarly, the measurement should not cause undue pain, trauma, or discomfort, unless it becomes necessary to endure these conditions in order to save the patient's life.

2.5. CONCLUSION

From the foregoing discussion it should be quite obvious that obtaining data from a living system greatly increases the complexity of instrumentation problems. Fortunately, however, new developments resulting in improved, smaller, and more effective measuring devices are continually being announced, thereby making possible measurements that had previously been considered impossible. In addition, greater knowledge of the physiology of the various systems of the body is emerging as man progresses in his monumental task of learning about himself. All of this will benefit the engineer, the technician, and the physician as time goes on by adding to the tools at their disposal in overcoming instrumentation problems.

· 3 ·

SOURCES OF
BIOELECTRIC
POTENTIALS

In carrying out their various functions, certain systems of the body generate their own monitoring signals, which convey useful information about the functions they represent. These signals are the bioelectric potentials associated with nerve conduction, brain activity, heartbeat, muscle activity, and so on. Bioelectric potentials are actually ionic voltages produced as a result of the electrochemical activity of certain special types of cells. Through the use of transducers capable of converting ionic potentials into electrical voltages, these natural monitoring signals can be measured and the results displayed in a meaningful way to aid the physician in his diagnosis and treatment of various diseases.

The idea of electricity being generated in the body goes back as far as 1786 when the Italian anatomy professor, Luigi Galvani, claimed to have found electricity in the muscle of a frog's leg. In the century that followed several other scientists discovered electrical activity in various animals and in man. But it was not until 1903, when the Dutch physician William Einthoven introduced the string galvanometer, that any practical application could be made of these potentials. The advent of the vacuum tube and amplification and, more recently, of solid-state technology has made possible better representation of the bioelectric potentials. These developments, combined with a large amount of physiological research activity, have opened many new avenues of knowledge in the application and interpretation of these important signals.

3.1. RESTING AND ACTION POTENTIALS

Certain types of cells within the body, such as nerve and muscle cells, are encased in a semipermeable membrane that permits some substances to pass through the membrane while others are kept out. Neither the exact structure of the membrane nor the mechanism by which its permeability is controlled is known, but the substances involved have been identified by experimentation.

Surrounding the cells of the body are the body fluids. These fluids are conductive solutions containing charged atoms, known as *ions*. The principal ions are sodium (Na^+), potassium (K^+), and chloride (Cl^-). The membrane of excitable cells readily permits entry of potassium and chloride ions but effectively blocks the entry of sodium ions. Since the various ions seek a balance between the inside of the cell and the outside, both according to concentration and electric charge, the inability of the sodium to penetrate the membrane results in two conditions. First, the concentration of sodium ions inside the cell becomes much lower than in the intercellular fluid outside. Since the sodium ions are positive, this would tend to make the outside of the cell more positive than the inside. Second, in an attempt to balance the electric charge, additional potassium ions, which are also positive, enter the cell, causing a higher concentration of potassium on the inside than on the outside. This charge balance cannot be achieved, however, because of the concentration imbalance of potassium ions. Equilibrium is reached with a potential difference across the membrane, negative on the inside and positive on the outside.

This membrane potential is called the *resting potential* of the cell and is maintained until some kind of disturbance upsets the equilibrium. Since measurement of the membrane potential is generally made from inside the cell with respect to the body fluids, the resting potential of a cell is given as negative. Research investigators have reported measuring membrane potentials in various cells ranging from -60 to -100 mV. Figure 3.1 illustrates in simplified form the cross section of a cell with its resting potential. A cell in the resting state is said to be *polarized*.

When a section of the cell membrane is excited by the flow of ionic current or by some form of externally applied energy, the membrane changes its characteristics and begins to allow some of the sodium ions to enter. This movement of sodium ions into the cell constitutes an ionic current flow that further reduces the barrier of the membrane to sodium

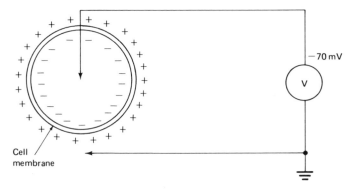

Figure 3.1. Polarized cell with its resting potential.

ions. The net result is an avalanche effect in which sodium ions literally rush into the cell to try to reach a balance with the ions outside. At the same time potassium ions, which were in higher concentration inside the cell during the resting state, try to leave the cell but are unable to move as rapidly as the sodium ions. As a result, the cell has a slightly positive potential on the inside due to the imbalance of potassium ions. This potential is known as the *action potential* and is approximately 20 mV positive. A cell that has been excited and that displays an action potential is said to be *depolarized;* the process of changing from the resting state to the action potential is called *depolarization*. Figure 3.2 shows the ionic movements associated with depolarization, and Figure 3.3 illustrates the cross section of a depolarized cell.

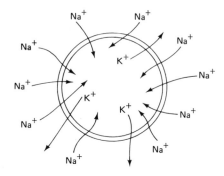

Figure 3.2. Depolarization of a cell. (Na$^+$ ions rush into the cell while K$^+$ ions attempt to leave.)

Once the rush of sodium ions through the cell membrane has stopped (a new state of equilibrium is reached), the ionic currents that lowered the barrier to sodium ions are no longer present and the membrane reverts back to its original, selectively permeable condition, wherein the passage of sodium ions from the outside to the inside of the cell is again blocked. Were this the only effect, however, it would take a long time for a resting

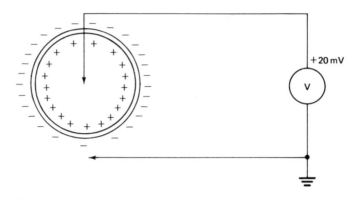

Figure 3.3. Depolarized cell during an action potential.

potential to develop again. But such is not the case. By an active process, called a *sodium pump,* the sodium ions are quickly transported to the outside of the cell, and the cell again becomes polarized and assumes its resting potential. This process is called *repolarization.* Although little is known of the exact chemical steps involved in the sodium pump, it is quite generally believed that sodium is withdrawn against both charge and concentration gradients supported by some form of high-energy phosphate compound. The rate of pumping is directly proportional to the sodium concentration in the cell. It is also believed that the operation of this pump is linked with the influx of potassium into the cell, as if a cyclic process involving an exchange of sodium for potassium existed.

Figure 3.4 shows a typical action potential waveform, beginning at the resting potential, depolarizing, and returning to the resting potential after repolarization. The time scale for the action potential depends on the type of cell producing the potential. In nerve and muscle cells, repolarization occurs so rapidly following depolarization that the action potential appears as a spike of as little as one millisecond total duration. Heart muscle, on the other hand, repolarizes much more slowly, with the action potential for heart muscle usually lasting from 150 to 300 msec.

Regardless of the method by which a cell is excited or the intensity of the stimulus (provided it is sufficient to activate the cell), the action potential is always the same for any given cell. This is known as the *all-or-nothing* law. The *net height* of the action potential is defined as the difference between the potential of the depolarized membrane at the peak of the action potential and the resting potential.

Following the generation of an action potential, there is a brief period of time during which the cell cannot respond to any new stimulus. This period, called the *absolute refractory period,* lasts about one millisecond in nerve cells. Following the absolute refractory period, there occurs a *rela-*

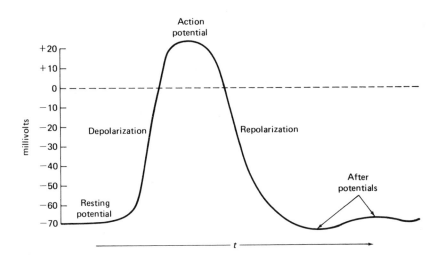

Figure 3.4. Waveform of the action potential. (Time scale varies with type of cell.)

tive refractory period, during which another action potential can be triggered, but a much stronger stimulation is required. In nerve cells, the relative refractory period lasts several milliseconds. These refractory periods are believed to be the result of afterpotentials that follow an action potential.

3.2. PROPAGATION OF ACTION POTENTIALS

When a cell is excited and generates an action potential, ionic currents begin to flow. This process can, in turn, excite neighboring cells or adjacent areas of the same cell. In the case of a nerve cell with a long fiber, the action potential is generated over a very small segment of the fiber's length but is propagated in both directions from the original point of excitation. In nature, nerve cells are excited only near their "input end" (see Chapter 10 for details). As the action potential travels down the fiber, it cannot reexcite the portion of the fiber immediately upstream, because of the refractory period that follows the action potential.

The rate at which an action potential moves down a fiber or is propagated from cell to cell is called the *propagation rate.* In nerve fibers the propagation rate is also called the nerve conduction rate, or conduction velocity. This velocity varies widely, depending on the type and diameter of the nerve fiber. The usual velocity range in nerves is from 20 to 140 meters per second. Propagation through heart muscle is slower, with an average rate from 0.2 to 0.4 meter per second. Special time-delay fibers between the atria and ventricles of the heart cause action potentials to propagate at an even slower rate, 0.03 to 0.05 meter per second.

3.3. THE BIOELECTRIC POTENTIALS

To measure bioelectric potentials, a transducer capable of converting ionic potentials and currents into electric potentials and currents is required. Such a transducer consists of two *electrodes,* which measure the ionic potential difference between their respective points of application. Electrodes are discussed in detail in Chapter 4.

Although measurement of individual action potentials can be made in some types of cells, such measurements are difficult because they require precise placement of an electrode inside a cell. The more common form of measured biopotentials is the combined effect of a large number of action potentials as they appear at the surface of the body, or at one or more electrodes inserted into a muscle, nerve, or some part of the brain.

The exact method by which these potentials reach the surface of the body is not known. A number of theories have been advanced that seem to explain most of the observed phenomena fairly well, but none exactly fits the situation. Many attempts have been made, for example, to explain the biopotentials from the heart as they appear at the surface of the body. According to one theory, the surface pattern is a summation of the potentials developed by the electric fields set up by the ionic currents that generate the individual action potentials. This theory, although plausible, fails to explain a number of the characteristics indicated by the observed surface patterns. A closer approximation can be obtained if it is assumed that the surface pattern is a function of the summation of the first derivatives (rates of change) of all the individual action potentials, instead of the potentials themselves. Part of the difficulty arises from the numerous assumptions that must be made concerning the ionic current and electric field patterns throughout the body. The validity of some of these assumptions is considered somewhat questionable. Regardless of the method by which these patterns of potentials reach the surface of the body or implanted measuring electrodes, they can be measured as specific bioelectric signal patterns that have been studied extensively and can be defined quite well.

The remainder of this chapter is devoted to a description of each of the more significant bioelectric potential waveforms. The designation of the waveform itself generally ends in the suffix *gram,* whereas the name of the instrument used to measure the potentials and graphically reproduce the waveform ends in the suffix *graph.* For example, the *electrocardiogram* (the name of the waveform resulting from the heart's electrical activity) is measured on an *electrocardiograph* (the instrument). Ranges of amplitudes and frequency spectra for each of the biopotential waveforms described below are included in Appendix B.

3.3.1. THE ELECTROCARDIOGRAM (ECG). The biopotentials generated by the muscles of the heart result in the *electrocardiogram,* abbreviated ECG (sometimes EKG, from the German *electrokardiogram*). To understand the origin of the ECG, it is necessary to have some familiarity with the anatomy of the heart. Figure 3.5 shows a cross section of the interior of the heart. The heart is divided into four chambers. The two upper chambers, the left and right *atria,* are synchronized to act together. Similarly, the two lower chambers, the *ventricles,* operate together. The right atrium receives blood from the veins of the body and pumps it into the right ventricle. The right ventricle pumps the blood through the lungs, where it is oxygenated. The oxygen-enriched blood then enters the left atrium, from which it is pumped into the left ventricle. The left ventricle pumps the blood into the arteries to circulate throughout the body. Because the ventricles actually pump the blood through the vessels (and therefore do most of the work), the ventricular muscles are much larger and more important than the muscles of the atria. In order for the cardiovascular system to function properly, both the atria and the ventricles must operate in a proper time relationship.

Each action potential in the heart originates near the top of the right atrium at a point called the *pacemaker* or *sinoatrial* (SA) *node.* The pacemaker is a group of specialized cells that spontaneously generate action potentials at a regular rate, although the rate is controlled by innervation. To initiate the heartbeat, the action potential generated by the pacemaker propagates in all directions along the surface of both atria. The wavefront of activation travels parallel to the surface of the atria toward the junction of the atria and the ventricles. The wave terminates at a point near the center of the heart, called the *atrioventricular node* (AV node). At this point, some special nerve fibers act as a "delay line" to provide proper timing between the action of the atria and the ventricles. Once the electrical excitation has passed through the delay line, action potentials are initiated in the powerful musculature of the ventricles. The wavefront in the ventricles, however, does not follow along the surface but is perpendicular to it and moves from the inside to the outside of the ventricular wall, terminating at the tip or *apex* of the heart. As indicated earlier, a wave of repolarization follows the depolarization wave by about 0.2 to 0.4 second. This repolarization, however, is not initiated from neighboring muscle cells but occurs as each cell returns to its resting potential independently.

Figure 3.6 shows a typical ECG as it appears when recorded from the surface of the body. Alphabetic designations have been given to each

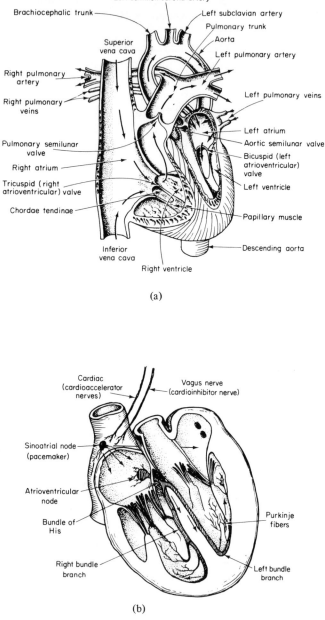

Figure 3.5. The heart: (a) internal structure; (b) conducting system. (From W.F. Evans, *Anatomy and Physiology, The Basic Principles,* Englewood Cliffs, N.J., Prentice-Hall, Inc., 1971, by permission.)

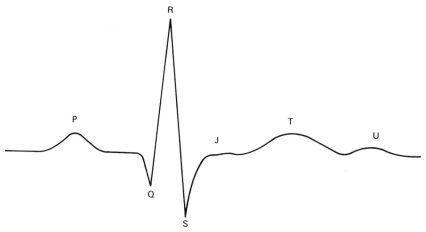

Figure 3.6. The electrocardiogram waveform.

of the prominent features. These can be identified with events related to the action potential propagation pattern. To facilitate analysis, the horizontal segment of this waveform preceding the *P wave* is designated as the *baseline* or the *isopotential line*. The *P wave* represents depolarization of the atrial musculature. *The QRS complex* is the combined result of the repolarization of the atria and the depolarization of the ventricles, which occur almost simultaneously. The *T* wave is the wave of ventricular repolarization, whereas the *U wave,* if present, is generally believed to be the result of afterpotentials in the ventricular muscle. The P-Q interval represents the time during which the excitation wave is delayed in the fibers near the AV node.

The shape and polarity of each of these features vary with the location of the measuring electrodes with respect to the heart, and a cardiologist normally bases his diagnosis on readings taken from several electrode locations. Measurement of the electrocardiogram is covered in more detail in Chapter 6.

3.3.2. THE ELECTROENCEPHALOGRAM (EEG). The recorded representation of bioelectric potentials generated by the neuronal activity of the brain is called the electroencephalogram, abbreviated EEG. The EEG has a very complex pattern much more difficult to recognize than the ECG. A typical sample of the EEG is shown in Figure 3.7. As can be seen, the waveform varies greatly with the location of the measuring electrodes on the surface of the scalp.

EEG potentials, measured at the surface of the scalp, actually represent the combined effect of neuronal potentials over a fairly wide region of

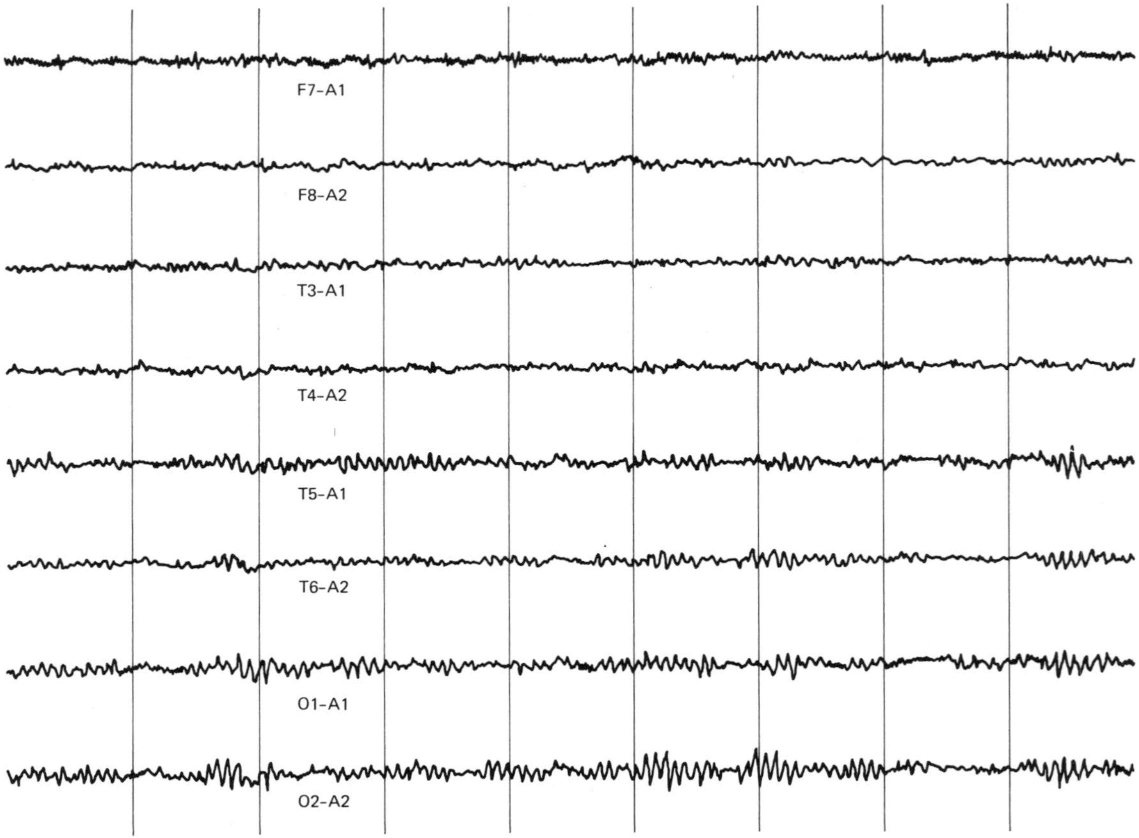

Figure 3.7. Typical human electroencephalogram. The eight tracings indicate regions of the scalp from which each channel of EEG was measured with respect to one of two reference ear electrodes (A1 and A2). (Figure 10.12 shows the layout of the electrodes.) (Courtesy Veterans Administration Hospital, Sepulveda, Calif.)

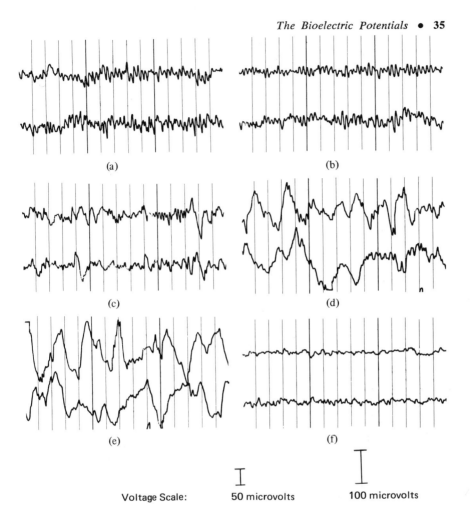

(a)

(b)

(c)

(d)

(e)

(f)

Voltage Scale: 50 microvolts   100 microvolts

Figure 3.8. Typical human EEG patterns for different stages of
sleep. In each case the upper record is from the left frontal region
of the brain and the lower tracing is from the right occipital region. (a) Awake and alert—mixed EEG frequencies; (b) Stage 1
—subject is drowsy and produces large amount of alpha waves;
(c) Stage 2—light sleep, occasional slow waves appear; (d) Stage
3—slow wave sleep; (e) Stage 4—deeper slow wave sleep; (f)
Paradoxical or rapid eye movement (REM) sleep. (Courtesy
Veterans Administration Hospital, Sepulveda, Calif.)

the cortex and from various points beneath. There are, however, certain
characteristic EEG waveforms that can be related to epileptic seizures and
sleep. The waveforms associated with the different stages of sleep are
shown in Figure 3.8. An alert, wideawake person usually displays an un-

synchronized high-frequency EEG. A drowsy person, particularly with his eyes closed, often produces a large amount of rhythmic activity in the range of 8 to 13 Hz. As the person begins to fall asleep, the amplitude and frequency of the waveform decrease; and in light sleep, a large–amplitude, low–frequency waveform emerges. Deeper sleep generally results in even slower and higher-amplitude waves. At certain times, however, a person, still sound asleep, breaks into an unsynchronized high-frequency EEG pattern for a time and then returns to the low-frequency sleep pattern. The period of high-frequency EEG that occurs during sleep is called *paradoxical sleep,* because the EEG is more like that of an awake, alert person than of one who is asleep. Another name is *rapid eye movement* (REM) sleep because associated with the high-frequency EEG is a large amount of rapid eye movement beneath the closed eyelids. This phenomenon is often associated with dreaming, although it has not been shown conclusively that dreaming is related to REM sleep.

Experiments have shown that the frequency of the EEG seems to be affected by the mental activity of a person. The wide variation among individuals and the lack of repeatability in a given person from one occasion to another make the establishment of specific relationships difficult, however.

The various frequency ranges of the EEG have arbitrarily been given Greek-letter designations because frequency seems to be the most prominent feature of an EEG pattern. Electroencephalographers do not agree on the exact ranges, but most classify the EEG frequency bands or rhythms approximately as follows:

Below $3\frac{1}{2}$ Hz	delta
From $3\frac{1}{2}$ Hz to about 8 Hz	theta
From about 8 Hz to about 13 Hz	alpha
Above 13 Hz	beta

Portions of some of these ranges have been given special designations, as have certain sub-bands that fall on or near the stated boundaries. Most humans seem to develop EEG patterns in the alpha range when they are relaxed with their eyes closed. This condition seems to represent a form of synchronization, almost like a "natural" or "idling" frequency of the brain. As soon as the person becomes alert or begins "thinking," the alpha rhythm disappears and is replaced with a "desynchronized" pattern, generally in the beta range. Much research is presently devoted to attempts to learn the physiological sources in the brain responsible for these phenomena, but so far nothing conclusive has resulted. Attempts are also under way to determine whether people, by having their own EEGs fed back to them either visually or audibly, can learn to control their EEG patterns. There is fairly good evidence that this can be done, but the experimentation reported thus far cannot be considered conclusive.

As indicated, the frequency content of the EEG pattern seems to be extremely important. In addition, phase relationships between similar EEG patterns from different parts of the brain are also of great interest. Information of this type may lead to discoveries of EEG sources and will, hopefully, provide additional knowledge regarding the functioning of the brain.

Another form of EEG measurement is the *evoked response*. This is a measure of the "disturbance" in the EEG pattern that results from external stimuli, such as a flash of light or a click of sound. Since these "disturbance" responses are quite repeatable from one flash or click to the next, the evoked response can be distinguished from the remainder of EEG activity, and from the noise, by averaging techniques. These techniques, as well as other methods of measuring EEG, are covered in Chapter 10.

3.3.3. ELECTROMYOGRAM (EMG). The bioelectric potentials associated with muscle activity constitute the *electromyogram* (EMG). These potentials may be measured at the surface of the body near a muscle of interest or directly from the muscle by penetrating the skin with needle electrodes. Since most EMG measurements are intended to obtain an indication of the amount of activity of a given muscle, or group of muscles, rather than of an individual muscle fiber, the pattern is usually a summation of the individual action potentials from the fibers constituting the muscle or muscles being measured. As with the EEG, EMG electrodes pick up potentials from all muscles within the range of the electrodes. This means that potentials from nearby large muscles may interfere with attempts to measure the EMG from smaller muscles, even though the electrodes are placed directly over the small muscles. Where this is a problem, needle electrodes inserted directly into the muscle are required.

As stated in Section 3.1, the action potential of a given muscle (or nerve fiber) has a fixed magnitude, regardless of the intensity of the stimulus that generates the response. Thus, in a muscle, the intensity with which the muscle acts does not increase the net height of the action potential pulse but does increase the rate with which each muscle fiber fires and the number of fibers that are activated at any given time. The amplitude of the measured EMG waveform is the instantaneous sum of all the action potentials generated at any given time. Because these action potentials occur in both positive and negative polarities at a given pair of electrodes, they sometimes add and sometimes cancel. Thus the EMG waveform appears very much like a random noise waveform with the energy of the signal a function of the amount of muscle activity and electrode placement. Typical EMG waveforms are shown in Figure 3.9. Methods and instrumentation for measuring EMG are described in Chapter 10.

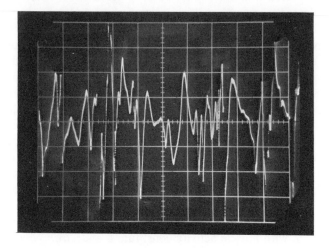

Figure 3.9. Typical electromyogram waveform. EMG of normal "interference pattern" with full strength muscle contraction producing obliteration of the baseline. Sweep speed is 10 milliseconds per cm; amplitude is 1 millivolt per cm. (Courtesy of the Veterans Administration Hospital, Portland, Oreg.)

3.3.4. OTHER BIOELECTRIC POTENTIALS. In addition to the three most significant bioelectric potentials (ECG, EEG, and EMG), several other electric signals can be obtained from the body, although most of them are special variations of EEG, EMG, or nerve-firing patterns. Some of the more prominent ones are

1. Electroretinogram (ERG). A record of the complex pattern of bioelectric potentials obtained from the retina of the eye. This is usually a response to a visual stimulus.
2. Electrooculogram (EOG). A measure of the variations in the corneal-retinal potential as affected by the position and movement of the eye.
3. Electrogastrogram (EGG). The EMG patterns associated with the peristaltic movements of the gastrointestinal tract.

· 4 ·

ELECTRODES

In observing the measurement of an electrocardiogram (ECG) or the result of some other form of bioelectric potential, as discussed in Chapter 3, a conclusion could easily be reached that the measuring electrodes are simply terminals or contact points with which voltages can be obtained at the surface of the body. Also, the electrolyte paste or jelly often used in such measurements might be considered to be applied only for the purpose of reducing the impedance of the skin in order to lower the overall input impedance of the system. This conclusion, however, is incorrect and does not satisfy the theory that explains the origin of these bioelectric potentials. It must be realized that the bioelectric potentials generated in the body are ionic potentials, produced by ionic current flow. Efficient measurement of these ionic potentials requires that they be converted into electronic potentials before they can be measured by conventional methods. It was the realization of this fact that led to the development of the modern noise-free, stable measuring devices now available.

Devices that convert ionic potentials into electronic potentials are called *electrodes*. The theory of electrodes and the principles that govern their design are inherent in an understanding of the measurement of bioelectric potentials. This same theory also applies to electrodes used in chemical transducers, such as those used to measure pH, P_{O_2}, and P_{CO_2} of the blood. This chapter deals first with the basic theory of electrodes and then with the various types used in biomedical instrumentation.

4.1. ELECTRODE THEORY

The interface of metallic ions in solution with their associated metals results in an electrical potential that is called the *electrode potential*. This potential is a result of the difference in diffusion rates of ions into and out of the metal. Equilibrium is produced by the formation of a layer of charge at the interface. This charge is really a double layer, with the layer nearest the metal being of one polarity and the layer next to the solution being of opposite polarity. Nonmetallic materials, such as hydrogen, also have electrode potentials when interfaced with their associated ions in solution. The electrode potentials of a wide variety of metals and alloys are listed in Table 4.1.

It is impossible to determine the absolute electrode potential of a single electrode, for measurement of the potential across the electrode and its ionic solution would require placing another metallic interface in the solution. Therefore all electrode potentials are given as relative values and must be stated in terms of some reference. By international agreement, the normal hydrogen electrode was chosen as the reference standard and arbitrarily assigned an electrode potential of zero volts. All the electrode potentials listed in Table 4.1 are given with respect to the hydrogen electrode. They represent the potentials that would be obtained across the stated electrode and a hydrogen electrode if both were placed in a suitable ionic solution.

Another source of an electrode potential is the unequal exchange of ions across a membrane that is semipermeable to a given ion when the membrane separates liquid solutions with different concentrations of that ion. An equation relating the potential across the membrane and the two concentrations of the ion is called the *Nernst equation* and can be stated as follows:

$$E = -\frac{RT}{nF} \ln \frac{C_1 f_1}{C_2 f_2}$$

where $R =$ the gas constant (8.315×10^7 ergs per mole per degree Kelvin)

$T =$ the absolute temperature in degrees Kelvin

$n =$ the valence of the ion (the number of electrons added or removed to ionize the atom)

TABLE 4.1. ELECTRODE POTENTIALS *

Electrode Reaction	E_0(volts)	Electrode Reaction	E_0(volts)
Li $\rightleftarrows$ Li+	−3.045	V $\rightleftarrows$ V³+	−0.876
Rb $\rightleftarrows$ Rb+	−2.925	Zn $\rightleftarrows$ Zn²+	−0.762
K $\rightleftarrows$ K+	−2.925	Cr $\rightleftarrows$ Cr²+	−0.74
Cs $\rightleftarrows$ Cs+	−2.923	Ga $\rightleftarrows$ Ga²+	−0.53
Ra $\rightleftarrows$ Ra²+	−2.92	Fe $\rightleftarrows$ Fe²+	−0.440
Ba $\rightleftarrows$ Ba²+	−2.90	Cd $\rightleftarrows$ Cd²+	−0.402
Sr $\rightleftarrows$ Sr²+	−2.89	In $\rightleftarrows$ In²+	−0.342
Ca $\rightleftarrows$ Ca²+	−2.87	Tl $\rightleftarrows$ Tl+	−0.336
Na $\rightleftarrows$ Na+	−2.714	Mn $\rightleftarrows$ Mn³+	−0.283
La $\rightleftarrows$ La³+	−2.52	Co $\rightleftarrows$ Co²+	−0.277
Mg $\rightleftarrows$ Mg²+	−2.37	Ni $\rightleftarrows$ Ni²+	−0.250
Am $\rightleftarrows$ Am³+	−2.32	Mo $\rightleftarrows$ Mo³+	−0.2
Pu $\rightleftarrows$ Pu³+	−2.07	Ge $\rightleftarrows$ Ge⁴+	−0.15
Th $\rightleftarrows$ Th⁴+	−1.90	Sn $\rightleftarrows$ Sn²+	−0.136
Np $\rightleftarrows$ Np³+	−1.86	Pb $\rightleftarrows$ Pb²+	−0.126
Bc $\rightleftarrows$ Bc²+	−1.85	Fe $\rightleftarrows$ Fe³+	−0.036
U $\rightleftarrows$ U³+	−1.80	D₂ $\rightleftarrows$ D+	−0.0034
Hf $\rightleftarrows$ Hf⁴+	−1.70	H₂ $\rightleftarrows$ H+	0.000
Al $\rightleftarrows$ Al³+	−1.66	Cu $\rightleftarrows$ Cu²+	+0.337
Ti $\rightleftarrows$ Ti²+	−1.63	Cu $\rightleftarrows$ Cu+	+0.521
Zr $\rightleftarrows$ Zr⁴+	−1.53	Hg $\rightleftarrows$ Hg₂²+	+0.789
U $\rightleftarrows$ U⁴+	−1.50	Ag $\rightleftarrows$ Ag+	+0.799
Np $\rightleftarrows$ Np⁴+	−1.354	Rh $\rightleftarrows$ Rh³+	+0.80
Pu $\rightleftarrows$ Pu⁴+	−1.28	Hg $\rightleftarrows$ Hg²+	+0.857
Ti $\rightleftarrows$ Ti³+	−1.21	Pd $\rightleftarrows$ Pd²+	+0.987
V $\rightleftarrows$ V²+	−1.18	Ir $\rightleftarrows$ Ir³+	+1.000
Mn $\rightleftarrows$ Mn²+	−1.18	Pt $\rightleftarrows$ Pt²+	+1.19
Nb $\rightleftarrows$ Nb³+	−1.1	Au $\rightleftarrows$ Au³+	+1.50
Cr $\rightleftarrows$ Cr²+	−0.913	Au $\rightleftarrows$ Au+	+1.68

* Reproduced by permission from Brown, J. H. V., J. E. Jacobs, L. Stark, *Biomedical Engineering*, F. A. Davis Company, 1971, Philadelphia, Penna.

F = the Faraday constant (96,500 coulombs)
C_1, C_2 = the two concentrations of the ion
f_1, f_2 = the respective activity coefficients of the ion on the two sides of the membrane

The activity coefficients, f_1 and f_2, depend on such factors as the charges of all ions in the solution and the distance between ions. The product, $C_1 f_1$, of a concentration and its associated activity coefficient is called the *activity* of the ion responsible for the electrode potential. From the Nernst equation it can be seen that the electrode potential across the membrane is proportional to the logarithm of the ratio of the activities of the subject ion on the two sides of the membrane. In a very dilute solution the activity coefficient f approaches unity, and the electrode potential becomes a function of the logarithm of the ratio of the two concentrations.

In electrodes used for the measurement of bioelectric potentials, the electrode potential occurs at the interface of a metal and an electrolyte, whereas in biochemical transducers both membrane barriers and metal-electrolyte interfaces are used. The sections that follow describe electrodes of both types.

4.2. BIOPOTENTIAL ELECTRODES

A wide variety of electrodes can be used to measure bioelectric events, but nearly all can be classified as belonging to one of three basic types:

1. *Microelectrodes.* Electrodes used to measure bioelectric potentials near or within a single cell.
2. *Skin surface electrodes.* Electrodes used to measure ECG, EEG, and EMG potentials from the surface of the skin.
3. *Needle electrodes.* Electrodes used to penetrate the skin to record EEG potentials from a local region of the brain or EMG potentials from a specific group of muscles.

All three types of biopotential electrodes have the metal-electrolyte interface described in the previous section. In each case, an electrode potential is developed across the interface, proportional to the exchange of ions between the metal and the electrolytes of the body. The double layer of charge at the interface acts as a capacitor. Thus the equivalent circuit of biopotential electrode in contact with the body consists of a voltage in series with a resistance-capacitance network of the type shown in Figure 4.1.

Since measurement of bioelectric potentials requires two electrodes, the voltage measured is really the difference between the instantaneous

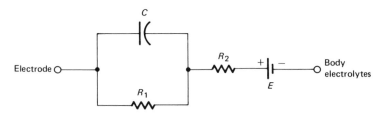

Figure 4.1. Equivalent circuit of biopotential electrode interface.

potentials of the two electrodes, as shown in Figure 4.2. If the two electrodes are of the same type, the difference is usually small and depends essentially on the actual difference of ionic potential between the two points

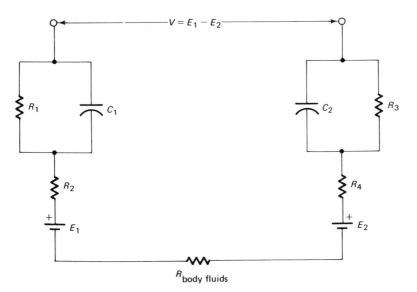

Figure 4.2. Measurement of biopotentials with two electrodes—equivalent circuit.

of the body from which measurements are being taken. If the two electrodes are different, however, they may produce a significant dc voltage that would cause current to flow through both electrodes as well as through the input circuit of the amplifier to which they are connected. The dc voltage due to the difference in electrode potentials is called the *electrode offset voltage*. The resulting current is often mistaken for a true physiological event. Even two electrodes of the same material may produce a small electrode offset voltage.

In addition to the electrode offset voltage, experiments have shown that the chemical activity that takes place within an electrode can cause voltage fluctuations to appear without any physiological input. Such variations may appear as noise on a bioelectric signal. This noise can be reduced by proper choice of materials or, in most cases, by special treatment, such as coating the electrodes by some electrolytic method to improve stability. It has been found that, electrochemically, the *silver-silver chloride electrode* is the most stable type of electrode. This type of electrode is prepared by electrolytically coating a piece of pure silver with silver chloride. The coating is normally done by placing a cleaned piece of silver into a bromide-free sodium-chloride solution. A second piece of silver is also placed in the solution, and the two are connected to a voltage source such that the electrode to be chlorided is made positive with respect to the other. The silver ions combine with the chloride ions from the salt to produce neutral silver chloride molecules that coat the silver electrode. Some variations in the process are used to produce electrodes with specific characteristics.

The resistance-capacitance networks shown in Figures 4.1 and 4.2 represent the impedance of the electrodes (one of their most important characteristics) as fixed values of resistance and capacitance. Unfortunately, the impedance is not constant. The impedance is frequency dependent because of the effect of the capacitance. Furthermore, both the electrode potential and the impedance are varied by an effect called *polarization*.

Polarization is the result of direct (dc) current passing through the metal-electrolyte interface. The effect is much like that of charging a battery with the polarity of the charge opposing the flow of current that generates the charge. Some electrodes are designed to avoid or reduce polarization. If the amplifier to which the electrodes are connected has an extremely high input impedance, the effect of polarization or any other change in electrode impedance is minimized.

Size and type of electrode are also important in determining the electrode impedance. Larger electrodes tend to have lower impedances. Surface electrodes generally have impedances of 2 to 10 kilohms, whereas small needle electrodes and microelectrodes have much higher impedances. For best results in reading or recording the potentials measured by the electrodes, the input impedance of the amplifier must be several times that of the electrodes.

4.2.1. MICROELECTRODES. Microelectrodes are electrodes with tips sufficiently small to penetrate a single cell in order to obtain readings from within the cell. The tip must be small enough to permit penetration without damaging the cell. This action is usually complicated by the difficulty of accurately positioning an electrode with respect to a cell.

Microelectrodes are generally of two types: metal and micropipette. Metal microelectrodes are formed by electrolytically etching the tip of a tungsten or stainless-steel wire to the desired size. Then the wire is coated almost to the tip with an insulating material. Some electrolytic processing can also be performed on the tip to lower the impedance. The metal-ion interface takes place where the metal tip contacts the electrolytes either inside or outside the cell.

The micropipette type of microelectrode is a glass micropipette with the tip drawn out to the desired size (usually around 1 micron diameter). The micropipette is filled with an electrolyte compatible with the cellular fluids. This type of microelectrode has a dual interface. One interface consists of a metal wire in contact with the electrolyte solution inside the micropipette, while the other is the interface between the electrolyte inside the pipette and the fluids inside or immediately outside the cell.

A commercial type of microelectrode is shown in Figure 4.3. In this

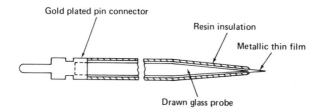

Figure 4.3. Commercial microelectrode with metal film on glass. (Courtesy of Transidyne General Corporation, Ann Arbor, Mich.)

electrode a thin film of precious metal is bonded to the outside of a drawn glass microelectrode. The manufacturer claims such advantages as lower impedance than the micropipette electrode, infinite shelf life, repeatable and reproducible performance, and easy cleaning and maintenance. The metal-electrolyte interface is between the metal film and the electrolyte of the cell.

Microelectrodes, because of their small surface areas, have impedances well up into the megohms. For this reason, amplifiers with extremely high impedances are required to avoid loading the circuit and to minimize the effects of small changes in interface impedance.

4.2.2. Body surface electrodes. Electrodes used to obtain bioelectric potentials from the surface of the body are found in many sizes and forms. Although any type surface electrode can be used to sense ECG, EEG, or EMG potentials, the larger electrodes are usually associated with ECG, since localization of the measurement is not important, whereas smaller electrodes are used in EEG and EMG measurements.

The earliest bioelectric potential measurements used *immersion electrodes,* which were simply buckets of saline solution into which the subject placed his hands and feet, one bucket for each extremity. As might be expected, this type of electrode (Figure 4.4) presented many difficulties, such as restricted position of the subject and danger of electrolyte spillage.

A great improvement over the immersion electrodes were the plate electrodes, which were first introduced about 1917. Originally these electrodes were separated from the subject's skin by cotton or felt pads soaked in a strong saline solution. Later a conductive jelly or paste (an electro-

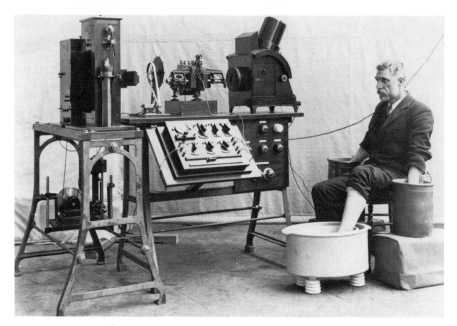

Figure 4.4. ECG measurement using immersion electrodes. Original Cambridge electrocardiograph (1912) built for Sir Thomas Lewis. Produced under agreement with Prof. Willem Einthoven, the father of electrocardiography. (Courtesy of Cambridge Instruments, Inc., Cambridge, Mass.)

lyte) replaced the soaked pads and the metal was allowed to contact the skin. Plate electrodes of this type are still in use today. Some examples are shown in Figure 4.5.

Another fairly old type of electrode still in use is the suction-cup elec-

Figure 4.5. Metal plate electrode. These plates are usually made of, or plated with, silver, nickel, or some similar alloy.

Figure 4.6. Suction cup electrode.

trode shown in Figure 4.6. In this type, only the rim actually contacts the skin.

One of the difficulties in using plate electrodes is the possibility of electrode slippage or movement. This also occurs with the suction-cup electrode after a sufficient length of time. A number of attempts were made to overcome this problem, including the use of adhesive backing and a surface resembling a nutmeg grater that penetrates the skin to lower the contact impedance and reduce the likelihood of slippage.

All the preceding electrodes suffer from a common problem. They are all sensitive to movement, some to a greater degree than others. Even the slightest movement changes the thickness of the thin film of electrolyte between metal and skin and thus causes changes in the electrode potential and impedance. In many cases, the potential changes are so severe that they completely block the bioelectric potentials the electrodes attempt to measure. The adhesive tape and "nutmeg grater" electrodes reduce this movement artifact by limiting electrode movement and reducing interface impedance, but neither is satisfactorily insensitive to movement.

Later, a new type of electrode, the *floating electrode,* was introduced in varying forms by several manufacturers. The principle of this electrode is to practically eliminate movement artifact by avoiding any direct contact of the metal with the skin. The only conductive path between metal and skin is the electrolyte paste or jelly, which forms an electrolyte bridge. Even with the electrode surface held at a right angle with the skin surface, performance is not impaired as long as the electrolyte bridge maintains contact with both the skin and the metal. Figure 4.7 shows a cross section

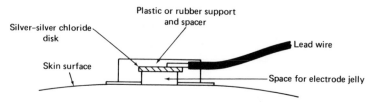

Figure 4.7. Diagram of floating type skin surface electrode.

of a floating electrode, and Figure 4.8 shows a commercially available configuration of the floating electrode.

Figure 4.8. Floating skin surface electrode. (Courtesy of Beckman Instruments, Inc., Fullerton, Calif.)

Floating electrodes are generally attached to the skin by means of two-sided adhesive collars (or rings), which adhere to both the plastic surface of the electrode and the skin. Figure 4.9 shows an electrode in position for biopotential measurement.

Special problems encountered in the monitoring of the ECG of astronauts during long periods of time, and under conditions of perspiration and considerable movement, led to the development of *spray-on* electrodes, in which a small spot of conductive adhesive is sprayed or painted over the skin, which had previously been treated with an electrolyte coating.

Various types of disposable electrodes have been introduced within the last few years to eliminate the usual requirement for cleaning and care after each use. Primarily intended for ECG monitoring, these electrodes can also be used for EEG and EMG as well. In general, disposable electrodes are of the floating type with simple snap connectors by which the leads, which are reusable, are attached. Although some disposable electrodes can be reused several times, their cost is usually low enough that cleaning for reuse is not warranted.

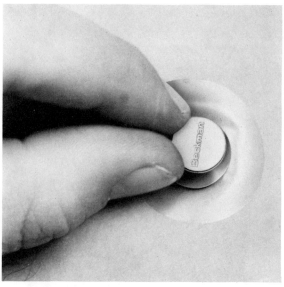

Figure 4.9. Application of floating type skin surface electrode. (Courtesy of Beckman Instruments, Inc., Fullerton, Calif.)

Special types of surface electrodes have been developed for other applications. For example, a special *ear clip electrode* (Figure 4.10) was

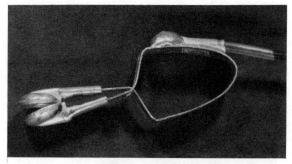

Figure 4.10. Ear clip electrode. (Courtesy of Sepulveda Veterans Administration Hospital.)

developed for use as a reference electrode for EEG measurements. Another special surface electrode, the *wick electrode* of Figure 4.11, consists of a metal contact that interfaces with a soft wick filled with the electrolyte. This type of electrode is used in applications where the pressure or weight of a standard surface electrode cannot be tolerated, such as in measuring potentials from the surface of the eye or from an internal organ. Scalp *surface electrodes* for EEG are usually small disks about 7 mm in diameter or small solder pellets that are placed on the cleaned scalp via an electrolyte paste. This type of electrode is shown in Figure 4.12.

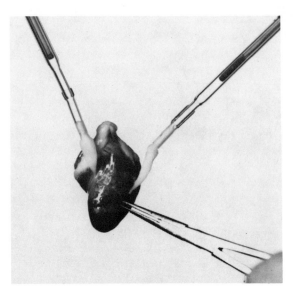

Figure 4.11. Wick electrodes. (Courtesy of Beckman Instruments, Inc., Fullerton, Calif.)

4.2.3. NEEDLE ELECTRODES. In order to reduce the interface impedance, and, consequently, movement artifact, some electroencephalographers use small subdermal needles to penetrate the scalp for EEG measurements. These needle electrodes, shown in Figure 4.13, are not inserted into the brain; instead they merely penetrate the skin. Generally, they are simply inserted through a small section of the skin just beneath the surface and parallel to it.

In animal research (and occasionally in man) longer needles are actually inserted into the brain in order to obtain localized measurement of potentials from a specific part of the brain. This process requires longer needles precisely located by means of a map or atlas of the brain. Sometimes a special instrument, called a *stereotaxic instrument,* is used to hold the animal's head and guide the placement of electrodes. Often these

Figure 4.12. EEG scalp surface electrode. (Courtesy of Sepulveda Veterans Administration Hospital.)

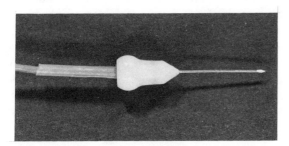

Figure 4.13. Subdermal needle electrode for EEG. (Courtesy of Sepulveda Veterans Administration Hospital.)

electrodes are implanted to permit repeated measurements over an extended period of time. In this case, a connector is cemented to the animal's skull and the incision through which the electrodes were implanted is allowed to heal.

In some research applications, simultaneous measurement from various depths in the brain along a certain axis is required. Special multiple-depth electrodes have been developed for this purpose. This type of electrode usually consists of a bundle of fine wires, each terminating at a different depth or each having an exposed conductive surface at a specific, but different, depth. These wires are generally brought out to a connector at the surface of the scalp and are often cemented to the skull.

Some needle electrodes consist merely of fine insulated wires, placed so that their tips, which are bare, are in contact with the nerve, muscle, or other tissue from which the measurement is made. The remainder of the wire is covered with some form of insulation to prevent shorting. Wire electrodes of copper or platinum are often used for EMG pickup from specific muscles. The wires are either surgically implanted or introduced by means of a hypodermic needle that is later withdrawn, leaving the wire electrode in place. With this type of electrode, the metal-electrolyte interface takes place between the uninsulated tip of the wire and the electrolytes of the body, although the wire is dipped into an electrolyte paste before insertion in some cases.

In some applications, the hypodermic needle is part of the electrode configuration and is not withdrawn. Instead the wires forming the electrodes are carried inside the needle, which creates the hole necessary for insertion, protects the wires, and acts as a grounded shield. A single wire inside the needle serves as a *unipolar electrode,* which measures the potentials at the point of contact with respect to some indifferent reference. If two wires are placed inside the needle, the measurement is called *bipolar* and provides a very localized measurement between the two wire tips.

Electrodes for measurement from beneath the skin need not actually take the form of needles, however. Surgical clips penetrating the skin of a mouse or rat in the spinal region provide an excellent method of measuring the ECG of an essentially unrestrained, unanesthetized animal. Conductive catheters permit the recording of the ECG from within the esophagus or even from within the chambers of the heart itself.

Needle electrodes and other types of electrodes that create an interface beneath the surface of the skin seem to be less susceptible to movement artifact than surface electrodes, particularly those of the older types. By making direct contact with the subdermal tissue or the intercellular fluids, these electrodes also seem to have lower impedances than surface electrodes of comparable interface area.

4.3. BIOCHEMICAL TRANSDUCERS

At the beginning of this chapter it was stated that an electrode potential is generated either at a metal-electrolyte interface or across a semipermeable membrane separating two different concentrations of an ion that can diffuse through the membrane. Both methods are used in transducers designed to measure the concentration of an ion or of a certain gas dissolved in blood or some other liquid. Also, as stated earlier, since it is impossible to have a single electrode interface to a solution, a second electrode is required to act as a reference. If both electrodes were to exhibit the same response to a given change in concentration of the measured solution, the potential measured between them would not be related to concentration and would, therefore, be useless as a measurement parameter. The usual method of measuring concentrations of ions or gases is to use one electrode (sometimes called the *indicator* or *active* electrode) that is sensitive to the substance or ion being measured and to choose the second or *reference* electrode of a type that is insensitive to that substance.

4.3.1. REFERENCE ELECTRODES. As stated in Section 4.1, the hydrogen-gas–hydrogen-ion interface has been designated as the reference interface and was arbitrarily assigned an electrode potential of zero volts. For this reason, it would seem logical that the hydrogen electrode should actually be used as the reference in biochemical measurements. Hydrogen electrodes can be built and are available commercially. These electrodes make use of the principle that an inert metal, such as platinum, readily absorbs hydrogen gas. If a properly treated piece of platinum is partially immersed in the solution containing hydrogen ions and is also exposed to hydrogen gas, which is passed through the electrode, an electrode potential is formed. The electrode lead is attached to the platinum.

Unfortunately, the hydrogen electrode is not sufficiently stable to serve as a good reference electrode. Furthermore, the problem of maintaining the supply of hydrogen to pass through the electrode during a measurement limits its usefulness to a few special applications. However, since measurement of electrochemical concentrations simply requires a change of potential proportional to a change in concentration, the electrode potential of the reference electrode can be any amount, as long as it is stable and does not respond to any possible changes in the composition of the solution being measured. Thus the search for a good reference electrode is essentially a search for the most stable electrode available. Two types of electrodes have interfaces sufficiently stable to serve as reference electrodes— the silver-silver chloride electrode and the calomel electrode.

The *silver-silver chloride electrode* used as a reference in electrochemical measurements utilizes the same type of interface described in Section 4.2 for bioelectric potential electrodes. In the chemical transducer, the ionic (silver chloride) side of the interface is connected to the solution by an electrolyte bridge, usually a dilute potassium chloride (KCl) solution. The electrode can be successfully employed as a reference electrode if the KCl solution is also saturated with precipitated silver chloride. The electrode potential for the silver-silver chloride reference electrode depends on the concentration of the KCl. For example, with a 0.01-mole * solution, the potential is 0.343 volt, whereas for a 1.0-mole solution the potential is only 0.236 volt.

The most popular type of reference electrode is the *calomel electrode,* shown in Figure 4.14. Calomel is another name for mercurous chloride,

Figure 4.14. Calomel reference electrode (Courtesy of Beckman Instruments, Inc., Fullerton, Calif.)

which is a chemical combination of mercury and chloride ions. The interface between mercury and mercurous chloride generates the electrode potential. By placing the calomel side of the interface in a potassium chloride (KCl) solution, an electrolytic bridge is formed to the solution from which the measurement is to be made. Like the silver-silver chloride electrode, the calomel electrode is very stable over long periods of time and serves well as a reference electrode in many electrochemical measurements. Also, like the silver-silver chloride electrode, the electrode potential

* A 0.01-mole solution of a substance is defined as 0.01 mole of the substance dissolved in 1 liter of solution. A mole is the quantity of the substance that has a weight equal to its molecular weight, usually in grams.

of the calomel electrode depends on the concentration of KCl. An electrode with a 0.01-mole solution of KCl has an electrode potential of 0.388 volt, whereas a saturated KCl solution (about 3.5 mole) has a potential of only 0.247 volt.

4.3.2. THE pH ELECTRODE. Perhaps the most important indicator of chemical balance in the body is the pH of the blood and other body fluids. The pH is directly related to the hydrogen ion concentration in a fluid. Specifically, it is the logarithm of the reciprocal of the H^+ ion concentration. In equation form,

$$pH = -\log_{10}[H^+] = \log_{10}\frac{1}{[H^+]}$$

The pH is a measure of the acid-base balance of a fluid. A neutral solution (neither acid nor base) has a pH of 7. Lower pH numbers indicate acidity, whereas higher pH values define a basic solution. Most human body fluids are slightly basic. The pH of normal arterial blood ranges between 7.38 and 7.42. The pH of venous blood is 7.35, due to the extra CO_2.

Because a thin glass membrane allows passage of only hydrogen ions in the form of H_3O^+, a glass electrode provides a "membrane" interface for hydrogen. Figure 4.15 shows a glass electrode used for measuring pH.

Figure 4.15. Glass electrode for pH measurement. (Couresty of Beckman Instruments, Inc., Fullerton, Calif.)

Inside the glass bulb is a highly acidic buffer solution. Measurement of the potential across the glass interface is achieved by placing a silver-silver chloride electrode in the solution inside the glass bulb and a calomel reference electrode in the solution in which the pH is being measured. In the measurement of pH and, in fact, any electrochemical measurement, each of the two electrodes required to obtain the measurement is called a half-cell. The electrode potential for a half-cell is sometimes called the *half-cell potential*. For pH measurement, the glass electrode with the silver-silver chloride electrode inside the bulb is considered one half-cell, while the calomel reference electrode constitutes the other half-cell.

Two configurations of probe assemblies, each consisting of two half-cells, a glass electrode for pH measurement, and a calomel reference electrode, are shown in Figure 4.16.

For in vivo measurement of pH, special configurations of pH electrodes,

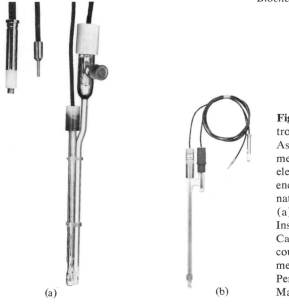

Figure 4.16. Probe electrodes for pH analysis. (a) Assembly for pH measurement consisting of glass pH electrode and calomel reference electrode; (b) Combination probe. [Figure 4.16 (a) is courtesy of Beckman Instruments, Inc., Fullerton, Calif. Figure 4.16 (b) is courtesy of Coleman Instruments, Division of the Perkin-Elmer Corporation, Maywood, Ill.]

(a)　　　　　　　　(b)

such as the stomach electrode shown in Figure 4.17, are available. This is a glass electrode fitted at the end of a tube for entry into the stomach via the mouth and esophagus. The glass electrode is quite adequate for

Figure 4.17. Stomach electrode for pH measurement. (Courtesy of Beckman Instruments, Inc., Fullerton, Calif.)

pH measurements in the physiological range but may produce considerable error at the extremes of the pH range (near pH $=0$ and pH $=13$ or 14). It is also subject to some deterioration after prolonged use but can be restored repeatedly by etching the glass in a 20-percent ammonium bifluoride solution.

The type of glass used for the membrane has much to do with the pH response of the electrode. Special hydroscopic glass that readily absorbs water provides the best pH response.

Modern pH electrodes have impedances ranging from 50 to 500 meg-ohms. Thus the input of the meter that measures the potential difference between the glass electrode and the reference electrode must have an extremely high input impedance. Most pH meters employ electrometer inputs.

4.3.3. BLOOD GAS ELECTRODES. Among the more important physiological chemical measurements are partial pressures of oxygen and carbon dioxide in the blood. The partial pressure of a dissolved gas is the contribution of that gas to the total pressure of all dissolved gases in the blood. The partial pressure of a gas is proportional to the quantity of that gas in the blood. The effectiveness of both the respiratory and cardiovascular systems is reflected in these important parameters.

The partial pressure of oxygen, P_{O_2}, can be measured both in vitro and in vivo. The basic principle is the same and is shown in Figure 4.18.

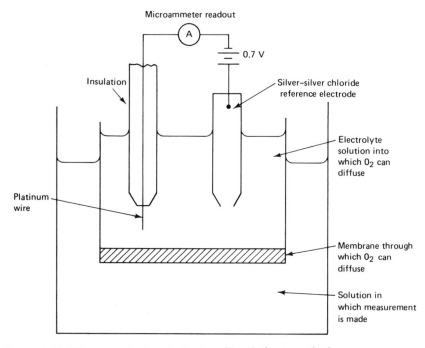

Figure 4.18. Diagram of P_{O_2} electrode with platinum cathode showing principle of operation.

A fine piece of platinum wire, embedded in glass for insulation purposes, with only the tip exposed, is placed in an electrolyte into which oxygen is allowed to diffuse. If a voltage of around 0.7 volt is applied between the platinum wire and a reference electrode (also placed into the electrolyte), with the platinum wire negative, reduction of the oxygen takes place at the platinum cathode. As a result, an oxidation-reduction current proportional to the partial pressure of the diffused oxygen can be measured. The electrolyte is generally sealed into the chamber that holds the platinum wire and the reference electrode by means of a membrane across which the dissolved oxygen can diffuse from the blood.

The platinum cathode and the reference electrode can be integrated into a single unit (the *Clark* electrode). This electrode can be placed in a cuvette of blood for in vitro measurements, or a microversion can be placed at the tip of a catheter for insertion into various parts of the heart or vascular system for direct in vivo measurements.

A similar integrated polarographic oxygen sensor has a rhodium cathode and a silver anode. Oxygen, which diffuses through a membrane, is consumed at the cathode in the same manner as in the previously described P_{O_2} electrode. A photograph of this P_{O_2} electrode is shown in Figure 4.19.

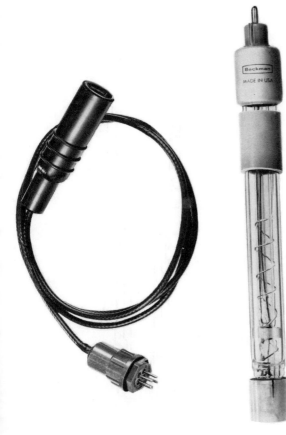

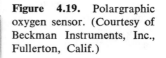

Figure 4.19. Polargraphic oxygen sensor. (Courtesy of Beckman Instruments, Inc., Fullerton, Calif.)

One of the problems inherent in this method of measuring P_{O_2} is the fact that the reduction process actually removes a finite amount of the oxygen from the immediate vicinity of the cathode. By careful design and use of proper procedures, modern P_{O_2} electrodes have been able to reduce this potential source of error to a minimum. Another apparent error in P_{O_2} measurement is a gradual reduction of current with time, almost like the polarization effect described for skin surface electrodes in Section 4.2.2. This effect, generally called *aging,* has also been minimized in modern P_{O_2} electrodes.

The measurement of the partial pressure of carbon dioxide, P_{CO_2}, makes use of the fact that there is a linear relationship between the logarithm of the P_{CO_2} and the pH of a solution. Since other factors also influence the pH, measurement of P_{CO_2} is essentially accomplished by surrounding a pH electrode with a membrane selectively permeable to CO_2. A modern, improved type of P_{CO_2} electrode is the *Severinghaus* electrode. A commercial P_{CO_2} electrode based on the Severinghaus principle is shown in Figure 4.20. In this electrode, the membrane permeable to the CO_2 is made of

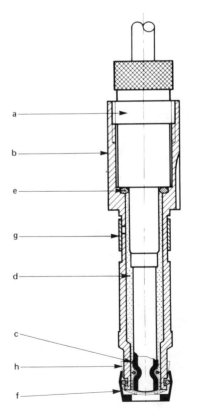

Figure 4.20. P_{CO_2} electrode based on the Severinghaus Principle. (Courtesy of Radiometer A/S, Copenhagen, Denmark.)

teflon, which is not permeable to other ions that might affect the pH. The space between the teflon and the glass contains a matrix consisting of thin cellophane, glass wool, or sheer nylon. This matrix serves as the support for an aqueous bicarbonate layer into which the CO_2 gas molecules can diffuse. One of the difficulties with older types of CO_2 electrodes is the length of time required for the CO_2 molecules to diffuse and thus obtain a reading. The principal advantage of the Severinghaus-type electrode is the more rapid reading that can be obtained because of the improved membrane and bicarbonate layer.

In some applications, measurements of P_{O_2} and P_{CO_2} are combined into a single electrode that also includes a common reference half-cell. Such a combination electrode is shown in diagram form in Figure 4.21.

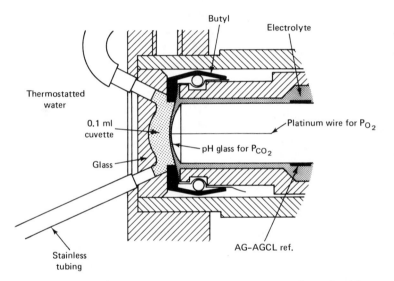

Figure 4.21. Combination P_{CO_2} and P_{O_2} electrode. (Courtesy of J.W. Severinghaus, M.D.)

4.3.4. SPECIFIC ION ELECTRODES. Just as the glass electrode provides a semipermeable membrane for the hydrogen ion in the pH electrode (see Section 4.3.2.), other materials can be used to form membranes that are semipermeable to other specific ions. In each case, measurement of the ion concentration is accomplished by measurement of potentials across a membrane that has the correct degree of permeability to the specific ion to be measured. The permeability should be sufficient to permit rapid establishment of the electrode potential. Both liquid and solid membranes are used for specific ions. As in the case of the pH electrode, a silver-silver chloride interface is usually provided on the electrode side of the membrane, and a standard reference electrode serves as the other half-cell in the solution.

Figure 4.22(a) shows an electrode for measurement of the sodium ion. Figure 4.22(b) shows three types of specific ion electrodes. The electrode at the left is a flow-through type of electrode in which the solution

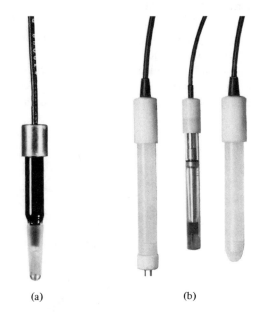

Figure 4.22. Specific ion electrodes: (a) sodium ion electrode. (Courtesy of Beckman Instruments, Inc., Fullerton, Calif.); (b) three specific ion electrodes. Left to right: Flow-through type liquid membrane electrode, combination solid state and reference electrode, liquid membrane electrode. (Courtesy of Orion Research, Inc., Cambridge, Mass.)

(a)   (b)

flows past a liquid membrane via a specified path. The electrode in the center is a combination of a specific ion electrode with a membrane of a solid material and a reference electrode. The electrode at the right is a liquid membrane electrode. The last two electrodes are of the conventional type, which are simply placed in the solution for measurement.

Figure 4.23 is a diagram showing the construction of a flow-through type of electrode. This is a liquid–membrane, specific ion electrode of the type shown at the left in Figure 4.22(b).

One of the difficulties encountered in the measurement of specific ions is the effect of other ions in the solution. In cases where more than one type of membrane could be selected for measurement of a certain ion, the choice of membrane actually used might well depend on other ions that may be expected. In fact, some specific ion electrodes can be used in measurement of a given ion only in the absence of certain other ions.

For measurement of divalent ions, a liquid membrane is often used for

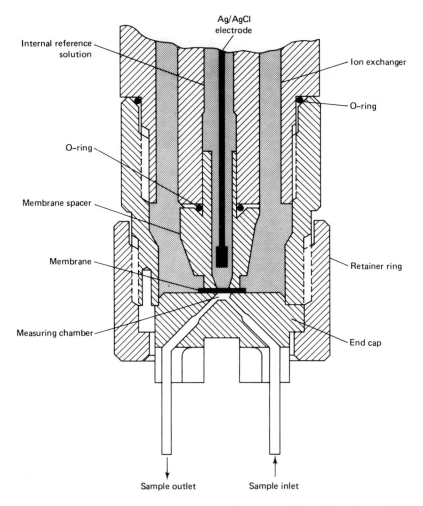

Figure 4.23. Diagram showing construction of flow-through liquid membrane specific ion electrode. (Courtesy of Orion Research, Inc., Cambridge, Mass.)

ion exchange. In this case, the exchanger is usually a salt of an organophosphoric acid, which shows a high degree of specificity to the ion being measured. A calcium-chloride solution bridges the membrane to the silver–silver chloride electrode. Electrodes with membranes of solid materials, such as the cupric ion electrode shown in Figure 4.24, are also used for measurement of divalent ions.

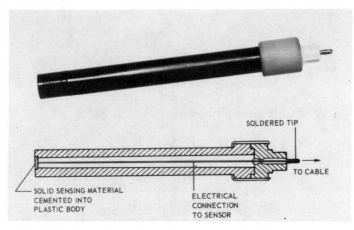

Figure 4.24. Solid state cupric (copper) ion electrode. (Courtesy of Beckman Instruments, Inc., Fullerton, Calif.)

· 5 ·

THE
CARDIOVASCULAR
SYSTEM

The heart attack, in its various forms, is the cause of many deaths in the world today. The use of engineering methods and the development of instrumentation have contributed substantially to progress made in recent years in reducing death from heart diseases.

There are many ways in which instrumentation is used both in medical research and in the hospital. Blood pressure, flow, and volume are measured by using engineering techniques. The electrocardiogram and the phonocardiogram are measured and recorded with electronic instruments. Intensive and coronary care units now installed in many hospitals rely on bioinstrumentation for their function. There are also cardiac assist devices, like the electronic pacemaker and defibrillator, which, although not measuring instruments per se, are electronic devices often used in conjunction with measurement systems.

In this chapter the cardiovascular system is discussed, not only from the point of view of basic physiology but also with the idea that it is an engineering system. In this way the important parameters can be illustrated in correct perspective. Included are the pump and flow characteristics, as well as the ancillary ideas of electrical activity and heart sounds.

63

The electrocardiogram has already been introduced in Chapter 3. The actual measurements and devices are discussed in Chapters 6 and 7.

5.1. THE HEART AND CARDIOVASCULAR SYSTEM

The heart may be considered as a two-stage pump, physically arranged in parallel but with the circulating blood passing through the pumps in a series sequence. The right half of the heart, known as the *right heart,* is the pump that supplies blood to the lungs for oxygenation, whereas the *left heart* supplies blood to the rest of the system. The circulatory path for blood flow through the lungs is called the *pulmonary circulation,* and the circulatory system that supplies oxygen and nutrients to the cells of the body is called the *systemic circulation.*

From an engineering standpoint, the systemic circulation is a high-resistance circuit with a large pressure gradient between the arteries and veins. Thus the pump constituting the left heart may be considered as a pressure pump. However, in the pulmonary circulation system, the pressure difference between the arteries and the veins is small, as is the resistance to flow, and so the right heart may be considered as a volume pump. The left heart is larger and of stronger muscle construction than the right heart because of the greater pressures required for the systemic circulation. The volume of blood delivered per unit of time by the two sides is the same when measured over a sufficiently long interval. The left heart develops a pressure head sufficient to cause blood to flow to all the extremities of the body.

The pumping action itself is performed by contraction of the heart muscles surrounding each chamber of the heart. These muscles receive their own blood supply from the *coronary arteries* which surround the heart like a crown (corona). The *coronary arterial system* is a special branch of the systemic circulation.

The analogy to a pump and hydraulic piping system should not be used too indiscriminately. The pipes, the arteries and the veins, are not rigid but flexible. They are capable of helping and controlling blood circulation by their own muscular action and their own valve and receptor system. Blood is not a pure Newtonian fluid; rather, it possesses properties that do not always comply with the laws governing hydraulic motion. In addition, the blood needs the help of the lungs for the supply of oxygen, and it interacts with the lymphatic system. Furthermore, many chemicals and hormones affect the operation of the system. Thus oversimplification could lead to error if carried too far.

The actual physiological system for the heart and circulation is illustrated in Figure 5.1, with the equivalent engineering-type piping diagram

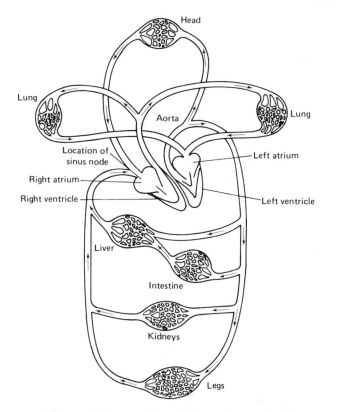

Figure 5.1. The cardiovascular system. (From K. Schmidt-Nielsen, *Animal Physiology,* 3rd ed., Prentice-Hall, Inc., 1970, by permission.)

shown in Figure 5.2. Referring to these figures, the operation of the circulatory system can be described as follows: Blood enters the heart on the right side through two main veins: the *superior vena cava,* which leads from the body's upper extremities, and the *inferior vena cava* leading from the body's organs and extremities below the heart. The incoming blood fills the storage chamber, the *right atrium.* In addition to the two veins mentioned, the *coronary sinus* also empties into the right atrium. The coronary sinus contains the blood that has been circulating through the heart itself via the coronary loop.

When the right atrium is full, it contracts and forces blood through the *tricuspid valve* into the *right ventricle,* which then contracts to pump the blood into the pulmonary circulation system. When ventricular pressure exceeds atrial pressure, the tricuspid valve closes and the pressure in the ventricle forces the semilunar *pulmonary valve* to open, thereby caus-

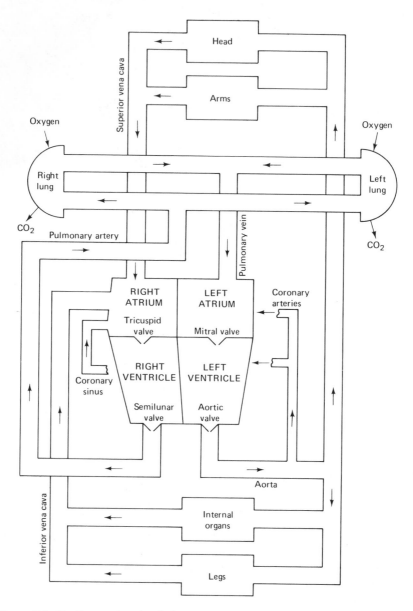

Figure 5.2. Cardiovascular circulation.

ing blood to flow into the pulmonary artery, which divides into the two lungs.

In the *alveoli* of the lungs, an exchange takes place. The red blood cells are recharged with oxygen and give up their carbon dioxide. Not shown on the diagram are the details of this exchange. The pulmonary artery *bifurcates* (divides) many times into smaller and smaller arteries, which become arterioles with extremely small cross sections. These arterioles supply blood to the alveolar capillaries in which the exchange of oxygen and carbon dioxide takes place. On the other side of the lung mass is a similar construction in which the capillaries feed into tiny veins, or *venules*. The latter combine to form larger veins, which in turn combine until ultimately all the oxygenated blood is returned to the heart via the pulmonary vein.

The blood enters the *left atrium* from the pulmonary vein, and from here it is pumped through the *mitral,* or *bicuspid valve,* into the left ventricle by contraction of the atrial muscles. When the left ventricular muscles contract, the pressure produced by the contraction mechanically closes the mitral valve, and the buildup of pressure in the ventricle forces the *aortic valve* to open, thus causing the blood to rush from the ventricle into the *aorta.* It should be noted that this action takes place synchronously with the right ventricle as it pumps blood into the pulmonary artery.

The heart's pumping cycle is divided into two major parts: systole and diastole. *Systole* (sĭs'tō·lē) is defined as the period of contraction of the heart muscles, specifically the ventricular muscle, at which time blood is pumped into the pulmonary artery and the aorta. *Diastole* (dī·ăs'tō·lē) is the period of dilation of the heart cavities as they fill with blood.

Once the blood has been pumped into the arterial system, the heart relaxes, the pressure in the chambers decreases, the outlet valves close, and in a short time the inlet valves open again to restart the diastole and initiate a new cycle in the heart. The mechanism of this cycle and its control will be discussed later.

After passing through many bifurcations of the arteries, the blood reaches the vital organs, the brain, and the extremities. The last stage of the arterial system is the gradual decrease in cross section and the increase in the number of arteries until the smallest type (arterioles) is reached. These feed into the capillaries, where oxygen is supplied to the cells and carbon dioxide is received from the cells. In turn, the capillaries join into venules, which become small veins, then larger veins, and ultimately form the superior and inferior vena cavae. The blood supply to the heart itself is from the aorta through the coronary arteries into a similar capillary system to the cardiac veins. This blood returns to the heart chambers by way of the coronary sinus as stated above.

Since continued reference has been made to the cardiovascular system in engineering terms, some numbers of interest should be mentioned. The heart beats at an average rate of about 75 beats per minute in a normal adult although this figure can vary considerably. The heart rate increases when a person stands up and decreases when he sits down, the range being from about 60 to 85. On the average, it is higher in women and generally decreases with age. In an infant, the heart rate may be as high as 140 beats per minute under normal conditions. The heart rate also increases with heat exposure and other physiological and psychological factors, which will be discussed later.

The heart pumps about 5 quarts of blood per minute, and since the volume of blood in the average adult is about 5 to 6 quarts, this corresponds to a complete turnover every minute during rest. With heavy exercise, the circulation rate is increased considerably. At any given time, about 75 to 80 percent of the blood volume is in the veins, about 20 percent in the arteries, and the remainder in the capillaries.

Systolic (maximum) blood pressure in the normal adult is in the range of 95 to 140 mm Hg, with 120 mm Hg being average. These figures are subject to much variation with age, climate, eating habits, and other factors. Normal diastolic blood pressure (lowest pressure between beats) ranges from 60 to 90 mm Hg, with 80 mm Hg being about average. This pressure is usually measured in the brachial artery in the arm. For comparison purposes with pressures of 130/75 in the aorta, 130/5 can be expected in the left ventricle, 9/5 in the left atrium, 25/0 in the right ventricle, 3/0 in the right atrium, and 25/12 in the pulmonary artery. These values are given as:

Systolic pressure/Diastolic pressure

5.2. THE HEART

The general behavior of the heart as the pump used to force the blood through the cardiovascular system has been discussed. Now a more detailed analysis of the anatomy of the heart, plus a discussion of the electrical excitation system necessary to produce and control the muscular contractions, should help to round out the background material needed for an understanding of cardiac dynamics. The electrocardiogram, as a record of biopotential events, has already been discussed in Chapter 3, but some repetition is necessary in order to consider the system as a whole and the relationships that exist between the electrical and mechanical events of the heart. Figure 5.3 is an illustration of the heart.

The heart is contained in the *pericardium,* a membranous sac consisting of an external layer of dense fibrous tissue and an inner serous layer that surrounds the heart directly. The base of the pericardium is attached to the central tendon of the diaphragm, and its cavity contains a thin serous liquid. The two sides of the heart are separated by the *septum,* or dividing wall of tissue. The septum also includes the *atrioventricular node* (AV node), which, as will be explained later, plays a role in the electrical conduction through the cardiac muscles.

Each of the four chambers of the heart is different from the others because of its functions. The *right atrium* is elongated and lies between the inferior (lower) and superior (upper) vena cava. Its interior is complex, the anterior (front) wall being very rough, whereas the posterior (rear) wall (which forms a part of the septum) and the remaining walls are smooth. At the junction of the right atrium and the superior vena cava is situated the *sinoatrial node* (SA node), which is the *pacemaker* or initiator of the electrical impulses that excite the heart. The *right ventricle* is situated below and to the left of the right atrium. They are separated by a fatty structure in which is contained the right branch of the coronary artery. This fatty separation is incapable of conducting electrical impulses; thus communication between the atria and the ventricles is accomplished only via the AV node and delay line.

Since the ventricle has to perform a pumping action, its walls are thicker than those of the atrium and its surfaces are ridged. Between the anterior wall of the ventricle and the septum is a muscular ridge that is part of the heart's electrical conduction system, known as the *bundle of His* (pronounced "hiss"). At the junction of the right and left atrium and the right ventricle on the septum is another node, the atrioventricular node. The bundle of His is attached to this node.

The right atrium and right ventricle are joined by a fibrous tissue known as the *atrioventricular ring,* to which are attached the three cusps of the tricuspid valve, which is the connecting valve between the two chambers.

The *left atrium* is smaller than the right atrium. Entry to it is through four pulmonary veins. The walls of the chamber are fairly smooth. It is joined to the left ventricle through the *mitral valve,* sometimes called the *bicuspid valve* since it consists of two cusps.

The *left ventricle* is considered the most important chamber, for this is the power pump for all the systemic circulation. Its walls are approximately three times as thick as the walls of the right ventricle because of this function. Conduction to the left ventricle is through the left bundle branch, which is in the ventricular muscle on the septum side.

As mentioned earlier, the outputs from the ventricles are through the aortic and pulmonary valves, respectively, for left and right ventricles.

Some aspects of the electrical activity of the heart have already been discussed in Section 3.3, but certain details will be elaborated on in the present context.

Excitation of the heart does not proceed directly from the central nervous system like most other muscle innervations but is initiated in the sinoatrial (SA) node, or pacemaker, a special group of excitable cells.

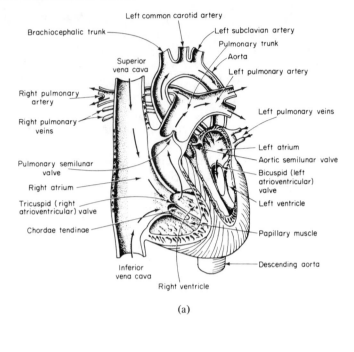

(a)

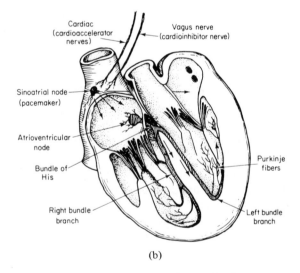

(b)

Figure 5.3. The heart: (a) internal structure; (b) conduction system. (From W.F. Evans, *Anatomy and Physiology, The Basic Principles,* Englewood Cliffs, N.J., Prentice-Hall, Inc., 1971, by permission.)

The electrical events that occur within the heart are reflected in the electro-cardiogram.

The SA node creates an impulse of electrical excitation that spreads across the right and left atria; the right atrium receives the earlier excitation due to its proximity to the SA node. This excitation causes the atria to contract and, a short time later, stimulates the atrioventricular (AV) node. The activated AV node, after a brief delay, initiates an impulse into the ventricles, through the bundle of His, and into the bundle branches that connect to the *"Purkinje fibers"* in the myocardium. The contraction re-sulting in the myocardium supplies the force to pump the blood into the circulatory systems.

The heart rate is controlled by the frequency at which the SA node gen-erates impulses. However, nerves of the sympathetic nervous system and the vagus nerve of the parasympathetic nervous system (See Chapter 10) cause the heart rate to quicken or slow down, respectively. Anatomically, the fibers of the sympathetic and vagus nerves enter the heart at the cardiac plexus under the arch of the aorta and are distributed quite profusely at and near both the SA and AV nodes. Vagal fibers are mostly found distributed in the atria, the bundle of His and branches, whereas the sympathetic fibers are found within the muscular walls of the atria and ventricles.

Although the effects of the sympathetic and vagus nerves are in oppo-sition to each other, if they both occur together in opposite directions the result is additive. That is, heart rate will increase from a combination of increased sympathetic activity concurrent with decreased vagal activity. The action of these nerves is called their *tone;* and by the various activities of each type of nerve, the rate of the heart, its coronary blood supply, and its contractability may be affected. The nerves affecting the rate of the heart in this way originate from the medullary centers in the brain and are con-trolled both by cardiac acceleration and by inhibition centers, each being sensitive to stimulation from higher centers of the brain. It is in this sequence that the heart rate is affected when a person is anxious, fright-ened, or excited. Heart attacks may be caused by this type of stimulation. The heart rate can also be affected, but in a more indirect way, by over-eating, respiration problems, extremes of body temperature, and blood changes.

One other effect that should be mentioned is that of the *pressoreceptors* or *baroreceptors* situated in the arch of the aorta and in the carotid sinus. Their function is to alter the vagal tone whenever the blood pressure within the aorta or cartoid sinus changes. When blood pressure rises, vagal tone is increased and the heart rate slows; when blood pressure falls, vagal tone is decreased and the heart rate increases.

A good enginering analog is illustrated in Figure 5.4, which shows the physiological system as a pump model. The pump is initially set to operate

CONTROL OF ARTERIAL BLOOD PRESSURE

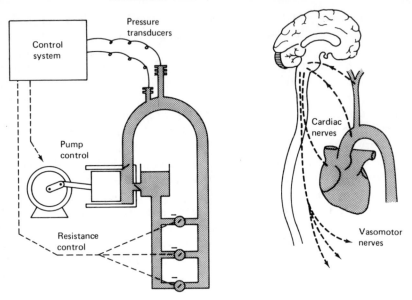

Figure 5.4. Control of Arterial Blood Pressure. (From R.F. Rushmer, *Cardiovascular Dynamics*, 3rd ed., W.B. Saunders Co., 1970, by permission.)

under predetermined conditions, as are the valves representing the resistance in the various organs. The pressure transducers sense the pressure continuously. With the pressure head set at some normal level, if one of the valves opens farther to obtain greater flow in that branch, the pressure head will decrease. This is picked up as a lower pressure by the sensors, which feed a signal to the controller, which, in turn, either closes other valves, speeds up the pump, or does both in order to try to maintain a constant pressure head.

5.3. BLOOD PRESSURE

In the arterial system of the body, the large pressure variations from systole to diastole are smoothed into a relatively steady flow through the peripheral vessels into the capillaries. This system, with some modifications, obeys the simple physical laws of hydrodynamics. As an analog, the potential (blood pressure) acting through the resistance of the arterial vascular pathways causes flow throughout the system. The resistance must not be so great as to impede flow, so that even the most remote capillaries receive sufficient blood and are able to return it into the venous system. On the other hand, the vessels of the system must be capable of damping out any large pressure fluctuations.

Since the system must be capable of maintaining an adequate pressure head while controlling flow, monitoring and feedback control loops are required. Demand on the system comes from various sources, such as from the gastrointestinal tract after a large meal or from the skeletal muscles during exercise. The result is vascular dilation at these particular points. If sufficient demands were to occur simultaneously so that increased blood flow were needed in many parts of the body, the blood pressure would drop. In this way, flow to the vital regions of heart and brain might be affected. Fortunately, however, the body is equipped with a monitoring system that can sense systemic arterial blood pressure and can compensate in the cardiovascular operations. Pressure is therefore maintained within a relatively narrow range, and the flow is kept within the normal range of the heart.

With regard to measurements, the events in the heart that relate to the blood pressure as a function of time should be understood. Figure 5.5 illustrates this point. The two basic stages of diastole and systole are shown with a more detailed time scale of phases of operation below. The blood pressure waves for the aorta, the left atrium, and the left ventricle are drawn to show time and magnitude relationships. Also, the correlated electrical events are shown at the bottom in the form of the electrocardiogram, and the basic relation of the heart sounds, which are discussed in Section 5.5, are shown in Figure 5.8.

Examining the aortic wave, it can be seen that, during systole, the ejection of blood from the left ventricle is rapid at first. As the rate of pressure change decreases, the rounded maximum of the curve is obtained. The peak aortic pressure during systole is a function of left ventricular stroke volume, the peak rate of ejection, and the distensibility of the walls of the aorta. In a diseased heart, ventricular contractability and rigid atherosclerotic arteries produce unwanted rises in blood pressure.

When the systolic period is completed the aortic valve is closed by the back pressure of blood (against the valve). This effect can be seen on the pressure pulse waveform as the *dicrotic notch*. When the valve is closed completely, the arterial pressure gradually decreases as blood pours into the countless peripheral vascular networks. The rate at which the pressure falls is determined by the pressure achieved during the systolic interval, the rate of outflow through the peripheral resistances, and the diastolic interval.

The form of the arterial pressure pulse changes as it passes through the arteries. The walls of the arteries cause damping and reflections; and as the arteries branch out into smaller arteries with smaller cross-sectional

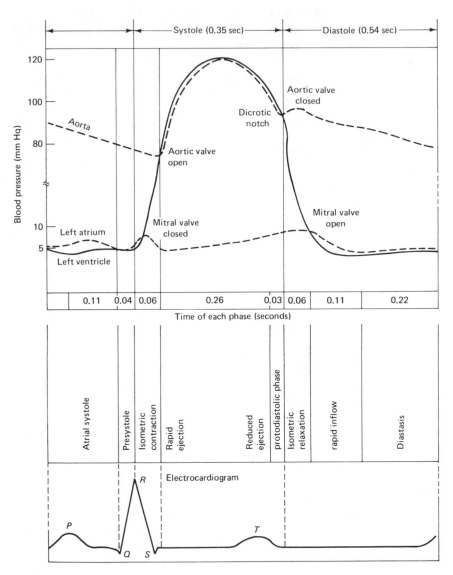

Figure 5.5. Blood pressure variations as a function of time.

areas, the pressures and volumes change, and hence the rate of flow also changes. The peak systolic pressure gets a little higher and the diastolic pressure flatter. The mean pressure in some arteries (e.g., the brachial artery) can be as much as 20 mm Hg higher than that in the aorta.

As the blood flows into the smaller arteries and arterioles, the pressure decreases and loses its oscillatory character. Pressure in the arterioles can vary from about 60 mm Hg down to 30 mm Hg. As the blood enters the venous system after flowing through the capillaries, the pressure is down to about 15 mm Hg.

In the venous system, the pressure in the venules decreases to approximately 8 mm Hg, and in the veins to about 5 mm Hg. In the vena cava, the pressure is only about 2 mm Hg. Because of these differences in pressure, measurement of arterial blood pressure is quite different from that of venous pressure. For example, a 2-mm Hg error in systolic pressure is only of the order of 1.5 percent. In a vein, however, this would be a 100-percent error. Also because of these pressure differences, the arteries have thick walls, while the veins have thin walls. Moreover, the veins have larger internal diameters. Since about 75 to 80 percent of the blood volume is contained in the venous system, the veins tend to serve as a reservoir for the body's blood supply.

A summary of the dimensions, blood flow velocities, and blood pressures at major points in the cardiovascular system is shown in Table 5.1. The figures are typical or average for comparative reference. The complete arterial and venous systems are shown in Figure 5.6.

TABLE 5.1. CARDIOVASCULAR SYSTEM—TYPICAL VALUES

Vessel	Number in Thousands	Diameter (mm)	Length (mm)	Mean Velocity (cm/sec)	Pressures (mm Hg)
Aorta	–	10.50	400	40.0	100
Terminal arteries	1.8	0.60	10	<10.0	40
Arterioles	40,000	0.02	2	0.5	40–25
Capillaries	$>$million	0.008	1	<0.1	25–12
Venules	80,000	0.03	2	<0.3	12–8
Terminal veins	1.8	1.50	100	1.0	<8
Vena cava	–	12.50	400	20.0	3–2

5.4. CHARACTERISTICS OF BLOOD FLOW

The blood flow at any point in the circulatory system is the volume of blood that passes that point during a unit of time. It is normally measured in milliliters per minute or liters per minute. Blood flow is highest in the pulmonary artery and the aorta where these blood vessels leave the heart. The flow at these points, called *cardiac output,* is between 3.5 and 5 liters per minute in a normal adult at rest. In the capillaries, on the other hand, the blood flow can be so slow that the travel of individual blood cells can be observed under a microscope.

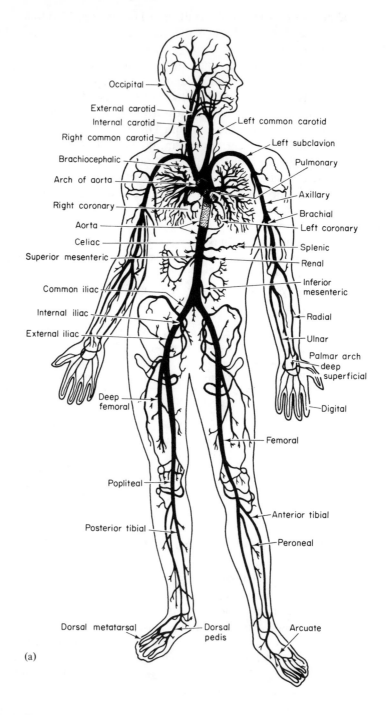

Occipital

External carotid

Internal carotid

Right common carotid

Left common carotid

Left subclavion

Brachiocephalic

Pulmonary

Arch of aorta

Right coronary

Axillary

Brachial

Aorta

Left coronary

Celiac

Splenic

Superior mesenteric

Renal

Common iliac

Inferior mesenteric

Internal iliac

Radial

External iliac

Ulnar

Palmar arch deep superficial

Deep femoral

Digital

Femoral

Popliteal

Anterior tibial

Posterior tibial

Peroneal

Dorsal metatarsal

Dorsal pedis

Arcuate

(a)

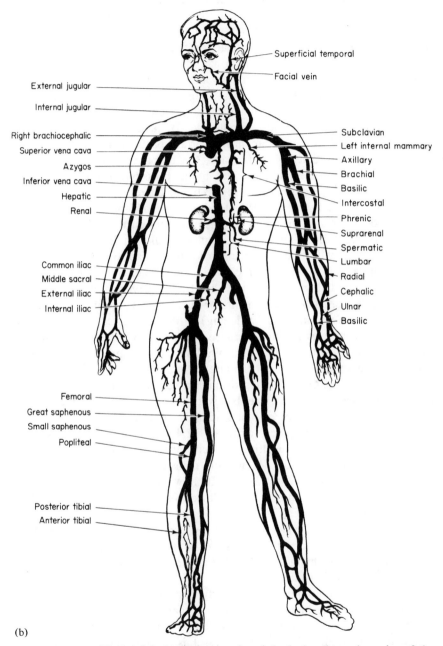

External jugular

Internal jugular

Right brachiocephalic

Superior vena cava

Azygos

Inferior vena cava

Hepatic

Renal

Common iliac

Middle sacral

External iliac

Internal iliac

Femoral

Great saphenous

Small saphenous

Popliteal

Posterior tibial

Anterior tibial

Superficial temporal

Facial vein

Subclavian

Left internal mammary

Axillary

Brachial

Basilic

Intercostal

Phrenic

Suprarenal

Spermatic

Lumbar

Radial

Cephalic

Ulnar

Basilic

(b)

Figure 5.6. (a) Major arteries of the body; (b) major veins of the body. (From W.F. Evans, *Anatomy and Physiology, The Basic Principles,* Englewood Cliffs, N.J., Prentice-Hall, Inc., 1971, by permission.)

From the cardiac output or the blood flow in a given vessel, a number of other characteristic variables can be calculated. The cardiac output divided by the number of heartbeats per minute gives the amount of blood that is ejected during each heartbeat, or the *stroke volume*. If the total amount of blood in circulation is known, and this volume is divided by the cardiac output, the *mean circulation time* is obtained. From the blood flow through a vessel, divided by the cross–sectional area of the vessel, the *mean velocity* of the blood at the point of measurement can be calculated.

In the arteries, blood flow is pulsatile. In fact, in some blood vessels a reversal of the flow can occur during certain parts of the heartbeat cycle. Because of the elasticity of their walls, the blood vessels tend to smooth out the pulsations of blood flow and blood pressure. Both pressure and flow are greatest in the aorta, where the blood leaves the heart.

Blood flow is a function of the blood pressure and of the flow resistance of the blood vessels in the same way as electrical current flow depends on voltage and resistance. The flow resistance of the capillary bed, however, can vary over a wide range. For instance, when exposed to low temperatures or under the influence of certain drugs (e.g., nicotine), the body reduces the blood flow through the skin by *vasoconstriction* (narrowing) of the capillaries. Heat, excitement, or local inflammation, among other things, can cause *vasodilation* (widening) of the capillaries, which increases the blood flow, at least locally. Because of the wide variations that are possible in the flow resistance, the determination of blood pressure alone is not sufficient to assess the status of the circulatory system.

The velocity of blood flowing through a vessel is not constant throughout the cross section of the vessel but is a function of the distance from the wall surfaces. A thin layer of blood actually adheres to the wall, resulting in zero velocity at this place, whereas the highest velocity occurs at the center of the vessel. The resulting "velocity profile" is shown in Figure 5.7. Some blood flow meters do not actually measure the blood flow but measure the mean velocity of the blood. If, however, the cross-sectional area of the blood vessel is known, these devices can be calibrated directly in terms of blood flow.

If the local blood velocity exceeds a certain limit, (as may happen when a blood vessel is constricted), small eddies can occur, and the *laminar flow* of Figure 5.7 changes to a *turbulent flow* pattern for which the flow rate is more difficult to determine.

The proper functioning of all body organs depends on an adequate blood supply. If the blood supply to an organ is reduced by a narrowing of the blood vessels, the function of that organ can be severely limited. When the blood flow in a certain vessel is completely obstructed (e.g., by a blood clot or *thrombus*), the tissue in the area supplied by this vessel may die. Such an obstruction in a blood vessel of the brain is one of the causes of the so-called *cerebrovascular accident (CVA)* or *stroke*. An ob-

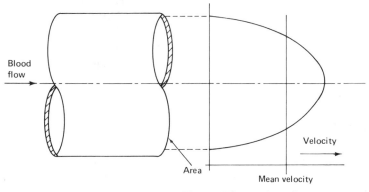

Blood
flow

Velocity

Area

Mean velocity

Figure 5.7. Laminar flow in a blood vessel.

struction of part of the coronary arteries that supply blood for the heart muscle is called a *myocardial (or coronary) infarct* or *heart attack,* whereas merely a reduced flow in the coronary vessels can cause a severe chest pain called *angina pectoris.* A blood clot in a vessel in the lung is called an *embolism.* Blood clots can also afflict the circulation in the lower extremities (*thrombosis*). Although the foregoing events afflict only a limited, though often vital, area of the body, the total circulatory system can also be affected. Such is the case if the cardiac output, the amount of blood pumped by the heart, is greatly reduced. This situation can be due to a mechanical malfunction such as a leaking or torn heart valve. It can also occur as *shock*—for example, after a severe injury when the body reacts with vasoconstriction of the capillaries, which reduces the blood loss but also prevents the blood from returning to the heart.

Most of these events have severe, and often fatal, results. Therefore it is of great interest to be able to determine the blood flow in such cases in order to provide an early diagnosis and begin treatment before irreparable tissue damage has occurred.

5.5. HEART SOUNDS

For centuries the medical profession has been aided in its diagnosis of certain types of heart disorders by the sounds and vibrations associated with the beating of the heart and the pumping of blood. The technique of listening to sounds produced by the organs and vessels of the body is called *auscultation,* and it is still in common use today. During his training the physician learns to recognize sounds or changes in sounds that he can associate with various types of disorders.

In spite of its widespread use, however, auscultation is rather subjective, and the amount of information that can be obtained by listening to the sounds of the heart depends largely on the skill, experience, and hearing ability of the physician. Thus different physicians may hear the same sounds differently, and perhaps interpret them differently.

The heart sounds heard by the physician through his stethoscope actually occur at the time of closure of major valves in the heart. This timing could easily lead to the false assumption that the sounds which are heard are primarily caused by the snapping together of the vanes of these valves. In reality, this snapping action produces almost no sound, because of the cushioning effect of the blood. The principal cause of heart sounds seems to be vibrations set up in the blood inside the heart by the sudden closure of the valves. These vibrations, together with eddy currents induced in the blood as it is forced through the closing valves, produce vibrations in the walls of the heart chambers and in the adjoining blood vessels.

With each heartbeat, the normal heart produces two distinct sounds that are audible in the stethoscope—often described as "lub-dub." The "lub" is caused by the closure of the *atrioventricular valves,* which permit flow of blood from the atria into the ventricles but prevent flow in the reverse direction. Normally this is called the *first heart sound,* and it occurs approximately at the time of the QRS complex of the electrocardiogram and just before ventricular systole. The "dub" part of the heart sounds is called the *second heart sound* and is caused by the closing of the *semilunar valves,* which release blood into the pulmonary and systemic circulation systems. These valves close at the end of systole, just before the atrioventricular valves reopen. This second heart sound occurs about the time of the end of the T wave of the electrocardiogram.

A *third heart sound* is sometimes heard, especially in young adults. This sound, which occurs from 0.1 to 0.2 second after the second heart sound, is attributed to the rush of blood from the atria into the ventricles, which causes turbulence and some vibration of the ventricular walls. This sound actually precedes atrial contraction, which means that the inrush of blood to the ventricles causing this sound is passive, pushed only by the venous pressure at the inlets to the atria. Actually, about 70 percent of blood flow into the ventricles occurs before atrial contraction.

A so-called *atrial heart sound,* which is not audible but may be visible on a graphic recording, occurs when the atria actually do contract, squeezing the remainder of the blood into the ventricles. The inaudibility of this heart sound is a result of low amplitude and low frequency of the vibrations.

Figure 5.8 shows the time relationships between the first, second, and third heart sounds with respect to the electrocardiogram, and the various pressure waveforms. Opening and closing times of valves are also shown. This figure should also be compared with Figure 5.5.

In abnormal hearts additional sounds, called *murmurs,* are heard be-

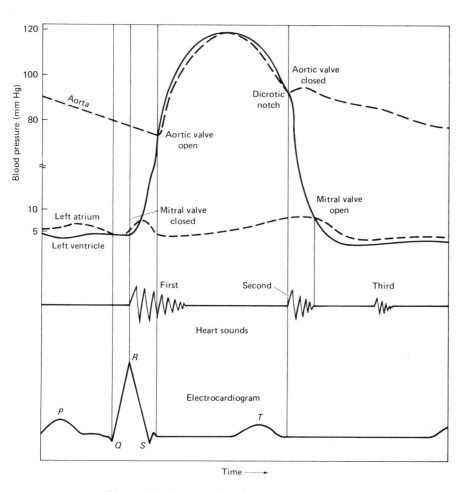

Figure 5.8. Relationship of heart sounds to function of the cardio-vascular system.

tween the normal heart sounds. Murmurs are generally caused either by improper opening of the valves (which requires the blood to be forced through a small aperture) or by regurgitation, which results when the valves do not close completely and allow some backward flow of blood. In either case, the sound is due to high-velocity blood flow through a small opening. Another cause of murmurs can be a small opening in the septum which separates the left and right sides of the heart. In this case, pressure differences between the two sides of the heart force blood through the opening, usually from the left ventricle into the right ventricle, bypassing the systemic circulation.

Normal heart sounds are quite short in duration, approximately one-tenth of a second for each, while murmurs usually extend between the normal sounds. Figure 5.9 shows a record of normal heart sounds and several types of murmurs.

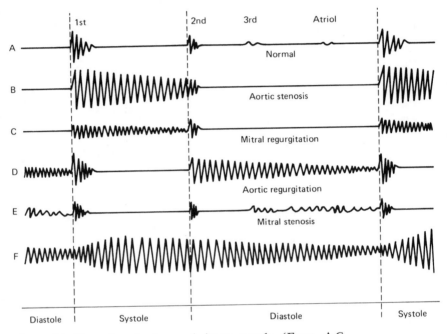

Figure 5.9. Normal and abnormal heart sounds. (From A.C. Guyton, *Textbook of Medical Physiology,* 4th ed., W.B. Saunders Co., 1971, by permission.)

There is also a difference in frequency range between normal and abnormal heart sounds. The first heart sound is composed primarily of energy in the 30– to 45–Hz range, with much of the sound below the threshold of audibility. The second heart sound is usually higher in pitch than the first, with maximum energy in the 50– to 70–Hz range. The third heart sound is an extremely weak vibration, with most of its energy at or below 30 Hz. Murmurs, on the other hand, often produce much higher pitched sounds. One particular type of regurgitation, for example, causes a murmur in the 100– to 600–Hz range.

Although auscultation is still the principal method of detecting and analyzing heart sounds, other techniques are often employed. For example, a graphic recording of heart sounds, such as shown in Figure 5.8, is called a *phonocardiogram*. Even though the phonocardiogram is a graphic record like the electrocardiogram, it extends to a much higher frequency range.

An entirely different waveform is produced by the vibrations of the heart against the thoracic cavity. The vibrations of the side of the heart as it thumps against the chest wall form the *vibrocardiogram,* whereas the tip or apex of the heart hitting the rib cage produces the *apex cardiogram.*

Sounds and pulsations can also be detected and measured at various locations in the systemic arterial circulation system where major arteries approach the surface of the body. The most common one is the *pulse,* which can be felt with the fingertips at certain points on major arteries. The waveform of this pulse can also be measured and recorded. In addition, when an artery is partially occluded so that the blood velocity through the constriction is increased sufficiently, identifiable sounds can be heard downstream through a stethoscope. These sounds, called the *Korotkoff sounds,* are used in the common method of blood pressure measurements and are discussed in detail in Chapter 6.

Another cardiovascular measurement worthy of note is the *ballisto-cardiogram.* Although not a heart sound or vibration measurement of the type described earlier, the ballistocardiogram is related to these measures in that it is a direct result of the dynamic forces of the heart as it beats and pumps blood into the major arteries. The beating heart exerts certain forces on the body as it goes through its sequence of motions. As in any situation involving forces, the body responds, but because of the greater mass of the body, these responses are generally not noticeable. However, when a person is placed on a platform that is free to move with these small dynamic responses, the motions of the body due to the beating of the heart and the corresponding blood ejection can be measured and recorded to produce the ballistocardiogram. Like the vibrocardiogram and the apex cardiogram, the ballistocardiogram provides information about the heart that cannot be obtained by any other measurement.

CARDIOVASCULAR
MEASUREMENTS

Having been given the overall concepts of the biomedical instrumentation system, the sources of bioelectric potentials, electrodes, and the basic concepts of the cardiovascular system, the reader should now have the background needed to understand practical systems for these measurements.

It is not by accident that the cardiovascular system has been chosen as the first of the major physiological measurement groupings to be studied. Instrumentation for obtaining measurements from this system has contributed greatly to advances in medical diagnosis.

Since such instrumentation includes devices to measure various types of physiological variables, such as electrical, mechanical, thermal, fluid, and auditory, this chapter is intended to provide a basis for studying all types of biomedical instrumentation. Each type of measurement is considered separately, beginning with the measurement of biopotentials that result in the electrocardiogram. Then the various methods, both direct and indirect, of measuring blood pressure, blood flow, cardiac output, and blood volume (plethysmography) are discussed. The final section is concerned with the measurement of heart sounds and vibrations.

6.1. ELECTROCARDIOGRAPHY

The *electrocardiogram* (ECG or EKG) is a graphic recording or display of the time-variant voltages produced by the myocardium during the cardiac cycle. Figure 6.1 shows the basic waveform of the nor-

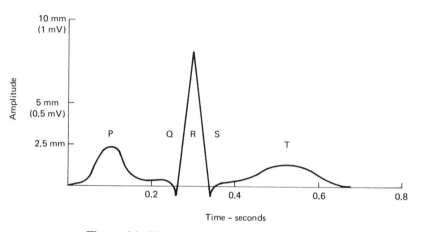

Figure 6.1. The electrocardiogram in detail.

mal electrocardiogram. The P, QRS, and T waves reflect the rhythmic electrical depolarization and repolarization of the myocardium associated with the contractions of the atria and ventricles. The electrocardiogram is used clinically in diagnosing various diseases and conditions associated with the heart. It also serves as a timing reference for other measurements.

A discussion of the ECG waveform has already been presented in Section 3.2, and will not be repeated here, except in the concept of measurement details. To the clinician, the shape and duration of each feature of the ECG are significant. The waveform, however, depends greatly upon the lead configuration used, as discussed below. In general, the cardiologist looks critically at the various time intervals, polarities, and amplitudes in order to arrive at his diagnosis.

Some normal values for amplitudes and durations of important ECG parameters are as follows:

Amplitudes:	P	wave	0.25 mV
	R	wave	1.60 mV
	Q	wave	25% of R wave
	T	wave	0.1 to 0.5 mV

Durations:

P-R	interval	0.12 to 0.20 second
Q-T	interval	0.35 to 0.44 second
S-T	segment	0.05 to 0.15 second
P	wave interval	0.11 second
QRS	interval	0.09 second

For his diagnosis, a cardiologist would typically look first at the heart rate. The normal value lies in the range of 60 to 100 beats per minute. A slower rate than this is called *bradycardia* (slow heart) and a higher rate, *tachycardia* (fast heart). He would then see if the cycles are evenly spaced. If not, an *arrhythmia* may be indicated. If the P-R interval is greater than 0.2 second, it can suggest blockage of the AV node. If one or more of the basic features of the ECG should be missing, a heart block of some sort might be indicated.

In healthy individuals the electrocardiogram remains reasonably constant, even though the heart rate changes with the demands of the body. It should be noted that the position of the heart within the thoracic region of the body, as well as the position of the body itself (whether erect or recumbent), influences the "electrical axis" of the heart. The *electrical axis* (which parallels the anatomical axis) is defined as the line along which the greatest electromotive force is developed at a given instant during the cardiac cycle. The electrical axis shifts continually through a repeatable pattern during every cardiac cycle.

Under pathological conditions, several changes may occur in the ECG. These include (a) altered paths of excitation in the heart, (b) changed origin of waves (ectopic beats), (c) altered relationships (sequences) of features, (d) changed magnitudes of one or more features, and (e) differing durations of waves or intervals.

As mentioned earlier, an instrument used to obtain and record the electrocardiogram is called an *electrocardiograph*. A brief history of electrocardiography provides a very interesting view of the development of instrumentation without the aid of electronic amplification.

Because the ECG voltages to be recorded are very small, sensitive photographic recorders had to be used before suitable electronic amplifiers became available. The first electrocardiogram was recorded in 1887 by Waller, who used a device called a *capillary electrometer,* introduced by Lippman in 1875. This device consisted of a mercury-filled glass capillary immersed in dilute sulfuric acid. The position of the meniscus which formed the dividing line between the two fluids changed when an electrical voltage was applied between mercury and acid. This movement was very small, but it could be recorded on a moving piece of light sensitive paper or film with the help of a magnifying optical projection system. The capillary electrometer, however, was cumbersome to operate and the inertia of the mercury column limited its frequency range.

The string galvanometer, which was introduced to electrocardiography by Einthoven in 1903, was a considerable improvement. It consisted of an extremely thin platinum wire or a gold-plated quartz fiber, about 5 microns thick, suspended in the air gap of a strong electromagnet. An electrical current flowing through the string caused movement of the string perpendicular to the direction of the magnetic field. The magnitude of the movement was small, but could be magnified several hundred times by an optical projection system for recording on a moving film or paper. The small mass of the moving fiber resulted in a frequency response sufficiently high for the faithful recording of the electrocardiogram. The sensitivity of the galvanometer could be adjusted by changing the mechanical tension on the string. To measure the sensitivity of the galvanometer, a standardization switch allowed a calibration voltage of 1 millivolt to be connected to the galvanometer terminals. Modern electrocardiographs, although they have a calibrated sensitivity, still retain this feature. The string galvanometer had dc response, and a difference in the contact potentials of the electrodes could easily drive the string off scale. A compensation voltage, adjustable in magnitude and polarity, was provided to center the shadow of the string on a ground glass screen prior to recording the electrocardiogram. In order to facilitate measurement of the time differences between the characteristic parts of the ECG waveform, time marks were provided on the film by a wheel with five spokes driven by a constant speed motor.

String-galvanometer electrocardiographs like the one shown in Figure 4.4 were used until about 1920 when they were replaced by devices incorporating electronic amplification. This allowed the use of less sensitive and more rugged recording devices. Early ECG machines incorporating amplification used the Dudell oscillograph as a recorder. The Dudell oscillograph was similar in design to the string galvanometer, but had the single string replaced by a hairpin shaped wire stretched between two fixed terminals and a spring-loaded support pulley. A small mirror cemented across the two legs of the hairpin wire was rotated when a current (the amplified ECG signal) flowed through the wire. The mirror was used to deflect a narrow light beam, throwing a small light spot on a moving film. While recording systems of this type are mechanically more rugged than the fragile string galvanometers, they still require photosensitive film or paper which has to be processed before the electrocardiogram can be read.

This and other disadvantages were overcome with the introduction of direct-writing recorders which use either ink, the transfer of pigments from a ribbon, or a special heat sensitive paper to make the recording immediately visible. Basically, the pen motor of such a recorder has a meter movement with a writing tip at the end of the indicator. Because this type of indicator naturally moves in a circular path, special measures are required to convert this motion to a straight line when a *rectilinear* rather than a *curvilinear* recording is desired.

The higher mass of moving parts used in direct-writing pen motors makes their frequency response inherently inferior to that of optical recording systems. Despite this handicap, modern direct-writing electrocardiographs have a frequency range extending to over 100 Hz, which is completely adequate for clinical ECG recordings. An improvement in performance over that of older direct-writing recorders can be partially attributed to the use of servo techniques in which the actual position of the pen is electrically sensed and the pen motor is included as part of a servo loop. For these reasons optical recording methods are seldom used in modern electrocardiographs.

The early string galvanometer had the advantage that it could easily be isolated from ground. Thus the potential difference between two electrodes on the patient could be measured with less electrical interference than can be done with a grounded system. Electronic amplifiers, however, are normally referenced to ground through their power supplies. This creates an interference problem (unless special measures are taken) when such amplifiers are used to measure small bioelectric potentials. The technique usually employed, not only in electrocardiography but also in the measurement of other bioelectric signals, is the use of a differential amplifier. The principle of the differential amplifier can be explained with the help of Figure 6.2.

A differential amplifier can be considered as two amplifiers with separate inputs (Figure 6.2a), but with a common output terminal which delivers the sum of the two amplifier output voltages. Both amplifiers have the same voltage gain, but one amplifier is inverting (output voltage is 180 degrees out of phase with respect to the input) while the other is noninverting (input and output voltages are in phase). If the two amplifier inputs are connected to the same input source, the resulting *common-mode gain* should be zero because the signals from the inverting and the noninverting amplifiers cancel each other at the common output. However, because the gain of the two amplifiers is not exactly equal, this cancellation is not complete. Rather, a small residual common-mode output remains. When one of the amplifier inputs is grounded and a voltage is applied only to the other amplifier input, the input voltage appears at the output amplified by the gain of the amplifier. This gain is called the *differential gain* of the differential amplifier. The ratio of the the differential gain to the common-mode gain is called the *common-mode rejection ratio* of the differential amplifier, which in modern amplifiers can be as high as 1,000,000:1.

When a differential amplifier is used to measure bioelectric signals that occur as a potential difference between two electrodes, as shown in Figure 6.2(b), the bioelectric signals are applied between the inverting and noninverting inputs of the amplifier. The signal is therefore amplified by the differential gain of the amplifier. For the interference signal, however, both inputs appear as though they were connected together to a common input

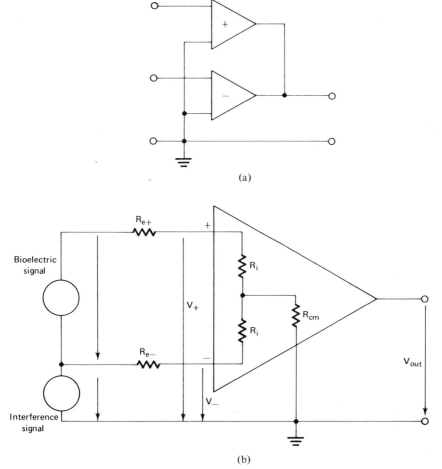

(a)

(b)

Figure 6.2. The Differential Amplifier: (a) represented as two amplifiers with separate inputs and common output; (b) as used for amplification of bioelectric signals (see text for explanation).

source. Thus, the common-mode interference signal is amplified only by the much smaller common-mode gain.

Figure 6.2(b) also illustrates another interesting point: The electrode impedances, R_{e+} and R_{e-}, each form a voltage divider with the input impedance of the differential amplifier. If the electrode impedances are not identical, the interference signals at the inverting and noninverting inputs of the differential amplifier may be different, and the desired degree of cancellation does not take place. Because the electrode impedances can never be made exactly equal, the high common-mode rejection ratio of a differential amplifier can only be realized if the amplifier has an input impedance much higher than the impedance of the electrodes to which it is connected. As indicated in the figure, this input impedance may not be the same for the differential signal as it is for the common-mode signal. The use of a differential amplifier also requires a third connection for the reference or ground input. Since the introduction of differential amplifiers into electrocardiography, an electrode attached to the *right leg* of the patient has been used for this purpose.

In electrocardiography, the amplitudes, polarities and even the timing and durations of the various features of the ECG are dependent to a large extent upon the location of the electrodes on the body. The standard locations for clinical electrode placement are on the left and right arms near the wrists, the left leg near the ankle, and several locations on the chest, called the *precordial positions.* In addition, a reference or ground electrode is generally placed on the right leg near the ankle, as indicated above. Each set of electrode locations from which the ECG is measured is called a *lead.* The standard lead configurations used in clinical cardiology are shown in Figure 6.3.

The three basic *limb leads,* which were originally established by Einthoven, are given as follows:

Lead I   Left Arm (LA) (Positive) to Right Arm (RA) (Negative)
Lead II   Left Leg (LL) (Positive) to Right Arm (RA) (Negative)
Lead III   Left Leg (LL) (Positive) to Left Arm (LA) (Negative).

In each of these lead positions, the QRS of a normal heart is such that the R wave is positive.

In working with electrocardiograms from these three basic limb leads, Einthoven postulated that at any given instant of the cardiac cycle, the frontal plane representation of the electrical axis of the heart is a two-dimensional vector. Further, the ECG measured from any one of the three basic limb leads is a time-variant single dimensional component of that vector. Einthoven also made the assumption that the heart (the origin of the vector) is near the center of an equilateral triangle, the apexes of which are the right and left shoulder and the crotch. By assuming that the ECG potentials at the shoulders are essentially the same as the wrists and that the potentials at the crotch differ little from those at either ankle, he let the points of this triangle represent the electrode positions for the three limb leads. This triangle, known as the *Einthoven Triangle* is shown in Figure 6.4.

The sides of the triangle represent the lines along which the three projections of the ECG vector are measured. Based on this, Einthoven showed that the instantaneous voltage measured from any one of the three limb lead positions is approximately equal to the algebraic sum of the other two, or that the vector sum of the projections on all three lines is equal to zero. For these statements to actually hold true, the polarity of the

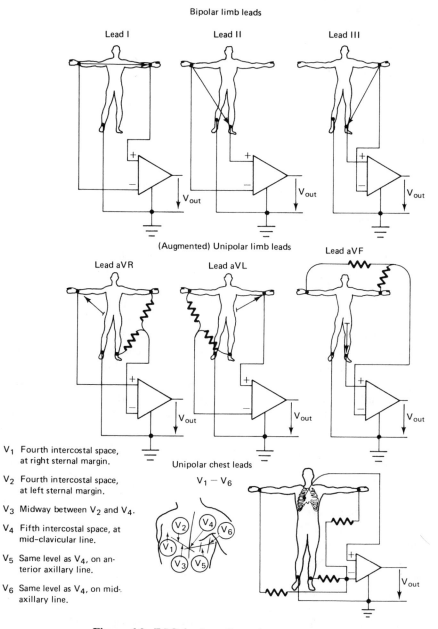

Figure 6.3. ECG lead configurations.

Lead II measurement must be reversed.

Of the three limb leads, Lead II produces the greatest R wave poten-

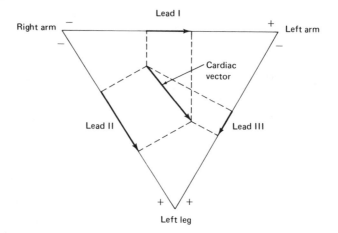

Figure 6.4. The Einthoven Triangle.

tial. Thus, when the amplitudes of the three limb leads are measured, the R wave amplitude of Lead II is equal to the sum of the R wave amplitudes of Leads I and III.

In addition to the three basic limb leads, a number of other leads are used routinely in clinical cardiology. These are also shown in Figure 6.3.

The basic limb leads are all *bipolar,* since each measurement is taken between two specific electrodes. A slightly different set of waveforms is obtained by measuring the potentials at each of the limb electrodes with respect to an *indifferent electrode* which is assumed to represent an average of the potentials of the three limb leads. This average can be approximated at a central terminal connected to each of the three limb electrodes through equal resistances. The resistance from the central terminal to any electrode is usually 5000 ohms. Measurements with respect to this terminal bear the designation *V* for "voltage." A *unipolar* measurement can then be made from any of the three limb electrodes or from an *exploring electrode* which can be placed at various points on the chest. The unipolar measurement from the left arm to the indifferent electrode is designated as Lead *VL,* while measurements from the right arm and left leg are called *VR* and *VF,* respectively (*F* stands for "foot").

Because of the loading effect of the resistance network required to provide the central terminal, the ECG potentials measured at any one of the three limb electrodes with respect to the central terminal are inconveniently small. To offset this difficulty, it was found that by disconnecting the measuring electrode from the network, the measured voltage can be increased by 50% without changing the waveform significantly. Unipolar leads obtained in this manner are called *augmented unipolar limb leads.*

They are designated *aVL, aVR* and *aVF*. Measurement of these leads is shown in the center row of diagrams in Figure 6.3. In reality, each of the augmented unipolar limb leads provides the voltage at one limb electrode with respect to the average of the potentials at the other two electrodes.

Twelve leads comprise the standard lead configuration used in clinical electrocardiography. These include the three standard bipolar limb leads (Leads I, II, and III), the three augmented unipolar limb leads (Leads *aVL, aVR* and *aVF*) and six positions on the chest with respect to a central terminal. The central terminal is the same as that described for the unipolar limb leads wherein the three limb electrodes are each connected to the central terminal through a 5000-ohm resistor. A single chest electrode (exploring electrode) is sequentially placed on each of the six predesignated points on the chest. These chest positions are called the *precordial unipolar* leads and are designated V_1 through V_6. These leads are diagrammed in the lower part of Figure 6.3.

In most clinical electrocardiographs, the twelve leads are measured one at a time and are selected by means of a selector switch on the instrument. Figure 6.5 shows a typical set of electrocardiogram readings as taken in a hospital. For convenience in comparing the various leads, the recordings taken from the twelve leads are mounted on a single sheet in a standard arrangement. This allows the cardiologist to analyze the complete set at one time.

As indicated above, the electrical axis of the heart is constantly in motion and its behavior can be represented as a three-dimensional time-variant vector. The bipolar limb leads and the augmented unipolar limb leads are essentially limited to measurements in the frontal plane. While the precordial leads do include some effect of the third dimension, they do not provide a true representation of the electrical activity in the front-to-back (saggital) direction that could be used in a three-dimensional analysis. When three-dimensional representation of the ECG is required, a special lead system is required in which electrodes are placed on the arms, one leg, the center of the chest and the center of the back. With such an arrangement, a set of *orthogonal* measurements can be obtained.

The building blocks of a modern electrocardiograph are shown on Figure 6.6(a), which also shows the controls that are typical for such an instrument. The patient is connected to the instrument via a detachable patient cable which ends in five lead wires. These lead wires are usually color coded and are attached to plate or floating electrodes for the four extremities and a suction cup electrode which serves as the exploratory electrode for recording the precordial leads. In the electrocardiograph, the wire from the right leg is connected to the chassis of the amplifier while the other four wires terminate at the *lead selector switch*. The lead selector

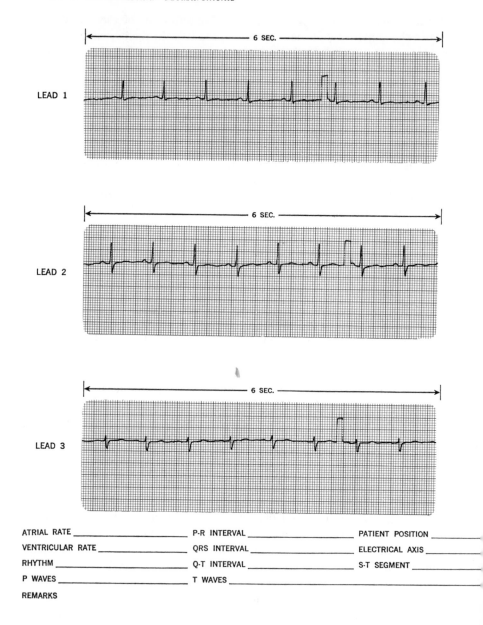

Figure 6.5. Typical patient ECG.

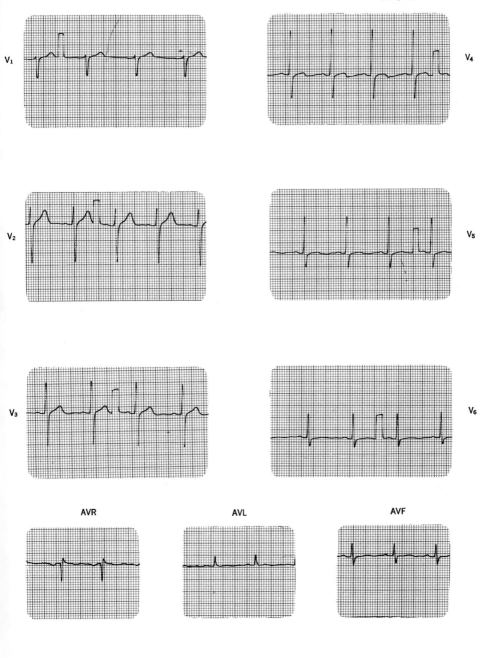

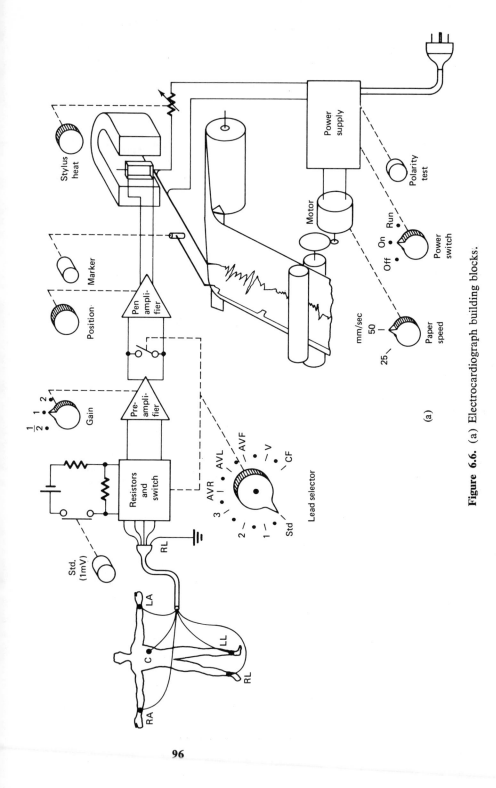

Figure 6.6. (a) Electrocardiograph building blocks.

switch circuit not only permits selection of the desired lead, but also includes the resistor combinations necessary for recording the augmented unipolar limb leads and the precordial leads.

By placing the lead selector switch in the appropriate position, the proper wires from the patient electrodes are selected and interconnected to allow the recording of any one of the 12 standard lead configurations. It is necessary, however, to change the position of the exploratory electrode between recordings when measuring the six precordial leads. The lead selector switch also contains a contact which blanks (shorts) the amplifier between successive lead positions to prevent erratic pen excursions during switching. Some types of electrocardiographs also stop the paper drive motor each time the selector is switched and restart it after a certain time delay to allow switching transients to disappear.

The first position of the lead selector switch shown in Figure 6.6(a) is marked STD. In this position, the pressing of a button marked STD (1MV) connects a 1mV dc voltage to the input of the amplifier to allow calibration of the overall gain of the instrument. Customarily, the gain is set so that the 1mV signal causes a pen deflection of 1 cm. Although modern electrocardiographs are not likely to show appreciable changes in gain, this *standardization* is usually still performed for every patient. The selected lead configuration is connected to the preamplifier, which has a differential input and is usually an ac amplifier. The preamplifier includes a switch to increase or decrease its gain by a factor of two to accommodate unusually large or small ECG signals. It also provides for a continuous gain adjustment, often accessible only by screwdriver, for the standardization of the overall gain.

The preamplifier drives a pen amplifier which is usually a dc type. The input to the pen amplifier is sometimes accessible through a separate connector to allow the electrocardiograph to be used as a graphic recorder with other medical instruments. The pen amplifier also contains a control marked POSITION which introduces an offset voltage to center the pen on the paper.

The *pen motor* is similar in design to a galvanometer or d'Arsonval meter, but is built much more ruggedly so it can transmit the necessary acceleration forces to the stylus. In modern design, the pen motor incorporates a transducer which provides a voltage proportional to either the position of the stylus or its velocity. This voltage is used as a feedback signal in the pen amplifier to improve the frequency response of the recorder.

When a heated stylus is used for the recording, a STYLUS HEAT control allows the width and darkness of the recorded trace to be varied. A marker stylus is often used to record markings on the electrocardiogram, identifying the lead configurations with a dash-dot code.

The recording paper is driven by a synchronous motor. A switch allows selection of a standard paper speed of 25 mm/sec or an alternate speed of 50 mm/sec when a very high heart rate is encountered.

The power switch of the modern electrocardiograph has an ON or STANDBY position in which only the amplifier is turned on while the paper motor and stylus heat are disconnected, and a RUN or RECORD position. In all older ECG machines, the power supply also has provisions for a *line polarity test,* which may consist of a neon lamp or a touch plate, sometimes in conjunction with a switch for polarity reversal. In such instruments, the chassis of the amplifier and the case of the machine are connected to the neutral conductor of the line cord, usually through a resistor of at least 100 kilohm. The purpose of this arrangement is to reduce line interference. In case the polarity of the line cord is reversed, by either miswiring a polarized receptacle or reversing a nonpolarized (two-pronged) power plug in the receptacle, severe ac interference results. Reversal of line cord polarity may also inject a current into the RL lead. The line polarity test, which is to be performed before a patient is connected, is intended to verify the correct polarity of the power line. Because of hazards which may result when this test is not performed conscientiously, modern ECG machines use less dangerous methods to reduce the line interference.

While Figure 6.6(a) shows the principal building blocks of an electrocardiograph, it does not show the circuit details that are found in modern devices of this type. Some of these features are shown in Figure 6.6(b). In order to increase the input impedance and thus reduce the effect of variations in electrode impedance, these instruments usually include a *buffer amplifier* for each patient lead. The transistors in these amplifiers are often protected by a network of resistors and neon lamps from overvoltages that may occur when the electrocardiograph is used during surgery in conjunction with high frequency devices for cutting and coagulation.

A more severe problem is the protection of the electrocardiograph from damage during defibrillation. The voltages that may be encountered in this case can reach several thousand volts. Thus special measures must be incorporated into the electrocardiograph to prevent burnout of components and provide fast recovery of the trace so as to permit the success of the countershock to be judged.

Some modern devices do not connect the right leg of the patient to the chassis, but utilize a so called "driven right leg lead." This involves a summing network to obtain the sum of the voltages from all other electrodes and a driving amplifier, the output of which is connected to the right leg of the patient. The effect of this arrangement is to force the reference connection at the right leg of the patient to assume a voltage equal to the sum of the voltages at the other leads. This arrangement increases the common-

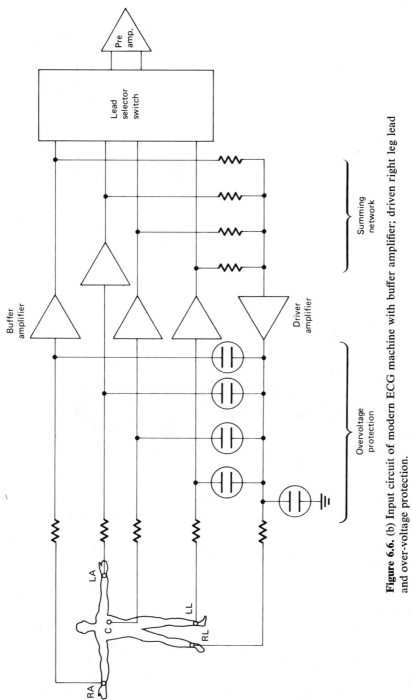

Figure 6.6. (b) Input circuit of modern ECG machine with buffer amplifier; driven right leg lead and over-voltage protection.

99

mode rejection ratio of the overall system and reduces interference. It also has the effect of reducing the current flow in the right leg electrode. Increased concern for the safety aspect of electrical connections to the patient have caused modern ECG designs to abandon the principle of a ground reference altogether and use isolated or floating input amplifiers as described in Chapter 16.

A modern electrocardiograph for clinical use is shown in Figure 6.6(c).

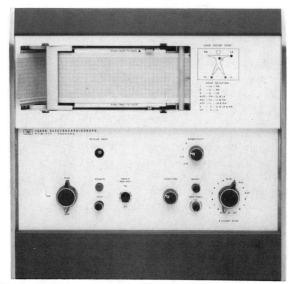

Figure 6.6. (c) An Electrocardiograph (Courtesy of Hewlett-Packard Company, Waltham, Mass.)

An electrocardiogram obtained with such a machine is a strip of paper several feet long. To convert this inconvenient output into a record suitable for interpretation and filing, such as the one shown in Figure 6.5, the strip has to be cut into pieces and pasted on a special form. Much of this time-consuming work is avoided in modern three-channel ECG machines. These devices record three leads simultaneously, switching automatically to the next group of three. A standard 12-lead electrocardiographic recording, including standardization, can be obtained in 10 seconds, ready for attachment to a file card.

With certain electrode configurations, the three leads that are recorded simultaneously on a three-channel recorder can be interpreted as three components of the cardiac vector recorded over time. In a *vectorcardiograph,* two of these vector components are shown in an X–Y display to obtain a presentation of the locus of the cardiac vector as projected on either the frontal, horizontal, or sagittal plane. When electrode positions corresponding to the three orthogonal planes are not directly accessible, a summation of voltages from several electrodes is sometimes used. The lead system most commonly employed for this purpose is called the *Frank Lead System*. Because of the frequency range necessary, a cathode-ray tube is normally used for the X–Y display, with timing marks introduced by blanking the beam. The vectorcardiogram consists of three loops, which correspond to the P, QRS, and T wave of the electrocardiogram. These three loops have one point (the "isoelectric" point, where all vector components are zero) in common. A recording of the vectorcardiogram can be obtained with an oscilloscope camera, the shutter of which is kept open during one complete cardiac cycle, or in some cases during a desired portion of the cycle.

In addition to its diagnostic function, the electrocardiogram is also a useful source for determining the heart rate. Since the electrocardiogram is recorded at a fixed rate on a chart or displayed on an oscilloscope with a fixed time scale, the heart rate can be determined directly from the ECG itself. Most monitoring devices installed in coronary care units also provide a lamp that flashes with each heartbeat to permit rapid estimation of heart rate. In the operating room, where the surgical staff may not be able to watch a light, an audible "beep" is often provided instead.

Where a direct readout of heart rate is desired, the electrocardiogram is used to trigger another instrument, called a *cardiotachometer*, which produces either a meter or digital readout of the heart rate or a chart recording that shows variations of heart rate over time. The heart rate indicated on a cardiotachometer is normally the average value measured over several heartbeats. In a *beat-to-beat cardiotachometer*, however, the heart rate is determined by measuring the interval between each pair of successive heartbeats. A recording from this device is capable of showing individual cardiac arrhythmias.

6.2. MEASUREMENT OF BLOOD PRESSURE

As one of the physiological variables that can be quite readily measured, blood pressure is considered a good indicator of the status of the cardiovascular system. A history of blood pressure measurements has saved many a person from an untimely death by providing warnings of dangerously high blood pressure (*hypertension*) in time to provide treatment.

In routine clinical tests, blood pressure is usually measured by means of an indirect method using a *sphygmomanometer* (from the Greek word, *sphygmos*, meaning pulse). This method is easy to use and can be automated. It has, however, certain disadvantages in that it does not provide a continuous recording of pressure variations and its practical repetition rate is limited. Furthermore, only systolic and diastolic arterial pressure readings can be obtained, with no indication of the details of the pressure waveform. The indirect method is also somewhat subjective, and often fails when the blood pressure is very low (as would be the case when a patient is in shock).

Methods for direct blood pressure measurement, on the other hand, do provide a continuous readout or recording of the blood pressure waveform and are considerably more accurate than the indirect method. They require, however, that a blood vessel be punctured in order to introduce the sensor. This limits their use to those cases in which the condition of the patient warrants invasion of the vascular system.

6.2.1. INDIRECT MEASUREMENTS. As stated earlier, the familiar indirect method of measuring blood pressure involves the use of a sphygmomanometer and a stethoscope. The sphygmomanometer consists of an inflatable pressure cuff and a mercury or aneroid manometer to measure the pressure in the cuff. The cuff consists of a rubber bladder inside an inelastic fabric covering that can be wrapped around the upper arm and fastened with either hooks or a Velcro® fastener. The cuff is normally inflated manually with a rubber bulb and deflated slowly through a needle valve. The stethoscope is described in detail in Section 6.5. A wall-mounted sphygmomanometer is shown in Figure 6.7. These devices are also manufactured as portable units.

Figure 6.7. Wall-mounted sphygmomanometer. (Courtesy of W.A. Baum, Inc., Copiague, N.Y.)

The sphygmomanometer works on the principle that when the cuff is placed on the upper arm and inflated, arterial blood can flow past the cuff only when the arterial pressure exceeds the pressure in the cuff. Further-more, when the cuff is inflated to a pressure that only partially occludes the brachial artery, turbulence is generated in the blood as it spurts through the tiny arterial opening during each systole. The sounds generated by this turbulence, called *Korotkoff* sounds, can be heard through a stetho-scope placed over the artery downstream from the cuff.

In order to obtain a blood pressure measurement with a sphygmoma-nometer and a stethoscope, the pressure cuff on the upper arm is first in-flated to a pressure well above systolic pressure. At this point no sounds can be heard through the stethoscope, which is placed over the brachial artery, for that artery has been collapsed by the pressure of the cuff. The pressure in the cuff is then gradually reduced. As soon as cuff pressure falls below systolic pressure, small amounts of blood spurt past the cuff and Korotkoff sounds begin to be heard through the stethoscope. The pressure of the cuff that is indicated on the manometer when the first Korotkoff sound is heard is recorded as the systolic blood pressure.

As the pressure in the cuff continues to drop, the Korotkoff sounds continue until the cuff pressure is no longer sufficient to occlude the vessel during any part of the cycle. Below this pressure the Korotkoff sounds dis-appear, marking the value of the diastolic pressure.

This familiar method of locating the systolic and diastolic pressure values by listening to the Korotkoff sounds is called the *auscultatory method* of sphygmomanometry. An alternate method, called the *palpatory method,* is similar except that the physician identifies the flow of blood in the artery by feeling the pulse of the patient downstream from the cuff instead of listening for the Korotkoff sounds. Although systolic pressure can easily be measured by the palpatory method, diastolic pressure is much more diffi-cult to identify. For this reason, the auscultatory method is more commonly used. Figure 6.8 shows a blood pressure measurement using the ausculta-tory method.

Because of the trauma imposed by direct measurement of blood pres-sure (described below) and the lack of a more suitable method for in-direct measurement, attempts have been made to automate the manual procedure. As a result, a number of automatic or semiautomatic systems have been developed. Most devices are of a type that ultilizes a pressure transducer connected to the sphygmomanometer cuff, a microphone placed beneath the cuff (over the artery), and a standard physiological recording system on which cuff pressure and the Korotkoff sounds are recorded. The basic procedure parallels the manual method. The pressure cuff is automatically inflated to about 220 mm Hg and allowed to deflate slowly. The microphone picks up the Korotkoff sounds from the artery near the surface, just below the compression cuff. The instrument either super-imposes the signal of the Korotkoff sounds on the voltage recording representing the falling cuff pressure or records the two separately. The

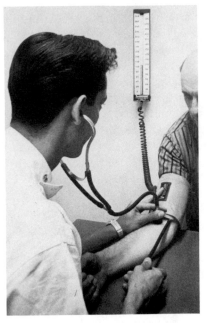

Figure 6.8. Measurement of blood pressure using sphygmomanometer. (Courtesy of W.A. Baum, Inc., Copiague, N.Y.)

pressure reading at the time of the first sound represents the systolic pressure; the diastolic pressure is the point on the falling pressure curve where the signal representing the last sound is seen. This instrument is actually only semiautomatic because the recording thus obtained must still be interpreted by the observer. False indications—caused, for instance, by motion artifacts—can often be observed on the recording. Fully automated devices use some type of signal-detecting circuit to determine the occurrence of the first and last Korotkoff sounds and retain and display the cuff pressure reading for these points, either electronically or with mercury manometers that are cut off by solenoid valves. These devices, by necessity, are more susceptible to false indications caused by artifacts. Methods other than those utilizing the Korotkoff sounds have been tried in detecting the blood pulse distal to the occlusion cuff. Among them is impedance plethysmography (see Chapter 6), which indicates directly the pulsating blood flow in the artery, and ultrasonic Doppler methods, which measure the motions of the arterial walls.

Many of the commercially available automatic blood pressure meters work well when demonstrated on a quiet, healthy subject but fail when used to measure blood pressure during activity or when used on patients in circulatory shock.

An example of an automatic blood pressure meter is the *programmed electrosphygmomanometer PE-300,* illustrated in Figures 6.9 and 6.10 in block diagram and pictorial form. This instrument is designed for use in

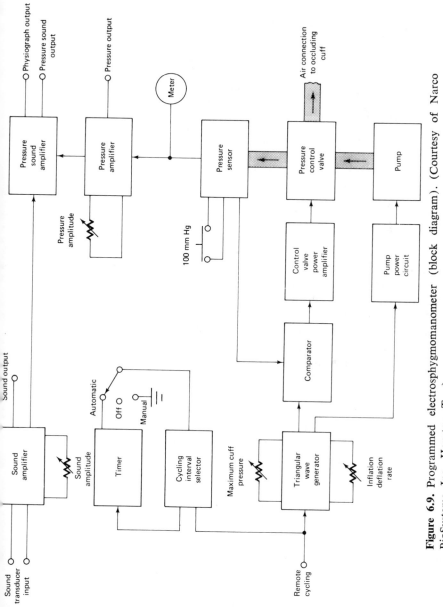

Figure 6.9. Programmed electrosphygmomanometer (block diagram). (Courtesy of Narco BioSystems, Inc., Houston, Tex.)

105

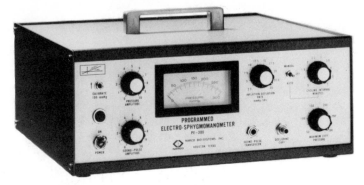

Figure 6.10. Electrosphygmomanometer. (Courtesy of Narco Bio-Systems, Inc., Houston, Tex.)

conjunction with an occluding cuff, microphone, or pulse transducer, and a recorder for the automatic measurement of indirect systolic and diastolic blood pressures from humans and many animal subjects.

The PE-300 incorporates a transducer-preamplifier that provides two output signals, a voltage proportional to the cuff pressure, and the amplified Korotkoff sounds or pulses. These signals can be monitored individually or with the sounds or pulses superimposed on the calibrated cuff-pressure tracing. The combined signal can be recorded on a graphic pen recorder.

The self-contained cuff inflation system can be programmed to inflate and deflate an occluding cuff at various rates and time intervals. Equal and linear rates of cuff inflation and deflation permit two blood pressure determinations per cycle. The PE-300 can be programmed for repeat cycles at adjustable time intervals for monitoring of blood pressure over long periods of time. Single cycles may be initiated by pressing a panel-mounted switch. Provision is also made for remote control via external contact closure. The maximum cuff pressure is adjustable, and the front panel meter gives a continuous visual display of the cuff pressure.

The block diagram of a similar device, developed for the Air Force by Allred and Johnson, is shown in Figure 6.11.

6.2.2. DIRECT MEASUREMENTS. In 1728 Hales first inserted a glass tube into the artery of a horse and crudely measured arterial pressure. Poiseuille substituted a mercury manometer for the piezometer tube of Hales, and Ludwig added a float and devised the *kymograph,* which allowed continuous, permanent recording of the blood pressure. It is only quite recently that electronic systems using strain gages as transducers have replaced the kymograph.

Regardless of the electrical or physical principles involved, direct measurement of blood pressure is usually obtained by one of three methods:

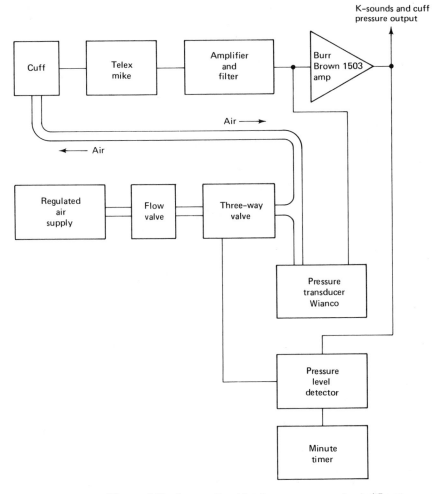

Figure 6.11. Automatic blood pressure system. (Courtesy of Allred and Johnson, U.S. Air Force)

1. percutaneous insertion
2. catheterization (vessel cutdown)
3. implantation of a transducer in a vessel or in the heart

Other methods, such as clamping a transducer on the intact artery, have also been used, but they are not common.

Percutaneous insertion and *catheterization* are both minor surgical techniques that involve invasion of the body. In the former, a catheter or needle is usually inserted into a blood vessel fairly close to the point of entry in the skin; the latter involves the guiding of a *catheter* through the artery or vein to the desired position, which may be in the heart itself.

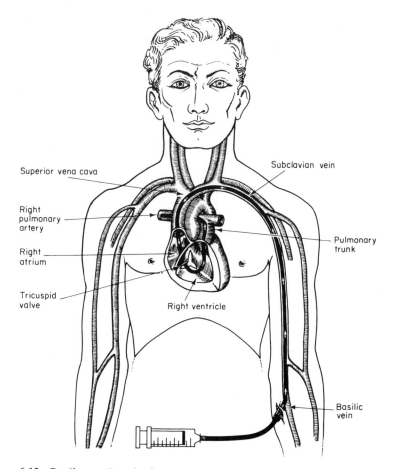

Figure 6.12. Cardiac catheterization. The tube is shown enter-
ing the basilic vein in this case. (From W.F. Evans, *Anatomy and
Physiology, The Basic Principles,* Englewood Cliffs, N.J., Prentice-
Hall, Inc., 1971, by permission.)

Figure 6.12 should give a general idea of both methods. Typically, for
percutaneous insertion, a local anesthetic is injected near the site of invasion. The vessel is occluded and a hollow needle is inserted at a slight
angle toward the vessel. When the needle is in place, a catheter is fed
through the hollow needle, usually with some sort of a guide. When the
catheter is securely in place in the vessel, the needle and guide are withdrawn. For some measurements, a type of needle attached to an airtight
tube is used, so that the needle can be left in the vessel and the blood pressure sensed directly by attaching a transducer to the tube. Other types
have the transducer built into the tip of the catheter. This latter type is

used in both percutaneous and full catheterization models. (See Figures 6.16 and 6.18.)

Catheterization was first developed in the late 1940s and has become a major diagnostic technique for analyzing the heart and other components of the cardiovascular system. Apart from obtaining blood pressures in the heart chambers and great vessels, this technique is also used to obtain blood samples from the heart for oxygen-content analysis and to detect the location of abnormal blood flow pathways. Also, catheters are used for investigations with injection of radiopaque dyes for X-ray studies, colored dyes for indicator dilution studies, and of vasoactive drugs directly into the heart and certain vessels. Essentially a *catheter* is a long tube that is introduced into the heart or a major vessel by way of a superficial vein or artery. The sterile catheter is designed for easy travel through the vessels.

Measurement of blood pressure with a catheter can be achieved in two ways. The first is to introduce a sterile saline solution into the catheter so that the fluid pressure is transmitted to a transducer outside the body (extracorporeal). A complete fluid pressure system is set up with provisions for checking against atmospheric pressure and for establishing a reference point. The frequency response of this system is a combination of both the frequency response of the transducer and the fluid column in the catheter. In the second method, pressure measurements are obtained at the source. Here, the transducer is introduced into the catheter and pushed to the point at which the pressure is to be measured, or the transducer is mounted at the tip of the catheter. This device is called a *catheter-tip blood pressure transducer.* For mounting at the end of a catheter, one manufacturer uses an unbonded resistance strain gage in the transducer, whereas another uses a variable inductance transducer. Each will be discussed later.

Implantation techniques involve major surgery and thus are normally only employed in research experiments. They have the advantage of having the transducer fixed in place in the appropriate vessel for long periods of time. The surgical techniques of a typical case are described in Section 12.4 and will not be discussed here. The type of transducer employed in that procedure is described later in this section.

Transducers can be categorized by the type of circuit element changed by pressure variations, such as capacitive, inductive, and resistive. The resistive types are most frequently used, so the other two types will be discussed only briefly.

In the *capacitance manometer,* a change in the distance between the plates of a capacitor changes its capacitance. In a typical application, one of the plates is a metal membrane separated from a fixed plate by some one-thousandth of an inch of air. Changes in pressure that change the distance between the plates thereby change the capacitance. If this element is contained in a high-frequency resonant circuit, the changes in capacitance vary the frequency of the resonant circuit to produce a form of frequency modulation. With suitable circuitry, blood pressure information can be obtained and recorded as a function of time.

An advantage of this type of transducer is that it can be long and thin in total contour and thus can be easily introduced into the bloodstream without deforming the contour of the recorded pressure waveform. Because of the stiff structure and the small movement of the membrane when pressure is applied, the volume displacement is extremely small and may be in the region of 10^{-6} cc per 100 mm Hg of applied pressure.

Disadvantages of this type of transducer are instability and a proneness to variations with small changes in temperature. Also, lead wires introduce errors in the capacitance, and this type of transducer is more difficult to use than resistance types.

A number of different devices use inductance effects. They measure the distortion of a membrane exposed to the blood pressure. In some of these types, two coils are used—a primary and secondary. When a spring-loaded core that couples the coils together magnetically is moved back and forth, the voltage induced into the secondary changes in proportion to the pressure applied.

A better-known method employs a *differential transformer,* described in detail in Chapter 9. In this device two secondary coils are wound oppositely and connected in series. If the spring-loaded core is symmetrically positioned, the induced voltage across one secondary coil opposes the voltage of the other. Movement of the core changes this symmetry, and the result is a signal developed across the combined secondary coils. The core can be spring-loaded to accept pressure from one side, or it can accept pressure from both sides simultaneously, thus measuring the difference of pressure between two different points.

The *physiological resistance transducer* is a direct adaptation of the strain gages used in industry for many years. The principle of a strain gage is that if a very fine wire is stretched, its resistance increases. (A detailed discussion of strain gages is given in Chapter 9.) If a voltage is applied to the resistance, the resulting current changes with the resistance variations according to Ohm's law. Thus the forces responsible for the strain can be recorded as a function of current. The method in which the blood pressure produces the strain is discussed in Section 6.2.3.

In order to obtain the degree of sensitivity required for blood pressure transducers, two or four strain gages are mounted on a diaphragm or membrane, and these resistances are connected to form a bridge circuit. Figure 6.13 shows such a circuit configuration.

In general, the four resistances are intially about equal when no pressure or strain is applied. The gages are attached to the pressure diaphragm in such a way that as the pressure increases, two of them stretch while the

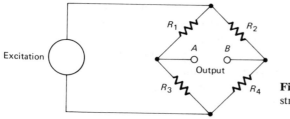

Figure 6.13. Resistance strain gage bridge.

other two contract. An excitation voltage is applied as shown. When pressure changes unbalance the bridge a voltage appears between terminals *A* and *B* proportional to the pressure. Excitation can be either direct current or alternating current, depending on the application.

Resistance-wire strain gages can be *bonded* or *unbonded* (see Chapter 9). In the bonded type, the gage is "bonded" to the diaphragm and stretches or contracts with bending. The unbonded type consists of two pairs of wires, coiled and assembled in such a way that displacement of a membrane connected to them causes one pair to stretch and the other to relax. The two pairs of wires are not bonded to the diaphragm material but are attached only by retaining lugs. Because the wires are very thin, it is possible to obtain relatively large signals from the bridge with small movement of the diaphragm.

Development of semiconductors that change their resistance in much the same manner as wire gages has led to the *bonded silicon element* bridge. Only small displacements (on the order of a few microns) of the pressure-sensing diaphragm, are needed for sizable changes of output voltage with low-voltage excitation. For example, with 10 volts excitation, a range of 300 mm Hg is obtained with a 3–micron deflection producing a 30–mV signal.

Semiconductor strain gage bridges are often temperature sensitive, however, and have to be calibrated for baseline and true zero. Therefore it is usually necessary to incorporate external resistors and potentiometers to balance the bridge initially, as well as for periodic correction.

In Chapter 9 the gage factor for a strain gage is defined as the amount of resistance change produced by a given change in length. Wire strain gages have gage factors on the order of 2 to 4, whereas semiconductor strain gages have gage factors ranging from 50 to 200. For silicon, the gage factor is typically 120. However, the use of semiconductors is restricted to those configurations that lend themselves to this technique.

When strain gages are incorporated in pressure transducers, the sensitivity of the transducer is expressed, not as a gage factor, but as a voltage change that results from a given pressure change. For example, the sensitivity of a pressure transducer can be given in microvolts per (applied) volt per millimeter of mercury.

6.2.3. SPECIFIC DIRECT MEASUREMENT TECHNIQUES. In Section 6.2.2 methods of direct blood pressure were classified in two ways, first by the clinical method by which the measuring device was coupled to the patient and, second, by the electrical principle involved. In the following discussion, the first category is expanded, with the electrical principles involved being used as subcategories where necessary. The four categories are as follows:

1. A catheterization method involving the sensing of blood pressure through a liquid column. In this method the transducer is external to the body, and the blood pressure is transmitted through a saline solution column in a catheter to this transducer. This method can use either an unbonded resistance strain gage to sense the pressure or a linear variable differential transformer. Externally these two devices are quite similar in appearance.

2. A catheterization method involving the placement of the transducer through a catheter at the actual site of measurement in the bloodstream (e.g., to the aorta), or by mounting the transducer on the tip of the catheter.

3. Percutaneous methods in which the blood pressure is sensed in the vessel just under the skin by the use of a needle or catheter.

4. Implantation techniques in which the transducer is more permanently placed in the blood vessel or the heart by surgical methods.

The most important aspects of the preceding methods will be discussed separately.

LIQUID COLUMN METHODS. Figure 6.14 shows a typical liquid-column type of blood pressure transducer. Figure 6.14(a) illustrates the "standard size" Statham Model P23AA pressure transducer, while Figure 6.14(b) shows the Model P23Db, which is a miniature transducer.

The electrical connection to the strain gage bridge is obtained from the cable at the head of the transducer. Pressure connections are made at the top through Luer fittings in the transparent dome, which offer connection to the catheter system and a means of flushing. The easily removable transparent dome is used so that air bubbles can be detected and eliminated. The transducer itself is of the unbonded wire strain gage type.

This type of blood pressure sensor is available in seven types. There are general-purpose models for arterial pressure (0 to 330 mm Hg) and venous pressure (0 to 50 mm Hg), and special models with differing sensitivities, volume displacement characteristics, and mechanical arrangements.

These transducers must be flushed to remove air bubbles and, also, during measurements, to prevent blood from clotting at the end of the catheter.

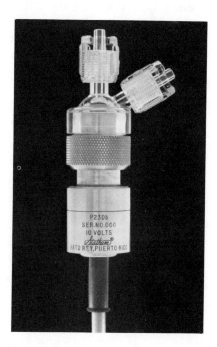

Figure 6.14. Fluid column blood pressure transducer. (a) Statham P23AA Standard; (b) Statham P23Db Miniature. (Courtesy of Statham Instruments, Inc., Oxnard, Calif.)

A frequency-response problem with this type of transducer results from the fact that the liquid column has a natural frequency of its own that can affect the frequency response of the system. Care in the choice of the particular transducer as to volume displacement and type of catheter, as well as in making sure that the frequency response of the complete system is adequate, is essential for good results.

Pressure transducers are normally mounted on the bed frame or on a stand near the patient's bed. It is important to keep the transducer at about the same height as the point at which the measurements are made in order to avoid errors due to hydrostatic pressure. If a differential pressure is desired, two transducers of this type may be used at two different points, and the difference in pressure may be obtained as the difference of the output signals.

The signal-conditioning and display instruments for these transducers come in a variety of forms. However, each basically consists of a method of excitation for the strain gage bridge, a way of zeroing or balancing the bridge, necessary amplification of the output signal, and a display device, such as an oscilloscope, recorder, meter, or digital readout. Most modern systems are designed for flexibility and matching so that many combinations are possible.

Another type of blood pressure transducer is the *linear variable differential transformer* (LVDT) device shown in an exploded view in Figure 6.15. Superficially these transducers look similar to the unbonded strain

Figure 6.15. LVDT blood pressure transducer—exploded view. (Courtesy of Biotronex Laboratory, Inc., Silver Springs, Md.)

gage type. Indeed, with respect to the plastic dome used for visibility, the two pressure fittings for attachment to the catheter and for flushing, and the cable coming out of the bottom, they are similar. Such transducers also come in a variety of models with a range of characteristics for venous or arterial pressure, for different sensitivities, and for alternate volume displacements. The various models also have different natural frequencies and frequency responses.

It should be noted from the exploded view that these units disassemble into three subassemblies—the dome and pressure fittings subassembly, the center portion consisting of a stainless-steel diaphragm and core assembly, and the linear variable differential transformer (LVDT) subassembly. There are two basic diaphragm and core assemblies with appropriate domes that are interchangeable in the coil-connector assembly. The first is used for venous and general-purpose clinical measurements and has a standard-size diaphragm with an internal fluid volume between the dome and diaphragm of less than 0.5 cc. The second design, with higher response characteristics for arterial pressure contours, has a reduced diaphragm area and an internal volume of approximately 0.1 cc.

The Biotronex BL-9630 transducer is a linear variable differential transformer in which the primary coil is excited by an ac carrier (5 to 20 volts peak to peak) in a range of 1500 to 15,000 Hz. Axial displacement of a movable iron core, attached to the diaphragm, cuts the magnetic lines of flux generated by the primary coil. Voltages induced in the secondary sensing coils are returned to the carrier amplifier, where they are differentially amplified and demodulated to remove the carrier frequency. The output of the carrier amplifier is a dc voltage proportional to diaphragm displacement. Linearity of the gage is better than ±1% of full-range. Ordinary jarring and handling will not harm the gage. A positive mechanical stop is provided to prevent damage by as much as a 100–percent overpressure. The LVDT offers much higher signal levels than conventional strain gage transducers for a given excitation voltage.

MEASUREMENT AT THE SITE. In order to avoid the problem inherent in measuring blood pressure through a liquid column, a "catheter-tip" manometer can be fed through the catheter to the site at which the blood pressure is to be measured. This process requires a small-diameter transducer that is fairly rigid but flexible.

One such transducer makes use of the variable inductance effect mentioned earlier. The tip is placed directly in the bloodstream so that the blood presses on a membrane surrounded by a protective cap. The membrane is connected to a magnetic slug that is free to move within a coil assembly and thus changes the inductance of the coil as a function of the pressure on the membrane.

Another type has a bonded strain gage sensor built into the tip of a cardiac catheter. The resistance changes in the strain gage are a result of pressure variations at the site itself, rather than through a fluid column. This gage can also be calibrated with a liquid system catheter at the same location.

PERCUTANEOUS TRANSDUCERS. An example of a percutaneous blood pressure transducer is shown in Figures 6.16 and 6.17. The former shows a Statham P23 transducer connected to a hypodermic needle that has been placed in a vessel of the arm, while Figure 6.17 is an exploded view showing the dome and stopcocks. The three-way stopcock dome permits flushing of the needle, administering of drugs, and withdrawing of blood samples (Figure 6.16). This transducer can measure arterial or venous pressures, or the pressures of other physiological fluids by direct attachment to a needle at the point of measurement. It can be used with a continuously self-flushing system without degradation of signal. The transparent plastic dome permits observation of air bubble formation and consequent ejection. It is designed for use with a portable blood pressure monitor, which provides bridge excitation, balancing, and amplification. The meter scale is calibrated directly into millimeters of mercury. This transducer also has the advantage that it can be connected to a standard intravenous infusion bottle.

Another type of percutaneous transducer is the Statham P866 shown

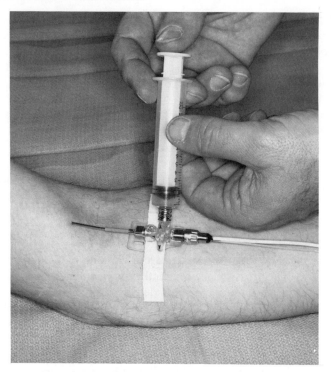

Figure 6.16. Percutaneous blood pressure measurement. Transducer in arm with three-way stop cock dome for administering drugs and withdrawing blood samples. (Courtesy of Statham Instruments, Inc., Oxnard, Calif.)

inserted into an arm in Figure 6.18. This has a thin-film gage sensor built into the tip of a No. 5 French catheter. It is capable of simultaneous pressure and phonocardiograph measurements.

IMPLANTABLE TRANSDUCERS. Figure 6.19 shows a type of transducer that can be implanted into the wall of a blood vessel or into the wall of the heart itself. A case study illustrating the use of this type is given in Section 12.4. This transducer is particularly useful for long-term investigations in animals.

Figure 6.17. Exploded view of P23 blood pressure transducer. (Courtesy of Statham Instruments, Inc., Oxnard, Calif.)

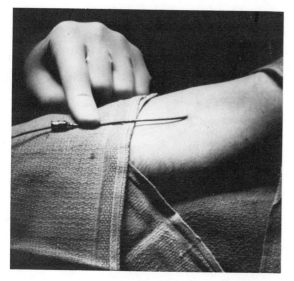

Figure 6.18. Thin film gage blood pressure sensor (P866 catheter tip). (Courtesy of Statham Instruments, Inc., Oxnard, Calif.)

Figure 6.19. Implantable pressure transducer. (Courtesy of Konigsberg Instruments, Pasadena, Calif.)

The transducer's body is made out of titanium, which has excellent corrosion-resistance characteristics, a relatively low thermal coefficient of expansion, and a low modulus of elasticity, which results in greater strain per unit stress. Four semiconductor strain gages are bonded to the inner surface of the pressure-sensing diaphragm. Transducers of this type come in a number of sizes (from 3 to 7 mm in diameter) for blood pressure measurement. A popular size is the 4.5 mm diameter. Larger sizes are available for pleural pressure. The thickness of the body is 1.2 to 1.3 mm in the various models.

The four semiconductors are connected in bridge fashion as shown in Figure 6.13. As blood pressure increases on the diaphragm, the inner surface is stressed. The strain gages are located so that two of them are strained in tension while two are in compression. When the bridge is excited, an output voltage proportional to the blood pressure can be obtained.

Additional resistors, connected externally to the bridge, provide temperature compensation, although these bridges are not extremely temperature sensitive. Since they operate in the bloodstream at a fairly constant temperature near 37°C, the temperature effects are not serious. (See Chapter 12.)

These transducers can be excited with ac or dc and easily lend themselves to telemetry circuits. In service, they have proven very reliable. Cases of chronic implants (in excess of 2 years) have been reported with no detrimental effect on the animal, the gage, or the wires. The wires are usually insulated with a plastic compound, polyvinylchloride, which is fairly impervious to body fluids.

6.3. BLOOD FLOW AND CARDIAC OUTPUT MEASUREMENTS

It would be desirable and convenient if blood flow at any point in the blood circulation system could be measured from the outside of the body. Unfortunately, doing so is not possible. In order to measure blood flow at a particular point in the circulatory system, a *blood flow sensor* must be brought to the point of measurement. This can be done by inserting a catheter through a peripheral vessel and guiding its tip to the point of interest, or the blood vessel can be excised and a transducer clamped around it. Both methods pose certain dangers and thus are not used routinely in clinical practice. Such methods are, however, widely utilized in medical research with animals. The determination of cardiac output by a tracer dilution method, on the other hand, which requires only the insertion of thin catheters for injection of the tracer substance and for the sampling of blood, is now routine in most hospitals. Visualization of blood flow using X rays and the injection of radiopaque dyes is also common practice.

Methods used in industry for flow measurements of other liquids, like the turbine flowmeter and the rotameter, are not very suitable for the measurement of blood flow because they require cutting the blood vessel. These methods also expose the blood to sharp edges, which are conducive to blood-clot formation.

Practically all blood flow meters currently used in clinical and research applications are based on one of the following physical principles:

1. Electromagnetic induction
2. Ultrasound transmission or reflection

3. Thermal convection
4. Radiographic principles
5. Indicator (dye or thermal) dilution

The blood flow meters described in this chapter actually measure the flow of blood in a given vessel. The blood flow in an entire section of the body (e.g., an arm), however, can be determined from volume changes of that section, measured with a plethysmograph, such as described in Section 6.4.

6.3.1. MAGNETIC BLOOD FLOW METERS. *Magnetic blood flow meters* are based on the principle of magnetic induction. When an electrical conductor is moved through a magnetic field, a voltage is induced in the conductor proportional to the velocity of its motion. The same principle applies when the moving conductor is not a wire but rather a column of conductive fluid that flows through a tube located in the magnetic field. Figure 6.20 shows how this principle is used in magnetic blood flow meters. A permanent magnet or electromagnet positioned around the blood vessel generates a magnetic field perpendicular to the direction of the blood flow. The voltage induced in the moving blood column is measured with stationary electrodes located on opposite sides of the blood vessel and perpendicular to the direction of the magnetic field.

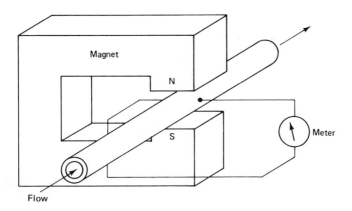

Figure 6.20. Magnetic blood flow meter, principle.

The most commonly used types of implantable magnetic blood flow probes are shown in Figures 6.21 through 6.23. The *slip-on* or *C type* is applied by squeezing an excised blood vessel together and slipping it through the slot of the probe. In some transducer models the slot is then closed by inserting a keystone-shaped segment of plastic as shown. Contact is provided by two slightly protruding platinum disks that touch the

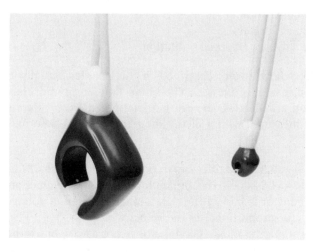

Figure 6.21. Samples of large and small lumen diameter blood flow probes. (Courtesy of Micron Instruments, Los Angeles, Calif.)

wall of the blood vessel. For proper operation, the orifice of the probe must fit tightly around the vessel. For this reason, probes of this type are manufactured in sets, with diameters increasing in steps of 0.5 or 1 mm from about 2 to 20 mm. The probes shown in Figure 6.21 can be implanted for chronic use. In contrast, Figure 6.22 shows a model, which is provided with a long handle, for use during surgery.

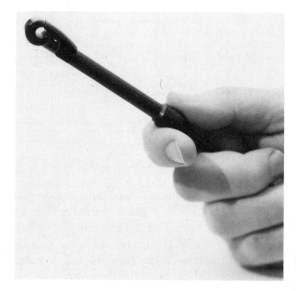

Figure 6.22. Blood flow probe—clip-on type for use during surgery. (Courtesy of Biotronex, Silver Springs, Md.)

In the *cannula-type transducer,* the blood flows through a plastic cannula around which the magnet is arranged. The contacts penetrate the walls of the cannula. This type of transducer requires that the blood vessel be cut and its ends slipped over the cannula and secured with a suture. A similar type of transducer (Figure 6.23) is also used to measure the blood

Figure 6.23. Extracorporeal blood flow probe. (Courtesy of Biotronex, Silver Springs, Md.)

flow in extracorporeal devices, such as dialyzers. Magnetic blood flow meters actually measure the mean blood velocity. Because the cross-sectional area at the place of velocity measurement is well defined with either type of transducer, these transducers can be calibrated directly in units of flow.

Magnetic blood flow transducers are also manufactured as catheter-tip transducers. For this type, the normal transducer design is essentially turned "inside out," with the electromagnet being located inside the catheter, which has the electrodes at the outside. For catheter transducers, however, the cross section of the blood vessel at the place of measurement is not defined, and these transducers, therefore, cannot be calibrated in flow units.

The output voltage of a magnetic blood flow transducer is very small, typically on the order of a few microvolts. In early blood flow meters, a constant magnetic field was used, which caused difficulties with electrode polarization and amplifier drift. In order to overcome these problems, all contemporary magnetic blood flow meters use electromagnets that are driven by alternating currents. The result, however, creates another problem: the change of the magnetic field causes the transducer to act like a transformer and induces error voltages that often exceed the signal levels by several orders of magnitude. Thus for recovering the signal in the presence of the error voltage, amplifiers with large dynamic range and phase-sensitive or gated detectors have to be used. Several different waveforms have been advocated for the magnet current in order to minimize the problem, as shown in Figure 6.24. With a sinusoidal magnet current, the induced voltage is also sinusoidal but is 90 degrees out of phase with the flow signal. With a suitable circuit, similar to a bridge, the induced voltage

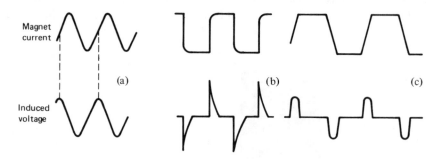

Figure 6.24. Waveforms used in magnetic blood flow meters and error signals induced by the current: (a) sine wave; (b) square wave; (c) trapezoidal wave.

can be partially balanced out. With the magnet current in the form of a square wave, the induced voltage should be zero once the spikes from the polarity reversal have passed. In practice, however, these spikes are often of extremely high amplitude, and the circuitry response tends to extend their effect. A compromise is the use of a magnet current having a trapezoidal waveform. None of the three waveforms used seems to have demonstrated a definite superiority.

The block diagram of a magnetic blood flow meter is shown in Figure 6.25. The oscillator, which drives the magnet and provides a control signal for the gate, operates at a frequency of between 60 and 400 Hz. The use of a gated detector makes the polarity of the output signal reverse when the flow direction reverses. The frequency response of this type of system is

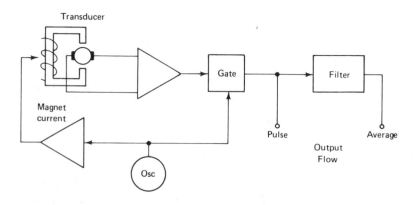

Figure 6.25. Magnetic blood flow meter, block diagram.

usually high enough to allow the recording of the flow pulses, while the mean or average flow can be derived by use of a low-pass filter. Figure 6.26 shows a single-channel magnetic blood flow meter that can be used with a variety of different transducers.

Figure 6.26. Magnetic blood flow meter. (Courtesy of Micron Instruments, Los Angeles, Calif.)

6.3.2. ULTRASONIC BLOOD FLOW METERS. In *ultrasonic blood flow meters,* a beam of ultrasonic energy is used to measure the velocity of flowing blood. This can be done in two different ways. In the *transit time ultrasonic flow meter,* a pulsed beam is directed through a blood vessel at a shallow angle and its transit time is then measured. When the blood flows in the direction of the energy transmission, the transit time is shortened. If it flows in the opposite direction, the transit time is lengthened.

More common are ultrasonic flow meters based on the Doppler principle (Figure 6.27). An oscillator, operating at a frequency of several megahertz, excites a piezoelectric transducer (usually made of barium

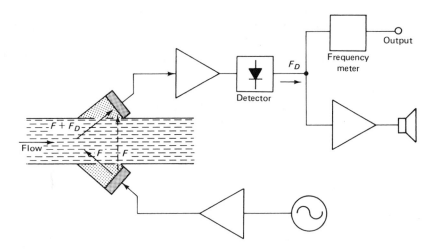

Figure 6.27. Ultrasonic blood flow meter, Doppler type.

titanate). This transducer is coupled to the wall of an exposed blood vessel and sends an ultrasonic beam with a frequency F into the flowing blood. A small part of the transmitted energy is scattered back and is received by a second transducer arranged opposite the first one. Because the scattering occurs mainly as a result of the moving blood cells, the reflected signal has a different frequency due to the Doppler effect. Its frequency is either $F + F_D$ or $F - F_D$, depending on the direction of the flow. The Doppler component F_D is directly proportional to the velocity of the flowing blood. A fraction of the transmitted ultrasonic energy, however, reaches the second transducer directly, with the frequency being unchanged. After amplification of the composite signal, the Doppler frequency can be obtained at the output of a detector as the difference between the direct and the scattered signal components.

With blood velocities in the range normally encountered, the Doppler signal is typically in the low-audio frequency range. Because of the velocity profile of the flowing blood, the Doppler signal is not a pure sine wave, but has more the form of narrow-band noise. Therefore, from a loud-speaker or earphone, the Doppler signal of the pulsating blood flow can be heard as a characteristic "swish—swish—." When the transducers are placed in a suitable mount (which defines the area of the blood vessel), a frequency meter used to measure the Doppler frequency can be calibrated directly in flow rate units. Unfortunately, Doppler flow meters of this simple design cannot discriminate the direction of flow. More complicated circuits, however, which use the insertion of two quadrature components of the carrier, are capable of indicating the direction of flow.

Transducers for ultrasonic flow meters can be implanted for chronic use. Some commercially available flow meters of this type incorporate a telemetry system to measure the blood flow in unrestrained animals.

A simple Doppler device, with the two transducers mounted in a hand-held probe, is now widely used to trace blood vessels close to the surface and to determine the location of vascular obstructions. In order to facilitate transmission of ultrasonic energy, the probe must be coupled to the skin with an aqueous jelly. Such devices can also be used to detect the motion of internal structures in the body—for example, the fetal heart (see Chapter 9).

6.3.3. BLOOD FLOW MEASUREMENT BY THERMAL CONVECTION. A hot object in a colder-flowing medium is cooled by thermal convection. The rate of cooling is proportional to the rate of the flow of the medium. This principle, often used to measure gas flow, has also been applied to the measurement of blood velocity. In one application, a thermistor in the bloodstream is kept at a constant temperature by a servo system. The electrical energy required to maintain this constant temperature is a measure of the flow rate. In another method an electric heater is placed between two thermocouples or thermistors that are located some distance apart along the axis of the vessel. The temperature difference between the upstream and the downstream sensor is a measure of the blood velocity. A device of the latter type is sometimes called *a thermostromuhr* (literally, from the German "heat current clock"). Thermal convection methods for blood flow determination, although among the oldest ones used for this purpose, have now been widely replaced by the other methods described in this chapter.

6.3.4. BLOOD FLOW DETERMINATION BY RADIOGRAPHIC METHODS. Blood is not normally visible on an X-ray image because it has about the same radio density as the surrounding tissue. By the injection of a contrast medium into a blood vessel (e.g., an iodated organic compound), the circulation pattern can be made locally visible. On a sequential record of the X-ray image (either photographic or on a video-tape recording), the progress of the contrast medium can be followed, obstructions can be detected, and the blood flow in certain blood vessels can be estimated. This technique, known as *cine* (or *video*) *angiography*, can be used to assess the extent of damage after a stroke or heart attack.

Another method is the injection of a radioactive isotope into the blood circulation, which allows the detection of vascular obstructions (e.g., in the lung) with an imaging device for nuclear radiation, such as a scanner or gamma camera (see Chapter 14).

Vascular obstructions in the lower extremities can sometimes be detected by measuring differences in the skin temperature caused by the reduced circulation. This can be accomplished by one of the various methods of skin surface temperature measurement described in Chapter 9.

6.3.5. MEASUREMENT BY INDICATOR DILUTION METHODS. *The indicator or dye dilution methods* are the only methods of blood flow measurement that really measure the blood flow and not the blood velocity. The indicator is a medium that mixes readily with blood and whose concentration in the blood after the mixing can be easily determined. It must be stable but should not be retained in the body, and it must not have toxic side effects. The most commonly used indicator is cardiogreen, a green indocyanine dye. This dye is used in an isotonic solution. Its concentration can be determined by measuring the light absorption with a densitometer (colorimeter), at a wavelength of about 800 mm, at which the dye has maximum absorption. Other indicators that have been used are radioactive isotopes, such as radio-iodated serum albumen, or simply an isotonic saline injected with a temperature lower than the body temperature. The isotope concentration can be determined with a scintillation counter. Safety considerations, however, limit the amount of isotope that can be injected and hence the accuracy of the measurement. Cold saline is the least harmful indicator. Its concentration can be determined with a thermistor thermometer. The warming of the blood-saline mixture by the surrounding blood vessel walls, however, limits its use to some extent.

The principle of the dilution method is shown in Figure 6.28. The upper left drawing shows a model of a part of the blood circulation under the (very simplified) assumption that the blood is not recirculated. The indicator is injected into the flow continuously, beginning at time t, at a constant infusion rate I (grams per minute). A detector measures the concentration downstream from the injection point. Figure 6.28(a) shows the output of a recorder that is connected to the detector. At a certain time after the injection, the indicator begins to appear, the concentration increases, and, finally, it reaches a constant value, C_0, (milligrams per liter). From the measured concentration and the known injection rate I (in milligrams per minute), the flow can be calculated as

$$F\text{(liters per minute)} = \frac{I\text{ (milligrams per minute)}}{C_0\text{ (milligrams per liter)}}$$

The earliest method for determining cardiac output, the *Fick method,* is based on this simple model. The indicator is the oxygen of the inhaled air that is "injected" into the blood in the lungs. The "infusion rate" is determined by measuring the oxygen content of the exhaled air and subtracting it from the known oxygen content of the inhaled room air. The oxygen metabolism only approximately resembles the model of the open circulation, because only part of the oxygen is consumed in the systemic circulation and the returning venous blood still contains some oxygen. Therefore the oxygen concentration in the returning venous blood has to be determined and subtracted from the oxygen concentration in the arterial blood leaving the lungs. The measurements are averaged over several minutes to reduce the influence of short-term fluctuations. An automated system is available that measures the oxygen concentration (by colorimetry) and the oxygen consumption, and that continuously calculates the cardiac output from these measurements.

When a dye or isotope is used as an indicator, the concentration does not assume a steady-state value but increases in steps whenever the recirculated indicator again passes the detector [points R in Figure 6.28(b)]. The recirculation often occurs before the concentration has reached a plateau. Consequently, a slightly different method is usually used, and the indicator is injected as a bolus instead of being infused at a constant rate. Figure 6.28(c) shows the concentration for this case, again under the assumption that it is an open system. The concentration increases at

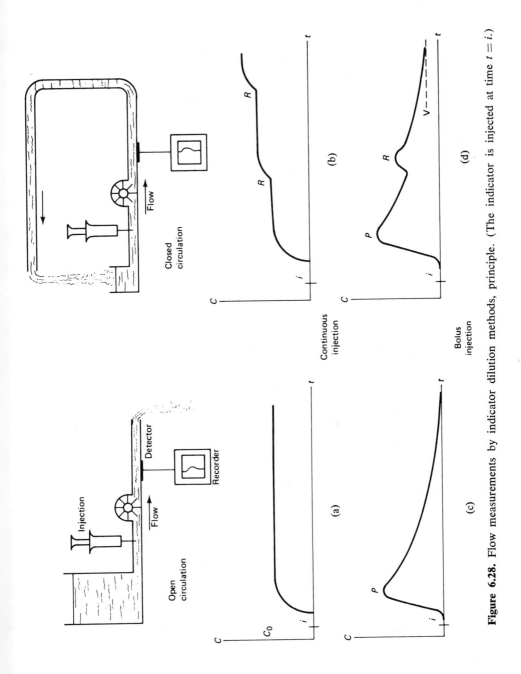

Figure 6.28. Flow measurements by indicator dilution methods, principle. (The indicator is injected at time $t = i$.)

first, reaches a peak P, and then decays as an exponential function. This "washout curve" is mainly a result of the velocity profile of the blood, which causes a "spread" of the bolus. In order to calculate the flow, the area under the concentration curve has to be determined. This is given by the integral:

$$\int_{t=i}^{t=\infty} C \, dt \left(\frac{\text{milligrams}}{\text{liter}} \times \text{minutes} \right)$$

From the value of this integral, and from the amount B of the injected indicator (in milligrams), the flow then can be calculated.

$$F \left(\frac{\text{liters}}{\text{minute}} \right) = \frac{B}{\int_{t=i}^{t=\infty} C \, dt} \left(\frac{\text{milligrams}}{\text{milligrams/liter} \times \text{minutes}} \right)$$

Because of the recirculation, the concentration does not show a monotonous decay as in Figure 6.28(b); instead, after some time, a "hump" (R in the figure) occurs. Normally the portion of the curve preceding the recirculation hump is exponential, and the decay curve can be determined, despite the recirculation, by exponential extrapolation (*Hamilton method*). This was originally done by manually replotting the curve on semilogarithmic paper, which resulted in a straight line for the exponential part of the curve. This line was then extended, the extended part replotted on the original plot, and the extrapolated curve integrated with a planimeter. This complicated, time-consuming procedure can now be performed by computers or programmable desk calculators, but it still requires that the curve be digitized manually. In order to avoid this step, a number of special-purpose analog computers have been developed that perform the calculation of cardiac output from the output signal of a densitometer directly on line. These computers utilize certain properties of the exponential decay function that can be explained with the help of Figure 6.29.

The exponentially decaying concentration curve $C(t)$ is assumed to have the value C_0 at time t_0. This curve is described by the equation

$$C = C_0 e^{-(t - t_0)/T}$$

The first derivative of the curve at the point $C_0(t_0)$ is

$$\frac{dc}{dt}(C_0) = -\frac{C_0}{T}$$

where

T is the time constant of the curve

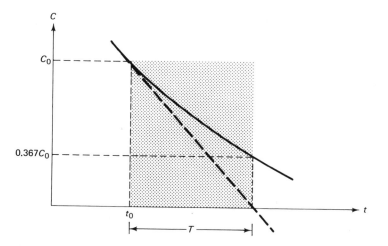

Figure 6.29. Properties of an exponential decay function.

The residual area A_0 under the curve from time t_0 on is given by

$$A_0 = \int_{t_0}^{\infty} C_0 e^{-(t - t_0)/T} \, dt = (C_0)(T)$$

With the equation for the first derivative of the curve resolved for T,

$$A_0 = \left[C_0 \frac{C_0}{dC/dt \, (C_0)} \right]$$

One commercial cardiac output computer, the block diagram of which is shown in Figure 6.30(a), uses this relationship. The concentration signal is integrated continuously, and at the same time a "prediction term"

$$C \left(\frac{C}{dc/dt} \right)$$

is calculated, and added to the integral. The sum is displayed on a meter or recorder. When the signal first goes through the upstroke of the curve, the reading fluctuates. As soon as the exponential part of the curve is reached, the output signal becomes steady, with the reading being proportional to the area under the extrapolated curve. The onset of recirculation is indicated by the reoccurrence of fluctuations of the signal. This method has the advantage that the absence of an exponential part of the curve is indicated by the absence of a steady output reading. In such a case, which can occur either because of a faulty technique or because of deformities in the vascular system (e.g., holes in the heart chamber walls), a cardiac

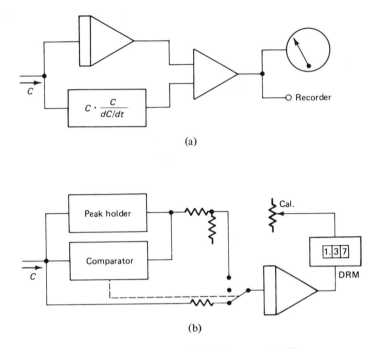

$$C \cdot \frac{C}{dC/dt}$$

(a)

(b)

Figure 6.30. Cardiac output computers, block diagram. (a) Lexington Instruments Corporation; (b) Columbus Instruments.

output reading obtained by an indicator dilution method is meaningless.

Another type of cardiac output computer uses a different property of the exponential decay curve, which can also be shown with the help of Figure 6.29. The residual area under the curve beginning at t_0 is also equal to the dotted rectangle with the sides C_0 and T. At the time $(t_0 + T)$, the concentration has decayed to $0.367C_0$. As shown in the block diagram in Figure 6.30(b), this computer begins the cardiac output determination by integrating the concentration signal. When the concentration reaches its maximum, the value at this point is stored in the capacitor of a "peak holder." From there onward, a comparator compares the actual concentration with the peak concentration. When the concentration has dropped to 85 percent of the maximum, it is presumed that the curve has entered its exponential phase, and this concentration value is used as C_0 for the further operation of the computer. The comparator next switches the input of the integrator from the concentration signal to the constant voltage representing C_0. At the same time, the gain of the integrator is increased by a factor of 2. The comparator continues to compare the concentration with the maximum until the concentration reaches 60 percent of C_0, which is 49.9 percent of the maximum. This is a point halfway between C_0 and

$0.367C_0$. When this point is reached, the determination is completed and the comparator turns the integrator off. The output of the integrator now represents a voltage. This corresponds to the sum of the area under the curve up to the point where C_0 was reached, plus the area of a rectangle with the sides $2C_0$ and $\frac{1}{2}$-T. A digital ratio meter (DRM) divides a calibration voltage by the output of the integrator and thus allows the direct reading of cardiac output. This computer, however, does not detect the absence of an exponential portion in the decay curve.

Most cardiac output computers can be used with any of the indicators mentioned above, depending on the detector used. The dye dilution method requires a flow-through densitometer and a constant rate withdrawal pump. Figure 6.31 shows a view of the computer (the block diagram of which was

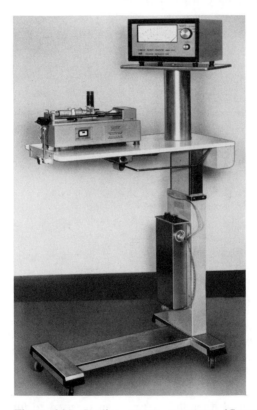

Figure 6.31. Cardiac output computer. (Courtesy of Lexington Instruments Corporation, Waltham, Mass.)

shown in Figure 6.30(a)), as a mobile unit complete with densitometer and withdrawal pump for cardiac output determinations at the bedside of a patient.

Prior to the cardiac output determination, these computers have to be zeroed, using a blood sample withdrawn from the patient. A measured volume of this sample is mixed with a known amount of indicator to determine the calibration factor of the densitometer-computer system. Great care must be taken in the preparation of the dye and the bolus injection to ensure a reasonable accuracy of this method, which, at its best, has an accuracy of about ±5 %.

6.4. PLETHYSMOGRAPHY

Related to the measurement of blood flow is the measurement of volume changes in any part of the body that result from the pulsations of blood occurring with each heartbeat. Such measurements are useful in the diagnosis of arterial obstructions as well as for pulse-wave velocity measurements. Instruments measuring volume changes or providing outputs that can be related to them are called *plethysmographs,* and the measurement of these volume changes, or phenomena related thereto, is called *plethysmography.*

A "true" plethysmograph is one that actually responds to changes in volume. Such an instrument consists of a rigid cup or chamber placed over the limb or digit in which the volume changes are to be measured, as shown in Figure 6.32. The cup is tightly sealed to the member to be measured so

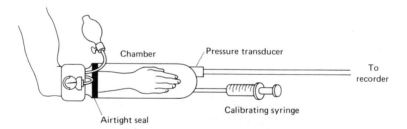

Figure 6.32. Plethysmograph. (Redrawn from A.C. Guyton, *Textbook of Medical Physiology,* 4th ed., W.B. Saunders Co., 1971, by permission.)

that any changes of volume in the limb or digit reflect as pressure changes inside the chamber. Either fluid or air can be used to fill the chamber.

Plethysmographs may be designed for constant pressure or constant volume within the chamber. In either case, some form of pressure or displacement transducer must be included to respond to pressure changes within the chamber and to provide a signal that can be calibrated to represent the volume of the limb or digit. (See description of pressure transducers in Section 6.2.2 and displacement transducers in Chapter 9.) The baseline pressure can be calibrated by use of a calibrating syringe.

This type of plethysmograph can be used in two ways (see Figure 6.32). If the cuff, placed upstream from the seal, is not inflated, the output signal is simply a sequence of pulsations proportional to the individual volume changes with each heartbeat.

The plethysmograph illustrated in Figure 6.32 can also be used to measure the total amount of blood flowing into the limb or digit being measured. By inflating the cuff (placed slightly upstream from the seal) to a pressure just above venous pressure, arterial blood can flow past the cuff, but venous blood cannot leave. The result is that the limb or digit increases its volume with each heartbeat by the volume of the blood entering during that beat. The output tracing for this measurement is shown in Figure 6.33. The slope of a line tangent to the peaks of these pulsations represents the overall rate at which blood enters the limb or digit. Note, however, that after a few seconds the slope tends to level off. The reason is

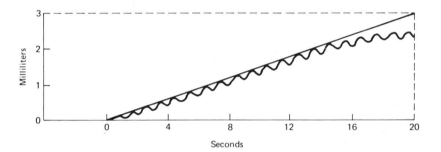

Figure 6.33. Blood volume record from plethysmograph. (From A.C. Guyton, *Textbook of Medical Physiology,* 4th ed., W.B. Saunders Co., 1971, by permission.)

a back pressure that builds up in the limb or digit from the accumulation of blood which cannot escape.

Another device that quite closely approximates a "true" plethysmograph is the *capacitance plethysmograph* shown in Figure 6.34. In this device, which is generally used on either the arm or leg, the limb in which the volume is being measured becomes one plate of a capacitor. The other plate is formed by a fixed screen held at a small distance from the limb by an insulating layer. Often a second screen surrounds the outside plate at a fixed distance to act as a shield for greater electrical stability. Pulsations of the blood in the arm or leg cause variations in the capacity of the capacitor, for the distance between the limb and the fixed screen varies with these pulsations. Some form of capacitance-measuring device is then used to obtain a continuous measure of the capacitance. Since the length of the

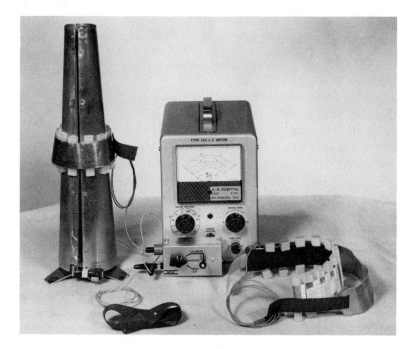

Figure 6.34. Capacitance plethysmograph. (Designed by one of the authors for V.A. Hospital, San Francisco.)

cuff is fixed, the variations in capacitance can be calibrated as volume variations. The device can be calibrated by using a special cone of known volume on which the diameter can be adjusted to provide the same capacitance reading as the limb on which measurements are made. Since the capacitance plethysmograph essentially integrates the diameter changes over a fairly large segment of the limb, its readings are reasonably close to those of a "true" plethysmograph. Also, as with the "true" plethysmograph, estimates of the total volume of blood entering an arm or leg over a given period of time can be made by placing an occluding cuff just upstream from the capacitance device and by pressurizing the cuff to a pressure greater than venous pressure but below arterial pressure.

Several devices, called plethysmographs, actually measure some variable related to volume rather than volume itself. One class of these "pseudoplethysmographs" measures changes in diameter at a certain cross section of a finger, toe, arm, leg, or other segment of the body. Since volume is related to diameter, this type of device is sufficiently accurate for many purposes.

A common method of sensing diameter changes is through the use of a *mercury strain gage,* which consists of a segment of small-diameter elastic tubing, just long enough to wrap around the limb or digit being measured. When the tube is filled with mercury, it provides a highly compliant strain gage that changes its resistance with changes in diameter. With each pulsation of blood that increases the diameter of the limb or digit, the strain gage elongates and, in stretching, becomes thinner, thus increasing its resistance. The major difficulty in using the mercury strain gage is its extremely low impedance. This drawback necessitates the use of a low-impedance bridge to measure small resistance variations and convert them into voltage changes that can be recorded. A mercury strain gage plethysmograph is shown in Figure 6.35. A difficulty that is common to all diameter-measuring pseudoplethysmographs is that of interpreting single-point diameter changes as volume changes.

An even more imitative type of "pseudoplethysmograph" is the *photoelectric plethysmograph.* This device operates on the principle that volume changes in a limb or digit result in changes in the optical density through and just beneath the skin over a vascular region. A photoelectric plethysmograph is shown in Figure 6.36. A light source in an opaque chamber illuminates a small area of the fingertip. Light scattered and transmitted

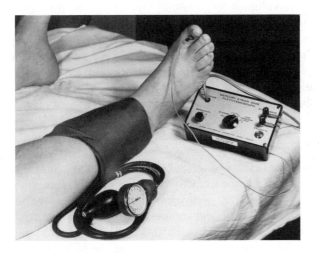

Figure 6.35. Mercury strain gage plethysmograph. (Courtesy of Parks Electronics Laboratory, Beaverton, Oreg.)

through the capillaries of the finger is picked up by the photocell, which is shielded from all other light. As the finger fills with blood (with each pulse), the blood density increases, thereby reducing the amount of light reaching the photocell. The result causes resistance changes in the photo-

Figure 6.36. Photoelectric plethysmograph. (Courtesy of Narco BioSystems, Inc., Houston, Tex.)

cell that can be measured on a bridge and recorded. Pulsations recorded in this manner are somewhat similar to those obtained by a true plethysmograph, but the photocell device cannot be calibrated to reflect absolute or even relative volumes. As a result, this type of measurement is primarily limited to detecting the fact that there are pulsations into the finger, indicating heart rate and determining the arrival time of the pulses. One serious difficulty experienced with this type of device is the fact that even the slightest movement of the finger with respect to the photocell or light source results in a severe amount of movement artifact. Furthermore, if the light source produces heat, the effect of the heat may change local circulation beneath the light source and photocell.

Another device for pseudoplethysmography is the *impedance plethysmograph* in which volume changes in a segment of a limb or digit are reflected as impedance changes. These impedance changes are due primarily to changes in the conductivity of the current path with each pulsation of blood. Impedance plethysmographic measurements can be made using a two-electrode or a four-electrode system. The electrodes are either conductive bands wrapped around the limb or digit to be measured or simple conductive strips of tape attached to the skin. In either case, the electrodes contact the skin through a suitable electrolyte jelly or paste to form an electrode interface and to remove the effect of skin resistance. In a two-electrode system, a constant current is forced through the tissue between the two electrodes and the resulting voltage changes are measured. In the four-electrode system, the constant current is forced through two outer, or current, electrodes, and the voltage between the two inner, or measurement, electrodes is measured. The internal body resistances between the electrodes form a physiological voltage divider. The advantage of the four-electrode system is a much smaller amount of current through the measuring electrodes, thus reducing the possibility of error due to changes in electrode resistance. Currents used for impedance plethysmography are commonly limited to the low-microampere range. The driving current is ac, sometimes a square wave, and usually of a high enough frequency (around 10 kHz or higher) to reduce the effect of skin resistance. At these frequencies the capacitive component of the skin electrode interface becomes a significant factor.

Several theories attempt to explain the actual cause of the measured impedance changes. One is that the mere presence of additional blood filling a segment of the body lowers the impedance of that segment. Tests reported by critics of this method, however, claim that the actual impedance difference between the blood-filled state and more "empty" state is not significant.

A second theory is that the increase in diameter due to additional blood in a segment of the body increases the cross-sectional area of the segment's conductive path and thereby lowers the resistance of the path. This may be true to some extent, but again the percentage of area change is very small.

Critics of impedance plethysmography argue that the measured impedance changes are actually changes in the impedance of the skin-electrode interface, caused by pressure changes on the electrodes that occur with each blood pulsation.

Whatever the reason, however, impedance plethysmography does produce a measure that closely approximates the output of a true plethysmograph. Its main difficulty is the problem of relating the output resistance to any absolute volume measurement. As with the photocell plethysmograph, detection of the presence of arterial pulsations, measurement of pulse rate, and determination of time of arrival of a pulse at any given point in the peripheral circulation can all be satisfactorily handled by impedance plethysmography. Also, the impedance plethysmograph can measure time-variant changes in blood volume.

A special form of impedance plethysmography is *rheoencephalography,* the measurement of impedance changes between electrodes positioned on the scalp. Although primarily limited to research applications, this technique provides information related to cerebral blood flow and is sometimes used to detect circulatory differences between the two sides of the head. Theoretically, such information might help in locating blockages in the internal carotid system, which supplies blood to the brain.

In certain rodents the tail is a convenient region for measurement of circulatory factors. For these measurements, a *caudal plethysmograph* is used. Caudal plethysmographs can utilize any of the previously described methods of sensing volume changes or the presence of blood pulsations. The same limitations encountered in human plethysmographic procedures are also found in caudal plethysmography. In addition, a special physiological factor must be considered in measuring blood pulsations from the tail of a rodent. Many animals use their tails as radiators in the control of body temperature. At low temperatures, very little blood actually flows through the vessels of the tail and plethysmographic measurements become very difficult.

If the animal is heated to a temperature at which the tail is used for cooling, however, sufficient blood flow for good plethysmographic measurements is usually found. Sometimes the necessary temperature for good caudal measurements is so near the point of overheating that traumatic effects are encountered.

6.5. MEASUREMENT OF HEART SOUNDS

In the early days of auscultation a physician listened to heart sounds by placing his ear on the chest of the patient, directly over the heart. It was probably during the process of treating some well-endowed, but bashful, young lady that someone developed the idea of transmitting heart sounds from the patient's chest to the physician's ear via a section of cardboard tubing. This was the forerunner of the stethoscope, which has become a symbol of the medical profession.

The *stethoscope* (from the Greek word, *stethos,* meaning chest, and *skopein,* meaning "to examine") is simply a device that carries sound energy from the chest of the patient to the ear of the physician via a column of air. There are many forms of stethoscopes, but the familiar configuration has two earpieces connected to a common bell or chest piece. Since the system is strictly acoustical, there is no amplification of sound, except for any that might occur through resonance and other acoustical characteristics.

Unfortunately, only a small portion of the energy in heart sounds is in the audible frequency range. Thus, since the dawning of the age of electronics, countless attempts have been made to convince the medical profession of the advantage of amplifying heart sounds, with the idea that if the sound level could be increased, a greater portion of the sound spectrum could be heard and greater diagnostic capability might be achieved. In addition, high-fidelity equipment would be able to reproduce the entire frequency range, much of which is missed by the stethoscope. In spite of these apparent advantages, *the electronic stethoscope* has never found favor with the physician. The principal argument is that doctors are trained to recognize heart defects by the way they sound through an ordinary stethoscope, and any variations therefrom are foreign and confusing. Nevertheless, a number of electronic stethoscopes are available commercially. An example is shown in Figure 6.37.

Instruments for graphically recording heart sounds have been more successful. As stated in Chapter 5, a graphic record of heart sounds is called a *phonocardiogram*. The instrument for producing this recording is called a *phonocardiograph*. Although instruments specifically designed for phonocardiography are rare, components suitable for this purpose are readily available.

The basic transducer for the phonocardiogram is a microphone having the necessary frequency response, generally ranging from below 5 Hz to above 1000 Hz. An amplifier with similar response characteristics is required, which may offer a selective low-pass filter to allow the high-frequency cutoff to be adjusted for noise and other considerations. In one instance, where the associated pen recorder is inadequate to reproduce higher frequencies, an integrator is employed and the envelope of frequencies over 80 Hz is recorded along with actual signals below 80 Hz.

The readout of a phonocardiograph is either a high-frequency chart recorder or an oscilloscope. Because most pen galvanometer recorders have an upper-frequency limitation of around 100 or 200 Hz, photographic or light galvanometer recorders are required for faithful recording of heart sounds. Although normal heart sounds fall well within the frequency range of pen recorders, the high-frequency murmurs that are often important in diagnosis require the greater response of the photographic device.

Some manufacturers of multiple-channel physiological recording systems claim the phonocardiogram as one of the measurements they offer. They have available as part of their system a microphone and amplifier suitable for the heart sounds, the amplifier often being the same one used for EMG (see Chapter 10). Some of these systems, however, have only a pen recorder output, which limits the high-frequency response of the recorded signal to about 100 or 200 Hz.

The presence of higher frequencies (murmurs) in the phonocardiogram indicates a possible heart disorder. For this reason, a spectral analysis of heart sounds can provide a useful diagnostic tool for discriminating between normal and abnormal hearts. This type of analysis, however, requires a digital computer with a high-speed analog-to-digital conversion capability and some form of Fourier transform software. A typical spectrum of heart sounds is shown in Figure 6.38.

Microphones for phonocardiograms are designed to be placed on the chest, over the heart. However, heart sounds are sometimes measured from other vantage points. For this purpose, special microphone transducers are placed at the tips of catheters in order to pick up heart sounds from within the chambers of the heart or from the major blood vessels near

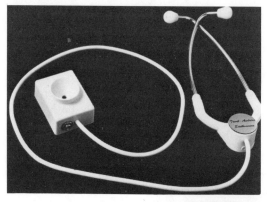

(a)

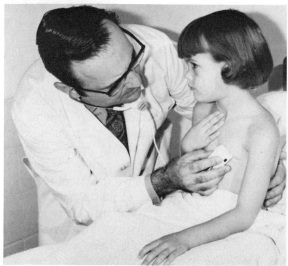

(b)

Figure 6.37. Electronic stethoscope: (a) the instrument; (b) in use on a patient. (Courtesy of Computer Medical Science Corporation, Tomball, Tex.)

the heart. Frequency-response requirements for these microphones are about the same as for phonocardiogram microphones. However, special requirements dictated by the size and configuration of the catheter must be considered in their construction. As might be expected, the difference in acoustical paths makes these heart sound patterns appear somewhat different from the usual phonocardiogram patterns.

The *vibrocardiograph* and the *apex cardiograph* which measure the vibrocardiogram and apex cardiogram respectively, also use microphones as transducers. However, since these measurements involve the low-frequency vibrations of the heart against the chest wall, the measurement is

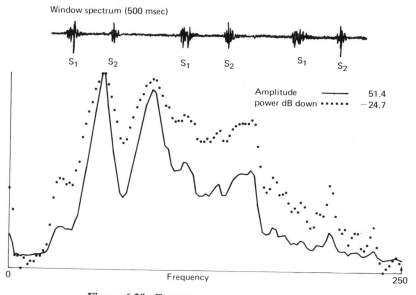

Figure 6.38. Frequency spectrum of heart sounds. (Courtesy of Computer Medical Science Corporation, Tomball, Tex.)

normally one of displacement or force rather than sound. Thus the microphone must be a good force transducer, with suitable low-frequency coupling from the chest wall to the microphone element. For the apex cardiogram, the microphone must be coupled to a point between the ribs. A soft rubber or plastic cone attached to the element of the microphone gives good results for this purpose.

Because the vibrocardiogram and the apex cardiogram do not contain the high-frequency components of the heart sounds, these signals can be handled by the same type of amplifiers and recorders as the electrocardiogram (see Section 6.1). Often these signals are recorded along with a channel of ECG data in order to maintain time reference. In this case, one channel of a multichannel ECG recorder is devoted to the heart vibration signal.

For recording the Korotkoff sounds from a partially occluded artery (see Chapter 5 and Section 6.2 in this chapter), a microphone is usually placed beneath the occluding cuff or over the artery immediately downstream from the cuff. The waveform and frequency content of these sounds are not as important as the simple identification of their presence, so these sounds generally do not require the high-frequency response specified for the phonocardiogram. Circuitry for identification of these sounds is included in certain automated, indirect blood pressure measuring devices (see Section 6.2.1.).

Measurement of the ballistocardiogram requires a platform mounted on a set of extremely flexible springs. When a person lies on the platform, the movement of his body in response to the beating of his heart and the ejection of blood causes similar movement of the platform. The amount of movement can be measured by any of the displacement or velocity transducers or accelerometers described in Chapter 9.

· 7 ·

PATIENT CARE
AND MONITORING

One area of biomedical instrumentation that is becoming increasingly familiar to the general public is that of patient monitoring. Here electronic equipment provides a continuous watch over the vital characteristics and parameters of the critically ill. In the coronary care and other intensive-care units in hospitals, thousands of lives have been saved in recent years because of the careful and accurate monitoring afforded by this equipment. Public awareness of this type of instrumentation has also been greatly increased by its frequent portrayal in television programs, both factual and fictional.

In hospitals that have engineering or electronics departments, patient–monitoring units, both fixed and portable, form a substantial part of the workload of the biomedical engineer or technician. Engineers and technicians are usually involved in the design of facilities for coronary or other intensive-care units, and they work closely with the medical staff to ensure that the equipment to be installed meets the needs of that particular hospital. Ensuring the safety of patients who may have conductive catheters or other direct electrical connections to their hearts is another function of these biomedical engineers and technicians. They also work with the contractors in the installation of the monitoring equipment and, when this job is completed, supervise equipment operation and maintenance. In addition, the engineering staff participates in the planning of improvements and additions, for there are many cases in which an intensive-care unit, even after careful design and installation, fails in some respect to meet the special needs of the hospital, and an in-house solution is required.

In this chapter the various elements that compose a patient-monitoring system are described, along with some of the problems often encountered with this type of equipment. Some specific systems are described in detail. Also discussed are pacemakers and defibrillators, which are generally associated with intensive coronary care.

7.1. THE ELEMENTS OF INTENSIVE-CARE MONITORING

The need for intensive-care and patient monitoring has been recognized for centuries. The 24–hour nurse for the critically ill patient has, over the years, become a familiar part of the hospital scene. But only in the last few years has equipment been designed and manufactured that is reliable enough and sufficiently accurate to be used extensively for patient monitoring. The nurse is still there, but her role has changed somewhat, for she now has powerful tools at her disposal for acquiring and assimilating information about the patients under her care. Thus she is able to render better service to a larger number of patients and is better able to react promptly and properly to an emergency situation. With the capability of providing an immediate alarm in the event of certain abnormalities in the behavior of a patient's heart, monitoring equipment makes it possible to summon a physician or nurse in time to administer emergency aid, often before permanent damage can occur. With prompt warning and by providing such information as the electrocardiogram record just prior to, during, and after the onset of cardiac difficulty, the monitoring system enables the physician to give the patient the correct drug rapidly. In some cases, even this process can be automated.

Physicians do not always agree among themselves as to which physiological parameters should be monitored. The number of parameters monitored must be carefully weighed against the cost, complexity, and reliability of the equipment. There are, however, certain parameters that provide vital information and that can be reliably measured at relatively low cost. For example, nearly all cardiac-monitoring units continuously measure the electrocardiogram from which the heart rate is easily derived. The electrocardiogram waveform is usually displayed and often recorded.

Temperature is also frequently monitored. On the other hand, there are some variables, such as blood pressure, in which the benefit of continuous monitoring is debatable in light of the problems associated with obtaining the measurement. Since continuous, direct, blood pressure monitoring requires catheterization of the patient, the traumatic experience of being catheterized may be more harmful to the patient than the lack of continuous pressure information. In fact, intermittent blood pressure measurements by means of a sphygmomanometer, either manual or automatic, might well provide adequate blood pressure information for most purposes. It is not the intent of this book to pass judgment on which measurements should be included but, rather, to familiarize the reader with the instrumentation used in patient care and monitoring.

Since patient-monitoring equipment is usually specified as a system, each manufacturer and each hospital staff has its own ideas as to what should be included in the unit. Thus a wide variety of configurations can be found in hospitals and in the manufacturers' literature. Since cardiac monitoring is the most extensively used type of patient monitoring today, it provides an appropriate example to illustrate the more general topic of patient monitoring.

The concept of intensive coronary care had little practicality until the development of electronic equipment that was capable of reliably measuring and displaying the electrical activity of the heart on a continuous basis. With such cardiac monitors, instant detection of potentially fatal arrhythmias finally became feasible. Combined with stimulatory equipment to reactivate the heart in the event of such an arrhythmia, a full system of equipment to prevent sudden death in such cases is now available.

In the intensive coronary care area, monitoring equipment is installed beside the bed of each patient to measure and display the electrocardiogram, heart rate, and other parameters being monitored from that patient. In addition, information from several bedside stations is usually displayed on a central console at the nurses' station.

Since many different arrangements are possible, a few systems in actual use are illustrated in Figures 7.1 through 7.5. Figure 7.1 shows a physician in a postsurgical intensive-care unit at a bedside terminal. Figure 7.2 shows a bedside terminal and a nurse's monitoring station. Figure 7.3 is a typical coronary-care ward showing a nurse at the central console monitoring eight patients. Figures 7.4 and 7.5 show other types of central monitoring situations.

As might be expected, many different room and facility layouts for intensive coronary care units are in use. One type that is quite popular is a U-shaped design in which six or eight cubicles or rooms with glass windows surround the nurses' central monitoring station. Although the

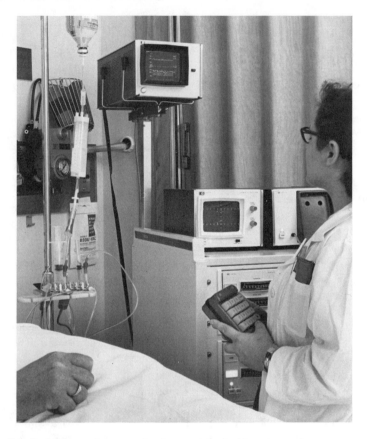

Figure 7.1. Bedside patient monitoring terminal. (Courtesy of Hewlett-Packard Company, Waltham, Mass.)

optimum number of stations per central console has not been established, a group of six or eight seems most efficient. For larger hospitals, monitoring of 16 to 24 beds can be accomplished by two or three central stations. The exact number depends on the individual hospital, its procedures, and the physical layout of the patient-care area. In certain areas in which recruitment of trained nurses is difficult, this factor could also be considered in selecting of the best design.

Although patient-monitoring systems vary greatly in size and configuration, certain basic elements are common to nearly all of them. A cardiac-care unit, for example, generally includes the following components:

1. Skin electrodes to pick up the ECG potentials. These electrodes are described in Chapter 4.

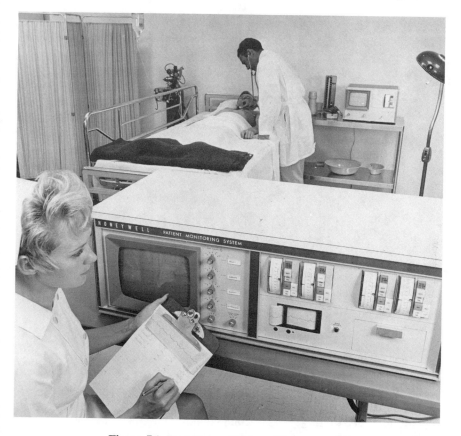

Figure 7.2. Bedside monitoring. (Courtesy of Honeywell Biomedical Electronic Products, Denver, Colo.)

2. Amplification equipment similar to that described in Chapter 6 for the electrocardiograph.
3. An oscilloscope display that permits direct observation of the ECG waveforms. The bedside monitors usually contain fairly small cathode-ray tube screens (2 to 5 inches in diameter), and each displays the ECG waveform from one patient. The central nurses' station generally has a larger oscilloscope screen on which electrocardiograms from several patients are displayed simultaneously.
4. A rate meter used to indicate the average number of heartbeats per minute and to provide a continuous indication of the heart rate. On most units, an audible beep or flashing light (or both) occurs with each heartbeat.
5. An alarm system, actuated by the rate meter, to alert the nurse or other

Figure 7.3. Coronary care central console unit. (Courtesy of Hewlett-Packard Company, Waltham, Mass.)

observer by audible or visible signals whenever the heart rate falls below or exceeds some adjustable preset range (e.g., 40 to 150 beats per minute).

In addition to these basic components, the following elements are useful and are often found in cardiac-monitoring systems:

1. A direct writeout device (an electrocardiograph) to obtain, on demand or automatically, a permanent record of the electrocardiogram seen on the oscilloscope. Such documentation is valuable for comparative purposes and is usually required in the event of an alarm condition. In combination with the tape loop described below, this written record provides a valuable diagnostic tool.

2. A memory-tape loop to record and play back the electrocardiogram for the 15 to 60 seconds just prior to an alarm condition. Recording of the ECG may continue until the system is reset. In this way, the electrical events associated with the heart immediately before, during, and following an alarm situation can be displayed if the nurse or some other observer was not present at the time of the occurrence.

3. Additional alarm systems triggered by ECG parameters other than the heart rate. These alarms may be activated by premature ventricular contractions or by widening of the QRS complex in the ECG (see Chap-

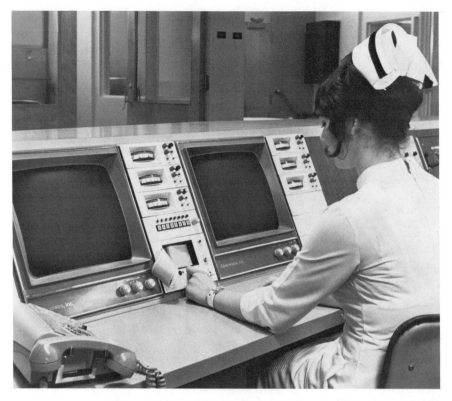

Figure 7.4. Multiple central monitoring unit. (Courtesy of Space-labs, Inc., Chatsworth, Calif.)

ter 3). Either situation may provide advance indication of a more serious problem.
4. Electrical circuits to indicate that an electrode has become disconnected or that a mechanical failure has occurred somewhere else in the monitoring system. Such a lead failure alarm permits instrumentation problems to be distinguished from true clinical emergencies.

Some of the features described in this paragraph can be seen in Figure 7.6, which is a closeup view of the front panel of the unit in Figure 7.2.

In addition to the components described above, some hospitals use computers or on-line computer terminals as an integral part of their patient-monitoring systems. In these situations, a special type of cathode-ray computer terminal may substitute for the more conventional patient-monitoring equipment. A terminal of this type is shown in Figure 7.7.

In this system all physiological signals are digitized and entered directly into the computer, which can monitor as many parameters as desired and

Figure 7.5. Eight channel nurses' station. (Courtesy of Spacelabs, Inc., Chatsworth, Calif.)

can watch for more subtle or complex conditions or for certain combinations of conditions that may indicate danger to the patient. Sudden changes in any of the parameters can also be detected. When appropriate, alarms are generated by the computer. From the buttons and controls at the terminal, the physician can request a printout or an oscilloscope plot of any desired parameter as an immediate aid to diagnosis or treatment. The more general topic of computers as a part of biomedical instrumentation is covered in Chapter 15.

Although not usually considered a part of the biomedical instrumentation system for patient monitoring, closed-circuit television is also used in some intensive-care areas in order to provide visual coverage in addition to monitoring the patients' vital parameters. Where television is employed, a camera is focused on each patient. The nurses' central station has either a bank of monitors, one for each patient, or a single monitor, which can be switched to any camera as desired.

Experience has shown that in spite of its value and increasing popu-

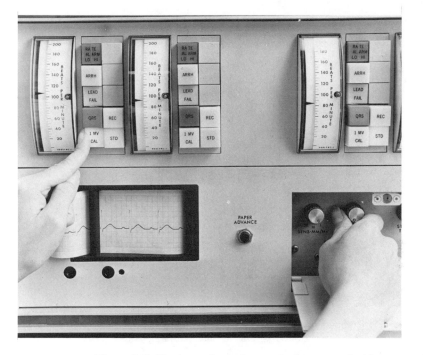

Figure 7.6. Close-up view of console front panel. (Courtesy of Honeywell Biomedical Electronic Products, Denver, Colo.)

larity, patient-monitoring equipment is not without its problems or limitations. Although many of the difficulties originally encountered in the development of such systems have been corrected, a number of significant problems remain. A few examples are given below.

EXAMPLE 1. Noise and movement artifact have always been a problem in the measurement of the electrocardiogram (see Chapter 6). Since heart rate meters, and subsequently the alarm devices, are usually trigged by the R wave of the ECG, and many systems cannot distinguish between the R wave and a noise spike of the same amplitude, movement or muscle interference may be counted as additional heartbeats. As a result, the rate meter shows a higher heart rate than that of the patient, and a high-rate false alarm is actuated. Unfortunately, repeated false alarms tend to cause the staff of the cardiac-care unit to lose confidence in the patient-monitoring equipment and either to ignore the alarms or turn them off altogether. Better electrodes that reduce patient movement artifact and more careful placement of electrodes to avoid areas of muscle activity can help to some extent. Electronic filtering to reduce the response of the system at frequencies at which interference might be

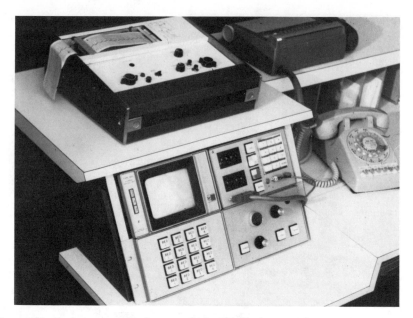

Figure 7.7. Computer terminal nurses station for patient monitoring system. (Courtesy of Latter Day Saints Hospital, Salt Lake City, Utah.)

expected is also partially effective. Some of the more sophisticated systems include circuitry that identifies additional characteristics of the ECG other than simply the amplitude of the R wave, thus further reducing the possibility of mistaking noise for the ECG signal. In spite of all these measures, the possibility of false alarm signals due to movement or muscle artifact is still a real problem.

EXAMPLE 2. The low-rate alarm can be falsely activated if the R wave of the ECG is of insufficient amplitude to trigger the rate meter. This can happen if the contact between the electrodes and the skin becomes disturbed because of improper application of the electrodes, excessive patient sweating, or drying of the electrode paste or jelly. In some cases, the indication on the oscilloscope approximates that which might appear if the heart stopped beating. Even though a false low-rate alarm indicates an equipment problem that requires attention, the danger of mistaking failure of the electrode connections for cardiac standstill can have serious consequences. To prevent this possibility, lead-failure alarms have been designed and are built into some patient-monitoring systems.

EXAMPLE 3. Because the ECG electrodes must remain attached to the skin for long periods of time during patient monitoring, inflammatory reactions at electrode locations are common. Special skin care and proper application of the electrodes can help minimize this problem.

Special electrode placement patterns are often used in patient monitoring applications. These patterns are generally intended to approximate the standard limb lead signals (see Chapter 6), while avoiding the actual placement of electrodes on the patient's arms and legs. Instead, the RA, LA, and RL electrodes are placed at appropriate positions on the patient's chest. Because many possible chest placement patterns provide suitable approximations of the three limb leads, no standard pattern has been determined, although some hospitals and manufacturers of patient monitoring equipment may specify a particular arrangement.

7.2. HOSPITAL SYSTEMS AND COMPONENTS

In addition to its use at the bedside, as discussed in Section 7.1 and illustrated in Figures 7.1 through 7.5, patient monitoring equipment is often found in other applications in the hospital. An important example is in the operating room. Figure 7.8 shows such a unit being used during surgery. The main features of this type of system are the large multichannel oscilloscope, the capability of obtaining a permanent ECG record on the chart recorder, and plug-in signal-conditioner modules that provide versatility and choice of measurement parameters.

The chart recorder has eight channels to be used as dictated by the specific requirements of the surgical team. The fluid writing system is pressurized and writes dry. The frequency response of such a unit is up to 40 Hz full scale. The trace is rectilinear, and the channel span is 40 mm graduated in 50 divisions. Two event channels are provided to relate information on the chart to specific events. The recorder has a large number of chart speeds selected by push buttons, ranging from 0.05 to 200 mm per second.

As mentioned earlier, a variety of plug-in modules are available for use with the unit. These biomedical signal conditioners include a universal unit for various bioelectric signals, an ECG unit, an EEG unit, a biotachometer, a transducer unit, an integrator, a differentiator, and an impedance unit. These units are all compatible. Figure 7.9 shows the front plates of some of them, whereas Figure 7.10 shows a complete biotachometer unit.

Each of the signal conditioners includes two separate submodules: a coupler and a medical amplifier. The coupler contains the circuitry and controls that are essential to its nameplate function. The amplifier or "back end" contains circuitry that is exactly the same for each channel. This "back end" can be obtained separately, thus reducing a possible

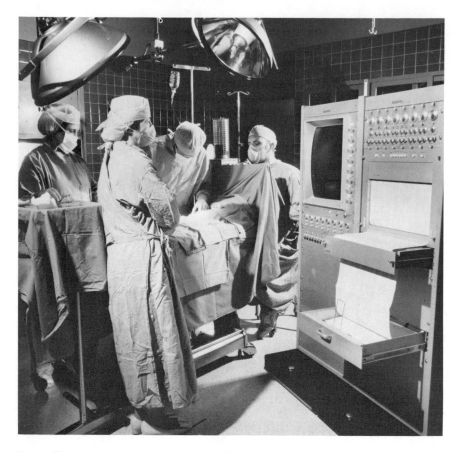

Figure 7.8. Surgical monitoring system. (Courtesy of Gould, Inc., Brush Instruments Division, Cleveland, Ohio.)

investment in idle circuitry. In the last few years many manufacturers of biomedical monitoring equipment have improved their systems to include such versatility.

The amplifier units are designed for broad-band amplification from dc to 10 kHz. The amplifier is a ground-isolation type that eliminates a potential shock hazard to the patient by isolating him to the extent that current through his body cannot exceed 2 μA (see Chapter 16).

Each amplifier provides a buffered output drive signal for peripheral monitoring equipment, such as tape recorders, oscilloscopes, and computers.

Each biomedical coupler unit has a high input impedance compatible with the function it performs, plus differential inputs with good common-mode rejection at 60 Hz. In general, the sensitivity varies with the par-

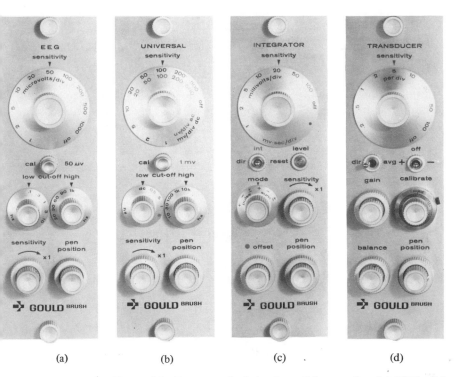

Figure 7.9. Front panel of signal conditioner units: (a) EEG; (b) universal; (c) integrator; (d) transducer. (Courtesy of Gould, Inc., Brush Instruments Division, Cleveland, Ohio.)

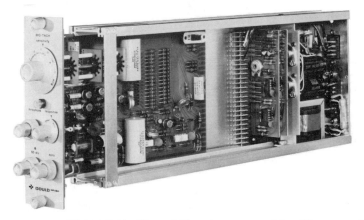

Figure 7.10. View of biotachometer module. (Courtesy of Gould, Inc., Brush Instruments Division, Cleveland, Ohio.)

ticular unit. However, the universal biomedical coupler, which can be used for phonocardiography, electromyography, electrocardiography, and other measurements involving low- or medium-voltage signals, has a measurement range on an ac setting of from 20 μV per division to 25 mV full scale, and, on dc, of from 1 mV per division to 25 volts full scale.

For greater versatility, a model similar to that shown in the operating room appears in Figure 7.11. An upright model with a swivel-top oscilloscope, it can be used equally well for research or for monitoring.

Figure 7.11. Upright surgical monitoring system. (Courtesy of Gould, Inc., Brush Instruments Division, Cleveland, Ohio.)

Most of the instruments illustrated in this book are from U.S. manufacturers; however, many other countries have become involved in patient monitoring. Although a sizable part of the equipment used elsewhere is manufactured in this country, a growing number of companies are manufacturing in other parts of the world. Since such equipment is often quite different from that made in the United States, it is useful to provide an example. Figure 7.12 shows a single-channel bedside station currently being used in hospitals in Australia.

The patient bedside unit consists of four or five modules placed in the assembly: an EEG preamplifier unit, an ECG preamplifier unit, a heart-rate alarm module, a single-channel monitorscope, and an automatic chart recorder. The recorder and monitorscope are similar to those described previously.

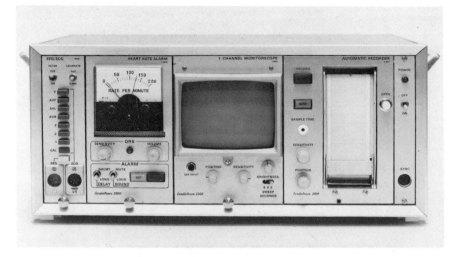

Figure 7.12. Australian bedside station. (Courtesy of Watson Victor, Ltd., Sydney, Australia.)

The modular unit for the central nursing station for a six-bed ward is shown in Figure 7.13. It is a duplicate of the bedside unit in terms of the heart-rate alarm system. The six electrocardiograms are monitored on three two-channel oscilloscopes. Printed records can be obtained by the use of a six-button automatic selector and the recorder. The system also has provisions for an automated sampling sequence with the elimination of vacant beds. Anytime a patient goes into "alarm," the selector automatically records the ECG from that patient.

7.3. PACEMAKERS

Although pacemakers are not actually measurement instruments, their operating principles and applications are so closely allied with patient care that the engineer or technician in the field should be familiar with their characteristics and use.

In Chapter 5 it was shown that the rhythmic activity of the heart is controlled by a neuronal "pacemaker" located on the surface of the heart near the top of the right atrium. Although this natural pacemaker (also called the sinoatrial or SA node) is capable of self-pacing and independent timing, it is normally controlled by both the parasympathetic and sympathetic nervous systems (see Chapter 10). When either the natural pace-

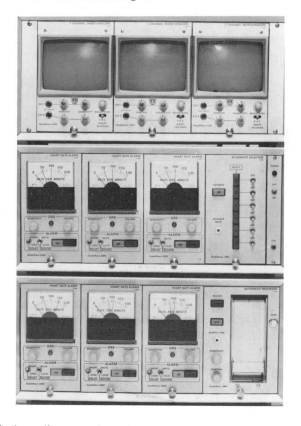

Figure 7.13. Australian central nursing station for six bed ward. (Courtesy of Watson Victor, Ltd., Sydney, Australia.)

maker or the controlling innervation to the heart becomes impaired to the point where the heart no longer provides sufficient circulation of the blood throughout the body, an artificial method of pacing the heart may be required. In such cases, periodic stimulation of the appropriate region of the myocardium substitutes for the natural pacing signals and triggers the heart at a rate adequate to maintain proper circulation.

Like the natural pacemaker, a device capable of generating artificial pacing signals and delivering them to the heart is also called a *pacemaker.* Artificial pacemakers come in a variety of forms. They can either be internally implanted for use by patients with permanent heart blocks who may require assisted pacing for the rest of their lives, or they may be worn externally for temporary requirements.

Internal pacemakers are surgically implanted beneath the skin, usually in the region of the abdomen. Internal leads connect to electrodes that are inserted directly into the myocardium. Since there are no external connections for applying power, the unit must be completely self-contained with a power source capable of continuously operating the pacemaker for a period of years. An internal pacemaker is shown in Figure 7.14, with a block diagram.

(a)

Figure 7.14. Internal pacemaker: (a) photograph of the unit; (b) block diagram. (Courtesy of Medtronic, Inc.)

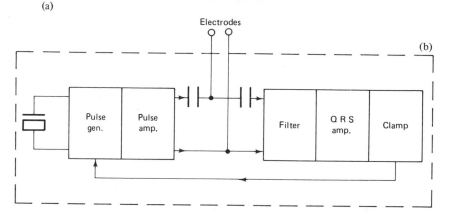

External pacemakers, which include all types of pacing units located outside the body, may either have electrodes attached directly to the myocardium with the connecting wires penetrating the skin, or the electrodes may be introduced into one of the chambers of the heart through a cardiac catheter, called the *pacing catheter,* as shown in Figure 7.15. When the pacing catheter is used, the leads are brought out through the catheter. Most external pacemakers in use today are small, portable, battery-operated units of the type shown in Figure 7.16. The pacemaker can be worn by an ambulatory patient, or it can be attached to the bed or arm of a patient confined to bed. Sometimes larger bedside units are used to provide the stimulating currents.

(a)

(b)

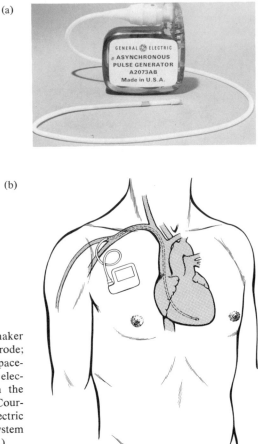

Figure 7.15. (a) Pacemaker used with catheter electrode; (b) Implanted standby pacemaker with catheter electrodes inserted through the right cephalic vein. (Courtesy of General Electric Company, Medical System Dept., Milwaukee, Wis.)

In some cases of acute heart block, when there is insufficient time to insert a catheter or otherwise place electrodes on or in the heart, pacing can be accomplished through the intact chest. In this method, also called external pacing, high-intensity pulses of current are applied through large electrodes placed on the chest. This form of pacing is rarely used today, however, because of the discomfort and possible burns caused by the large current levels required and because of improved techniques of catheter pacing.

All pacemakers, both internal and external, consist of circuitry for generating the stimulating pulses and for controlling their intensity and

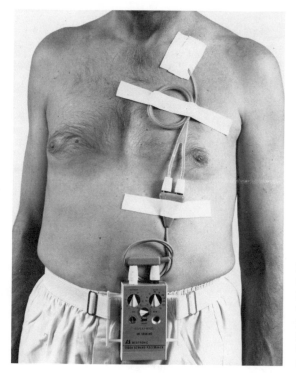

Figure 7.16. Portable external pacemaker. Patient is being temporarily paced with an external demand pacemaker and transvenous pacing catheter. (Courtesy of Medtronic, Inc., Minneapolis, Minn.)

rate. In addition, the pacemaker must include a suitable power source and electrodes for delivering the pulses to the heart. The pulse generator must be capable of producing a continuous sequence of fixed, short-duration pulses of current with a relatively long, controllable interval between pulses. The pulses are usually rectangular with a duration of from 2 to 5 msec. Pulse rates normally range from about 40 or 50 per minute to over 120 per minute. The actual rates that can be obtained from a given pacemaker, and the intensity of the output pulses, depend on the type of pacemaker (internal or external) and the needs of the patient. In the design of internal pacemakers, where every effort is made to conserve battery power, pulses are kept as low in intensity and as short as possible without jeopardizing the pacemaker's performance.

Pulses applied directly to the heart are generally in the range of 5 to 15 volts or below 20 mA, depending on whether voltage or current is controlled. If pacing must be done through the intact chest wall, however, amplitudes ten times as great are required.

Both *unipolar electrodes,* in which one electrode is placed on or in the heart and the reference electrode is located somewhere away from the heart, and *bipolar electrodes,* in which both electrodes are on or in the

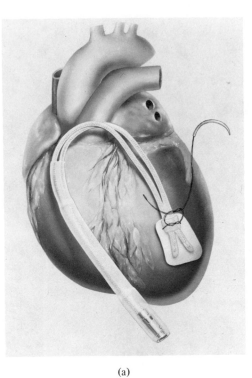

Figure 7.17. (a) Pacing electrodes attached to the myocardium; (b) Myocardial electrodes with pacemaker generator implanted in abdomen. (Courtesy of General Electric Company, Medical Systems Department, Milwaukee, Wis.)

(a)

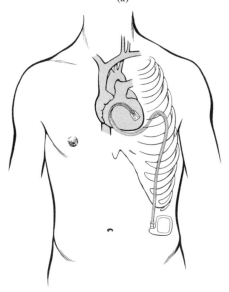

(b)

heart, are used in pacing. Either type can be used when electrodes are attached to the myocardium or with a pacing catheter. Attachment of electrodes to the heart is shown in Figure 7.17.

Large skin surface electrodes are used when pacing must be performed through the intact chest wall. By having a greater surface contact area on the electrodes, the danger of burns from the pacing current is reduced.

The power source used depends on whether the pacemaker is internal or external. External pacemakers are either battery powered or they receive power from the ac power line. Because of the need to isolate patients with direct connections to their hearts from any possible source of power line leakage current (see Chapter 16), and for the sake of portability, battery-powered units are preferred.

Efficiency of design in an external pacemaker is not as crucial as that of an internal pacemaker, because batteries are easily accessible for replacement and the size of the unit is not critical. When the entire unit is surgically implanted, as is the case with the internal pacemaker, however, every effort must be made to avoid the need for power supply replacement any more often than is absolutely necessary. For this reason, circuitry for internal pacemakers must be carefully designed to minimize power drain. Long-life mercury batteries, capable of operating the pacemaker up to 5 years, are most commonly used. These batteries, as might be expected, are still the limiting factor in the useful life of the pacemaker. Although manufacturers go to extremes in manufacture and testing in order to ensure long life, such as quality checking each cell by X rays to detect possible sources of failures, the average life of a pacemaker battery is still only 2 to 3 years.

The use of nuclear power for pacemaker operation is currently under development and seems promising. It is estimated that incorporation of nuclear batteries will extend the useful life of the pacemaker to at least 10 years. Attempts are also underway to harness the natural motion of the heart and the great vessels for the generation of power to operate the pacemaker. Efforts to apply power to an internal pacemaker from an external energy source by use of radio-frequency transmission or inductive coupling have not proven successful, although a possible breakthrough with the latter type has been claimed.

Since the natural "pacemaker" in the heart, under neuronal control, is constantly able to change its rate to meet the needs of the body, it would seem desirable to provide the artificial pacemaker with automatic control of its rate. This desire, however, has not yet been achieved. External pacemakers can be manually adjusted to any rate within their range; and by observing the patient and his electrocardiogram, the physician or nurse can correct for an obvious error in rate.

With internal pacemakers, the problem is complicated by the difficulty in gaining access to the pacemaker. Most internal pacemakers have fixed rates at some value around 70 beats per minute. Dual-rate units are also available with a slow (60 to 70 beats per minute) rate for resting and a faster (about 85 beats per minute) rate for periods of activity. The unit is switched by means of an external control using an induction coil or a permanent magnet. Units with up to four switchable rates have been implanted.

The switchable rate units described are intended to be switched from one rate to the other as required throughout the life of the pacemaker. Another type of internal pacemaker has a continuous rate adjustment controlled by a needle that must penetrate the skin. This adjustment is usually made once after implantation and is left at that setting unless an emergency situation dictates readjustment.

External pacemakers also permit adjustment of the intensity of the stimulation pulses to meet the needs of the individual patient. Internal pacemakers, on the other hand, have a fixed intensity, usually at least three times the heart's normal threshold level to ensure continued operation as the batteries wear down. Again there is one exception: an internal pacemaker with an adjustment that can be made by penetrating the skin with a needle to set the level after implantation.

All the pacemakers described so far operate continuously at the rate to which they are set. There are patients, however, who can usually generate a normal heart rhythm but who require the assistance of a pacemaker intermittently. In this situation, a special *demand pacemaker* is implanted. The demand pacemaker includes a sensing device that is triggered by the R wave of the ECG. If, for any reason, the R wave does not occur within a preset "standby" interval of time, the pacemaker triggers the heartbeat. Thus stimulating pulses are applied to the heart only when required.

7.4. DEFIBRILLATORS

As discussed in Chapters 3 and 5, the heart is able to perform its important pumping function only through precisely synchronized action of the heart muscle fibers. The rapid spread of action potentials over the surface of the atria causes these two chambers of the heart to contract together and pump blood through the two atrioventricular valves into the ventricles. After a critical time delay, the powerful ventricular muscles are synchronously activated to pump blood through the pulmonary and systemic circulatory systems. A condition in which this necessary synchronism is lost is known as *fibrillation*. During fibrillation the normal rhythmic contractions of either the atria or the ventricles are replaced by rapid irregular twitching of the muscular wall. Fibrillation of atrial muscles is called *atrial fibrillation,* whereas fibrillation of the ventricles is known as *ventricular fibrillation.*

Under conditions of atrial fibrillation, the ventricles can still function normally, but they respond with an irregular rhythm to the nonsynchronized bombardment of electrical stimulation from the fibrillating atria. Since most of the blood flow into the ventricles occurs before atrial contraction, there is still blood for the ventricles to pump. Thus even with atrial fibrillation circulation may still be maintained, although not as efficiently. The sensation produced, however, by the fibrillating atria and irregular ventricular action can be quite traumatic for the patient.

Ventricular fibrillation is far more dangerous, for under this condition the ventricles are unable to pump blood; and if the fibrillation is not corrected, death will usually occur within a few minutes. Unfortunately, fibrillation, once begun, is not self-correcting. Hence a patient susceptible to ventricular fibrillation must be watched continuously so that the medical staff can respond immediately if an emergency occurs. This is the primary reason for cardiac monitoring, which was discussed earlier.

Although mechanical methods (heart massage) for defibrillating patients have been tried over the years, the most successful method of defibrillating is the application of an electric shock to the area of the heart. Since the heart muscle fibers respond to electrical excitation, if sufficient current to contract all musculature of the heart simultaneously is applied for a brief period and then released, all the heart muscle fibers enter their refractory periods together, after which normal heart action may resume. The discovery of this phenomenon led to the rather widespread use of defibrillation by applying a brief (0.25 to 1 second) burst of 60-Hz ac at an intensity of around 6 amperes to the chest of the patient. This application of an electrical shock to resynchronize the heart is sometimes called *countershock.* If the patient did not respond, the burst was repeated until defibrillation had occurred. This method of countershock was known as *ac defibrillation,* and a large number of ac defibrillators appeared on the market and found their way into most hospitals.

A number of disadvantages can be accrued to ac defibrillation, however. Successive attempts to correct ventricular fibrillation are often required. Moreover, ac defibrillation cannot be successfully used to correct atrial defibrillation. In fact, attempts to correct atrial fibrillation by this method often result in the more serious ventricular fibrillation. Another problem is that in some hospitals the application of the direct load of an ac defibrillation on the power line caused a momentary voltage drop that interfered with other equipment and made control of the defibrillating current difficult.

About 1960 a number of experimenters began working with direct currents for defibrillation. Various schemes and waveforms were tried until, in late 1962, Dr. Bernard Lown of the Harvard School of Public Health and Peter Bent Brigham Hospital developed a new method of dc defibrillation that has found common use today. In this method a capacitor is charged to a high dc voltage and, at a crucial part of the ECG cycle (if there is an ECG), the capacitor is discharged within a few milliseconds across the chest of the patient. It was found that this method of defibrillation is not only more successful in correcting ventricular fibrillation than the ac method, but it can also be used successfully for correcting atrial fibrillation and other types of arrhythmias. The dc method requires fewer repetitions and is less likely to harm the patient.

In the Lown method the energy discharged by the capacitor ranges between 100 and 400 watt-seconds or joules. Duration of the discharge pulse is usually between 2.5 and 5 msec. The waveform is approximately that shown in Figure 7.18. An inductance is used to damp the waveform

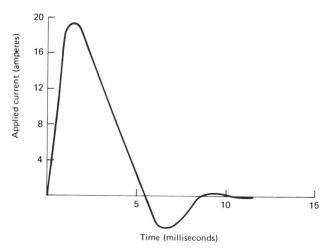

Figure 7.18. Lown defibrillator discharge waveform.

and to eliminate the sharp current spike that would otherwise occur at the beginning of the discharge.

Even with the Lown waveform there is danger of damage to the myocardial surface or the skin at the point of application, for voltages of 6000 or 7000 volts are used. To reduce this risk, a tapered delay-line discharge system with two sections of capacitance and inductance is used in one type of commercially available defibrillator. With this method, a dual-peak waveform of longer duration can be used at a much lower voltage. Effective defibrillation can be achieved with from 80 to 140 watt-seconds. The waveform from a tapered delay-line defibrillator is shown in Figure 7.19.

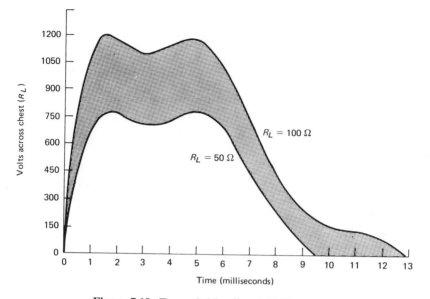

Figure 7.19. Tapered delay line defibrillator waveform. (Courtesy of Travenol Laboratories, Inc., Zenith Radio Corporation, Deerfield, Ill.)

To avoid the possibility of ventricular fibrillation resulting from the application of the dc pulse for atrial defibrillation or correction of some other form of arrhythmia, the capacitor discharge must be synchronized with the electrocardiogram. The optimum time for discharge is during or immediately following the downward slope of the R wave. The time near the occurrence of the T wave should particularly be avoided, for this is the period of the heart's greatest susceptibility to ventricular fibrillation. Most modern defibrillators include provisions for synchronizing the discharge pulse with the ECG. A typical defibrillator is shown in Figure 7.20, with a typical circuit shown in Figure 7.21.

In order to prevent burns from the large current discharge applied through the skin for defibrillation, a special type of electrode is used. These electrodes, often called *paddles,* usually measure 3 to 4 inches in diameter. For application of the defibrillation pulse, a pair of these electrodes is held against the chest of the patient, as shown in Figure 7.22. To protect the person applying the electrodes from electric shock, special insulated handles are provided. A thumb switch, located in one (or both) of the handles is used to apply the impulse when the paddles are in place. This device prevents the patient, or anyone else, from receiving the shock prematurely. In some earlier systems, a foot switch was used instead. The possibility of someone accidentally stepping on the foot switch before the paddles are in place, in the excitement of an emergency, makes the thumb switches in

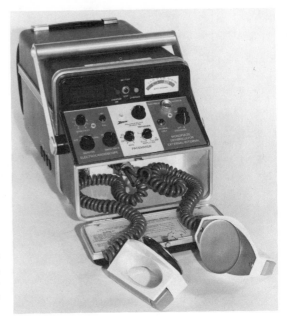

Figure 7.20. Monopulse dc defibrillator with paddles. This portable unit incorporates a defibrillator, electrocardioscope and pacemaker. (Courtesy of Travenol Laboratories, Inc., Zenith Radio Corporation, Deerfield, Ill.)

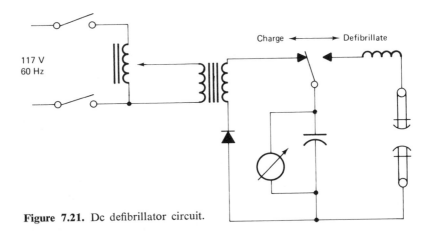

Figure 7.21. Dc defibrillator circuit.

the handles preferable. When the discharge is to be automatically synchronized with the ECG waveform, closing of the thumb switches simply permits the defibrillator to discharge the next time the appropriate portion of the ECG cycle occurs.

With a dc defibrillator there is no direct loading of the power line, for the burst of energy is provided by the discharge of one or more capacitors. Thus interference with other equipment is greatly reduced. Because the

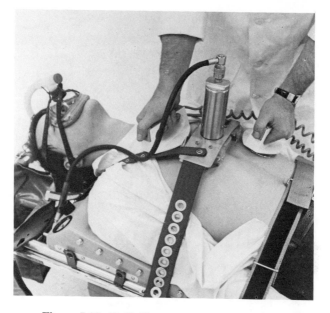

Figure 7.22. Defibrillator paddles applied to the chest of a patient, who is undergoing cardio-pulmonary resuscitation. A facemask maintains respiration while a pneumatic "thumper" applies periodic pressure to the sternum to maintain blood circulation. (Courtesy Travenol Laboratories, Inc., Zenith Radio Corporation, Deerfield, Ill.)

capacitor must recharge after each use, however, several seconds must be allowed between applications for recharging.

Defibrillation is sometimes required during cardiac surgery when the heart is exposed. Here internal defibrillation is accomplished by direct application of the discharge current to the heart. A much smaller charge, usually at a voltage less than one-tenth of that used for external defibrillation, is required when the discharge is applied directly to the heart.

· 8 ·

MEASUREMENTS IN
THE RESPIRATORY
SYSTEM

The exchange of gases in any biological process is termed *respiration.* To sustain life, the human body must take in oxygen, which combines with carbon, hydrogen, and other elements furnished by food material to produce heat and energy for the performance of work. As a result of this process of *metabolism,* which takes place in the cells, a certain amount of water is produced along with the principal waste product, carbon dioxide (CO_2). The entire process of taking in oxygen from the environment, transporting the oxygen to the cells, removing the carbon dioxide from the cells, and exhausting this waste product into the atmosphere must be considered within the definition of respiration.

In the human body, the tissue cells generally are not in direct contact with their external environment. Instead the cells are bathed in fluid. This tissue fluid can be considered as the *internal environment* of the body. The cells absorb oxygen from this fluid and discharge waste materials, including carbon dioxide, into the fluid. The circulatory blood is the medium by which oxygen is brought to the internal environment. Carbon dioxide is carried from the tissue fluids by the same mechanism. The exchange of

170

gases between the blood and the external environment takes place in the *lungs* and is termed *external respiration.*

The function of the lungs is to oxygenate the blood and to eliminate carbon dioxide in a controlled manner. During inspiration fresh air enters the respiratory tract, becomes humidified and heated to body temperature, and is mixed with the gases already present in the region comprising the trachea and bronchi (see Figure 8.1). This gas is then mixed further with the gas residing in the alveoli as it enters these small sacs in the wall of the lungs. Oxygen diffuses from the alveoli to the pulmonary capillary blood supply, whereas carbon dioxide diffuses from the blood to the alveoli. The oxygen is carried from the lungs and distributed among the various cells of the body by the blood circulation system, which also returns the carbon dioxide to the lungs. The entire process of inspiring and expiring air, exchange of gases, distribution of oxygen to the cells, and collection of CO_2 from the cells forms what is known as the *pulmonary function.* Tests for assessing the various components of the process are called *pulmonary function tests.*

Unfortunately, no single laboratory test or even a simple group of tests is capable of completely measuring pulmonary function. In fact, the field of instrumentation for obtaining pulmonary measurements is quite complex. However, tests and instrumentation for the measurement of respiration can be divided into two categories. The first includes tests designed to measure the mechanics of breathing and the physical characteristics of the lungs; the second category is involved with diffusion of gases in the lungs, the distribution of oxygen, and the collection of carbon dioxide.

This chapter begins with a brief presentation of the physiology of the respiratory system; then the tests and instrumentation associated with each of the two categories of measurements described above are covered. Because of the complexity of the field, it is almost impossible to cover all tests or all types of instrumentation used in either category. However, an attempt has been made to include the most meaningful ones, as well as those with which the biomedical engineer or technician is most likely to become associated.

8.1. THE PHYSIOLOGY OF THE RESPIRATORY SYSTEM

Air enters the lungs through the air passages, which include the *nasal cavities, pharynx, larynx, trachea, bronchi,* and *bronchioles,* as shown in Figure 8.1.

The lungs are elastic bags located in a closed cavity, called the *thorax*

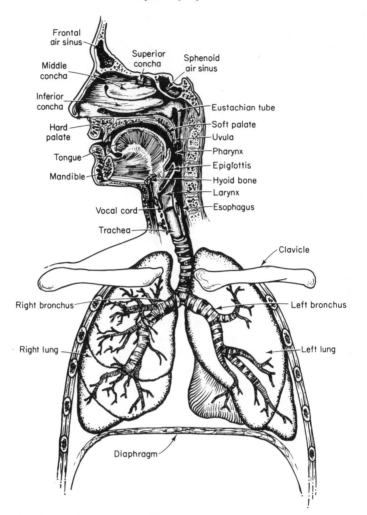

Figure 8.1. The respiratory tract. (From W.F. Evans, *Anatomy and Physiology, The Basic Principles,* Prentice-Hall, Inc., 1971, by permission.)

or *thoracic cavity*. The right lung consists of three lobes (upper, middle, and lower), and the left lung has two lobes (upper and lower).

The *larynx,* sometimes called the "voice box" (because it contains the vocal cords), is connected to the bronchi through the *trachea,* sometimes called the "windpipe." Above the larynx is the *epiglottis,* a cover that closes whenever a person swallows, so that food and liquids are directed to the esophagus (tube leading to the stomach) and into the stomach rather than into the larynx and trachea.

The trachea is about 1.5 to 2.5 cm in diameter and approximately 11 cm long, extending from the larynx to the upper boundary of the chest. Here it bifurcates (forks) into the right and left main stem *bronchi*. Each bronchus enters into the corresponding lung and divides like the limbs of a tree into smaller branches. The branches are of unequal length and at different angles, with over 20 of these nonsymmetrical bifurcations normally present in the human body. Farther along these branchings, where the diameter is reduced to about 0.1 cm, the air-conducting tubes are called *bronchioles*. As they continue to decrease in size to about 0.05 cm in diameter, they form the *terminal bronchioles,* which branch again into the *respiratory bronchioles,* where some alveoli are attached as small air sacs in the walls of the lung. After some additional branching, these air sacs increase in number, becoming the *pulmonary alveoli.* The alveoli are each about 0.02 cm in diameter. It is estimated that, all told, some 300 million alveoli are found in the lungs. (See Figure 8.2.)

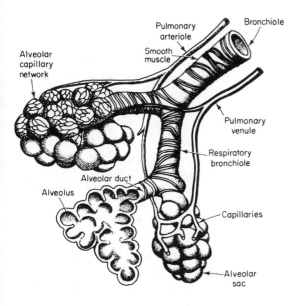

Figure 8.2. Alveoli and capillary network. (From W.F. Evans, *Anatomy and Physiology, The Basic Principles,* Prentice-Hall, Inc., 1971, by permission.)

Beyond about the tenth stage of branching, the bronchioles are embedded within alveolar lung tissue; and with the expansion and relaxation of the lung, their diameters are greatly affected by the lung size or lung volume. Up to this point the diameter of the air sacs is more affected by the *pleural pressure,* or the pressure inside the thorax.

The lungs are covered by a thin membrane, called the *pleura,* which passes from the lung at its root onto the interior of the chest wall and upper surface of the diaphragm. The two membranous sacs so formed are called the *pleural cavities,* one on each side of the chest, between the lungs and the thoracic boundaries. These so-called cavities are potential ones only, for the pleura covering the lung and that lining the chest are in contact in the healthy condition. Fluid or blood, as well as air, may collect in this potential space to create an actual space in certain diseases. The part of the pleural membrane lining the thoracic wall is called the *parietal* pleura, whereas that portion covering and firmly adherent to the surface of the lungs themselves is called the *pulmonary pleura* or *visceral pleura.* A small amount of fluid, just wetting the surfaces between the pleura, allows the lungs and the lobes of the lungs to slide over each other and on the chest wall easily with breathing.

Breathing is accomplished by musculature that literally changes the volume of the thoracic cavity and, in so doing, creates negative and positive pressures that move air into and out of the lungs. Two sets of muscles are involved: those in and near the diaphragm that cause the diaphragm to move up and down, changing the size of the thoracic cavity in the vertical direction, and those that move the rib cage up and down to change the lateral diameter of the thorax.

The *diaphragm* is a special dome- or bell-shaped muscle located at the bottom of the thoracic cavity, which, when contracted, pulls downward to enlarge the thorax. This action is the principal force involved in inspiration. At the same time as the diaphragm moves downward, a group of external intercostal muscles lifts the rib cage and sternum. Because of the shape of the rib cage, this lifting action also increases the effective diameter of the thoracic cavity. The resultant increase in thoracic volume creates a negative pressure (vacuum) in the thorax. Since the thorax is a closed chamber and the only opening to the outside is from the inside of the lungs, the negative pressure is relieved by air entering the lungs. The lungs themselves are passive and expand only because of the internal pressure of air in the lungs, which is greater than the pressure in the thorax outside the lungs.

Normal expiration is essentially passive, for, on release of the inspiratory muscles, the elasticity of the lungs and the rib cage, combined with the tone of the diaphragm, reduces the volume of the thorax, thereby developing a positive pressure that forces air out of the lungs. In forced expiration a set of abdominal muscles pushes the diaphragm upward very powerfully while the internal intercostal muscles pull the rib cage downward and apply pressure against the lungs to help force air out.

During normal inspiration the pressure inside the lungs, the *intra-alveolar pressure,* is about -3 mm Hg, whereas during expiration the pressure becomes about $+3$ mm Hg. The ability of the lungs and thorax to expand during breathing is called the *compliance,* which is expressed as the volume increase in the lungs per unit increase in intra-alveolar pressure. The resistance to the flow of air into and out of the lungs is called *airway resistance.*

As described in Chapter 5, blood from the body tissues and their capillaries is brought via the superior and inferior vena cava into the right atrium of the heart, which in turn empties into the right ventricle. The right ventricle pumps the blood into and through the lungs in a pulsating fashion, with a systolic pressure of about 20 mm Hg and a diastolic pressure of 1 to 4 mm Hg. By perfusion, the blood passes through the pulmonary capillaries, which are in the walls of the air sacs, wherein oxygen is taken up by the red blood cells and hemoglobin. The compound formed by the oxygen and the hemoglobin is called *oxyhemoglobin.* At the same time, carbon dioxide is removed from the blood into the alveoli.

From the pulmonary capillaries, the blood is carried through the pulmonary veins to the left atrium. From here it enters the left ventricle, which pumps the blood out into the aorta at pressures of 120/80 mm Hg. It is then distributed to all the organs and muscles of the body. In the tissues, the oxyhemoglobin gives up its oxygen, while carbon dioxide diffuses into the blood from the tissue and surrounding fluids. The blood then flows from the capillaries into the venous system back into the superior and inferior vena cava.

The interchange of the oxygen from the lungs to the blood and the diffusion of carbon dioxide from the blood to the lungs take place in the capillary surfaces of the alveoli. The alveolar surface area is about 80 square meters, of which more than three-quarters is capillary surface.

In order to understand some of the terminology used in conjunction with the tests and instrumentation involved in respiratory measurements, definition of a few medical terms is necessary. Additional definitions are included in the glossary in Appendix A.

Hypoventilation is a condition of insufficient ventilation by an individual to maintain his normal P_{CO_2} level, whereas *hyperventilation* refers to abnormally prolonged, rapid, or deep breathing. Hyperventilation is also the condition produced by overbreathing. *Dyspnea* is the sensation of inadequate or distressful respiration, a condition of abnormal breathlessness. *Hypercapnia* is an excess amount of CO_2 in the system, and *hypoxia* is a shortage of oxygen. Both hypercapnia and hypoxia can result from inadequate ventilation.

8.2. TESTS AND INSTRUMENTATION FOR THE MECHANICS OF BREATHING

The mechanics of breathing concern the ability of a person to bring air into his lungs from the outside atmosphere and to exhaust air from the lungs. This ability is affected by the various components of the air passages, the diaphragm and associated muscles, the rib cage and associated musculature, and the characteristics of the lungs themselves. Tests can be performed to assess each of these factors, but no one measurement has been devised that can adequately and completely evaluate the performance of the breathing mechanism. This section describes a number of the most prominent measurements and tests that are used clinically and in research in connection with the mechanics of breathing. In addition, the instrumentation required for these tests and measurements is described and discussed. In some cases, one instrument can be used for the performance of several tests.

8.2.1. LUNG VOLUMES AND CAPACITIES. Among the basic pulmonary tests are those designed for determination of lung volumes and capacities. These parameters, which are a function of an individual's physical characteristics and the condition of his breathing mechanism, are given in Figure 8.3.

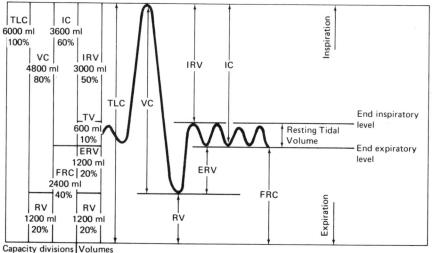

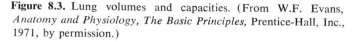

Figure 8.3. Lung volumes and capacities. (From W.F. Evans, *Anatomy and Physiology, The Basic Principles,* Prentice-Hall, Inc., 1971, by permission.)

The *tidal volume* (TV), or normal depth of breathing, is the volume of gas inspired or expired during each normal, quiet, respiration cycle.

Inspiratory reserve volume (IRV) is the extra volume of gas that a person can inspire with maximal effort after reaching the normal end inspiratory level. The *end inspiratory level* is the level reached at the end of a normal, quiet inspiration.

The *expiratory reserve volume* (ERV), is that extra volume of gas that can be expired with maximum effort beyond the end expiratory level. The *end expiratory level* is the level reached at the end of a normal, quiet expiration.

The *residual volume* (RV) is the volume of gas remaining in the lungs at the end of a maximal expiration.

The *vital capacity* (VC) is the maximum volume of gas that can be expelled from the lungs by forceful effort after a maximal inspiration. It is actually the difference between the level of maximum inspiration and the residual volume, and it is measured without respect to time. The vital capacity is also the sum of the tidal volume, inspiratory reserve volume, and expiratory reserve volume.

The *total lung capacity* (TLC) is the amount of gas contained in the lungs at the end of a maximal inspiration. It is the sum of the vital capacity and residual volume. Total lung capacity is also the sum of the tidal volume, inspiratory reserve volume, expiratory reserve volume, and residual volume.

The *inspiratory capacity* (IC) is the maximum amount of gas that can be inspired after reaching the end expiratory level. It is the sum of the tidal volume and the inspiratory reserve volume.

The functional residual capacity, often referred to by its abbreviation, FRC, is the volume of gas remaining in the lungs at the end expiratory level. It is the sum of the residual volume and the expiratory reserve volume. The FRC can also be calculated as the total lung capacity minus the inspiratory capacity, and it is often regarded as the baseline from which other volumes and capacities are determined, for it seems to be more stable than the end inspiratory level.

Typical values for these volumes and capacities in a normal 20–30 year old male are given in Appendix C. All volumes and capacities are about 20 to 25 percent less in females than in males and are generally much greater in athletes than in less-active people. These values also vary, of course, with body size and weight.

In addition to the static volumes and capacities given above, several dynamic measures are used to assess the breathing mechanism. These measures are important because breathing is, in fact, a dynamic process, and the rate at which gases can be exchanged with the blood is a direct function of the rate at which air can be inspired and expired.

A measure of the overall output of the respiratory system is the *respiratory minute volume.* This is a measure of the amount of air inspired during one minute at rest. It is obtained by multiplying the tidal volume by the number of respiratory cycles per minute.

A number of forced breathing tests are used to assess the muscle power associated with breathing and the resistance of the airway. Among them is the *forced vital capacity* (FVC), which is really a vital capacity measurement taken as quickly as possible. By definition, the FVC is the total amount of air that can forcibly be expired as quickly as possible after taking the deepest possible breath. If the measurement is made with respect to the time required for the maneuver, it is called a *timed vital capacity* measurement. A measure of the maximum amount of gas that can be expelled in a given number of seconds is called the *forced expiratory volume* (FEV). This is usually given with a subscript indicating the number of seconds over which the measurement is made. For example, FEV_1 indicates the amount of air that can be blown out in one second following a maximum inspiration, while FEV_3 is the maximum amount of air that can be expired in 3 seconds. Sometimes FEV is given as a percentage of the forced vital capacity.

Since forced vital capacity measurements are often encumbered by patient hesitation and the inertia of the instrument, a measure of the *maximum midexpiratory flow* rate is taken. This is flow measurement over the middle half of the forced vital capacity (from the 25-percent level to the 75-percent level). The corresponding FEV measurement is called $FEV_{25\%-75\%}$.

Another important flow measurement is the *maximal expiration flow* (MEF) rate, which is the rate during the first liter expired after 200 ml have been exhausted at the beginning of the FEV. It differs from the *peak flow,* which is the maximum rate of airflow attained during a forced expiration.

Another useful measurement for assessing the integrity of the breathing mechanism is the *maximal breathing capacity* (MBC) or *maximal voluntary ventilation* (MVV). This is a measure of the maximum amount of air that can be breathed in and blown out over a sustained interval, such as 15 or 20 seconds. A ratio of the maximal breathing capacity to the vital capacity is also of clinical interest.

The results of many of the preceding tests are generally reported as percentages of predicted normal values. In the presentation of various respiratory volumes, the term *BTPS* is often used, indicating that the measurements were made at body temperature and ambient pressure, with the gas saturated with water vapor. Sometimes, in order to use these values in the reporting of metabolism, they must be converted to standard temperature and pressure and dry measurement conditions, indicated by the term *STPD.*

With each breath, most of the air enters the lungs to fill the alveoli. However, a certain amount of air is required to fill the various cavities of the air passages. This air is called the *dead space air,* and the space it occupies is called the *dead space.* The amount of air that actually reaches the alveolar interface with the bloodstream with each breath is the tidal volume minus the volume of the dead space. The respiratory minute volume can be broken down into the alveolar *ventilation per minute* and the *dead space ventilation per minute.*

8.2.2. MECHANICAL MEASUREMENTS. The volume and capacity measurements just described, particularly the forced measurements, are a good indication of the compliance of the lungs and rib cage and the resistance of the air passages. However, direct measurement of these parameters is also possible and is often used in the measurement of pulmonary function.

Determination of *compliance,* which has been defined as the volume increase in the lungs per unit increase in lung pressure, requires measurement of an inspired or expired volume of gas and of intrathoracic pressure. Compliance is actually a static measurement. However, in practice, two types of compliance measurement, static and dynamic, are made. *Static compliance* is determined by obtaining a ratio of the difference in lung volume at two different volume levels and the associated difference in intraalveolar pressure. To measure dynamic compliance, tidal volume is used as the volume measurement, while intrathoracic pressure measurements are taken during the instants of zero airflow that occur at the end inspiratory and expiratory levels with each breath (refer to Figure 8.3). The lung compliance varies with the size of the lungs; a child has a smaller compliance than an adult. Furthermore, the volume-pressure curve is not linear. Hence compliance does not remain constant over the breathing cycle but tends to decrease as the lungs are inflated. Fortunately, over the tidal volume range, in which dynamic compliance measurements are usually performed, the relationship is approximately linear and a constant compliance is assumed. Compliance values are given as liters per centimeter H_2O.

Resistance of the air passages is generally called *airway resistance,* which is a pneumatic analog of hydraulic or electrical resistance and, as such, is a ratio of pressure to flow. Thus for the determination of airway resistance, intra-alveolar pressure and airflow measurements are required. As was the case with compliance, airway resistance is not constant over the respiratory cycle. As the pressure in the thoracic cavity becomes more negative, the airways are widened and the airway resistance is lowered. Conversely, during expiration, when the pressure in the thorax becomes positive, the airways are narrowed and resistance is increased. The intra-alveolar pressure is given in centimeters H_2O and the flow in liters per second; the airway resistance is expressed in centimeters H_2O per liter per second. Most airway resistance measurements are made at or near the functional residual capacity (end expiratory) level.

From the preceding discussion it can be seen that in order to obtain compliance and airway resistance determinations, volume, intra-alveolar pressure, intrathoracic pressure, and instantaneous airflow measurements are required. The methods for measurement of volume for these determinations are no different from those used for the volume and capacity measurements discussed earlier.

8.2.3. INSTRUMENTATION FOR MEASURING THE ME-CHANICS OF BREATHING. As shown in the previous sections, all the parameters dealing with the mechanics of breathing can be derived from measurements of lung volumes at various levels and conditions of breathing, pressures within the lungs and the thorax with respect to outside air pressure, and instantaneous airflow. The complexity of pulmonary measurements lies not in the variety of measurements required but rather in gaining access to the sources of these measurements and in providing suitable conditions to make the measurements meaningful.

The most widely used laboratory instrument for respiratory volume measurements is the *recording spirometer,* an example of which is shown in Figure 8.4. All lung volumes and capacities that can be determined by measuring the amount of gas expired under a given set of conditions or during a specified time interval can be obtained by use of the spirometer. Included are the timed vital capacity and forced expiratory volume measurements. The only volume and capacity measurements that cannot be obtained with a spirometer are those requiring measurement of the gas that cannot be expelled from the lungs under any conditions. Such measurements include the residual volume, functional residual capacity, and total lung capacity.

The standard spirometer consists of a movable bell inverted over a chamber of water. Inside the bell, above the water line, is the gas that is to be breathed. The bell is counterbalanced by a weight to maintain the gas inside at atmospheric pressure so that its height above the water is proportional to the amount of gas in the bell. A breathing tube connects the mouth of the patient with the gas under the bell. Thus, as the patient breathes into the tube, the bell moves up and down with each inspiration and expiration in proportion to the amount of air breathed in or out. Attached to the bell or the counterbalancing mechanism is a pen that writes on an adjacent drum recorder, called a *kymograph.* As the kymograph rotates, the pen traces the breathing pattern of the patient.

Various bell volumes are available, but 9 and 13.5 liters are most common. A well-designed spirometer offers little resistance to the airflow, and the bell has little inertia. Various paper speeds are available for the kymograph, with 32, 160, 300, and 1920 mm per minute most common. The compact spirometer shown in Figure 8.4 is a widely used instrument for pulmonary function testing. It is used both in the physician's office and in the hospital ward. Its 9-liter capacity is often considered adequate for recording the largest vital capacities, for extended-period oxygen uptake determinations, and even for spirography during mild exercise. However, many physcians prefer the larger size (13.5 liters) because of the extra capacity. The principle of operation is stmilar for both. Easily removable flutter valves and a CO_2 absorbent container permit minimized breathing resistance during tests for maximal respiratory flow rates. This instrument is equally suitable for clinical spirography, for cardiopulmonary function testing, and for metabolism determinations. The instrument directly records basal minute volume, exercise ventilation, or maximum breathing capacity. The ventilation equivalent for oxygen may be calculated directly from the spirogram slope lines for ventilation and oxygen uptake.

In addition to the standard type of spirometer just described, and illustrated in Figure 8.4, several other types are available. A few are described briefly below.

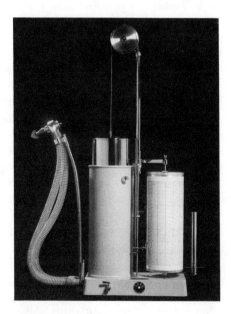

Figure 8.4. Spirometer. (Courtesy of Warren E. Collins, Inc., Braintree, Mass.)

The *waterless spirometers* are instruments that provide electrical outputs proportional to both volume and flow. They have a high dynamic response and do not impose a load on the patient's lungs. The volume and flow signals are obtained independently from two linear transducers. A standard amplifier-type recorder with high input impedance can be used to obtain the spirogram.

A *broncho-spirometer* is a dual spirometer that, with an appropriate input device, measures the volumes and capacities of each lung individually. The input device is a double-lumen tube that divides for entry into the airway to each lung and thus provides isolation for differential measurement. Its main function is the preoperative evaluation of oxygen consumption of the individual's lungs.

Spirometers with various types of readouts are available, including instruments with built-in computation ability to provide direct readout of the calculated parameters. Digital readouts are also available on various types of spirometers.

The procedure in making ventilatory tests using a spirometer is as follows: The patient is seated and the mouthpiece is set to the proper height. The nose is blocked with a clip so that all breathing is through the mouth, and quiet breathing is observed for a short time. The kymograph is set at slow speed to measure vital capacity (VC).

The subject makes a maximal inspiration, and when his lungs are filled, he is told to breath out as rapidly as possible until no more air can be expired from the lungs. The tracing shows the inspiratory and expiratory curves, as well as the vital capacity. The process is repeated three times, and the highest vital capacity is recorded. This process may then be repeated with the patient inspiring the maximum amount of air following maximal expiration.

The vital capacity may also be recorded in a two-stage test. First, the inspiratory capacity (maximum inspiration) from the functional residual capacity (FRC) state is measured. Then, by measuring maximum expiration from the FRC, expiratory reserve volume (ERV) is obtained. By adding the inspiratory capacity to the expiratory reserve volume, the vital capacity is determined. Obstructive airway disease may be detected if the two types of vital capacity tests are performed and there is a significant difference.

Maximum inspiratory and expiratory rates of flow can be determined from the slopes of the inspiration and expiration curves. Typical flow rates are 350 to 450 liters per minute and 350 to 500 liters per minute, respectively.

The static vital capacity is not a good indication of respiratory function. It is only when the volume expired per unit time is measured that useful information concerning airway obstruction may be obtained. However, with a spirometer and a high-speed kymograph, timed vital capacity and forced expiratory volume measurements can be made as the subject produces a forced expiration after a maximum inspiration.

Figure 8.5 is a typical *spirogram,* showing the tests and measurements required to calculate vital capacity, FEV_1, and maximum voluntary ventilation (MVV). The trace reads from right to left along the time base and from bottom to top for inspiratory-to-expiratory conditions. The vital

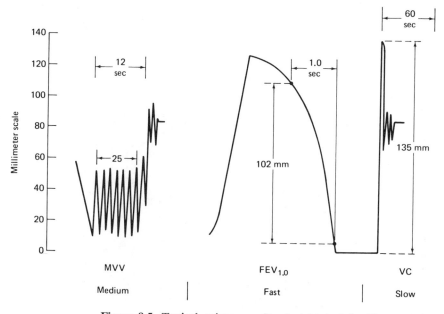

Figure 8.5. Typical spirogram. Read right to left. (See text for explanation.)

capacity was taken at a slow speed of 32 mm per minute. After breathing normally at rest to provide a baseline, the patient exhales completely, after which he takes a maximum inhalation. He then holds his breath, the kymograph is changed to a fast speed (1920 mm per minute), and the patient is told to exhale as forcefully and completely as possible. This produces the vital capacity at slow speed and the FEV at high speed. The FEV_1 may be obtained by calculation. For the MVV determination, the patient, after resting, takes a few resting breaths and is then asked to breathe in and out as rapidly as possible for about 10 seconds. After he rests again, the test is repeated. The maximal reading of the three tests is taken as the value of MVV. Although some instruments are calibrated for direct readout, others require that the height of the tracings be converted to liters by using a calibration factor for the instrument, called the *spirometer factor*. A table may be prepared to facilitate these calculations.

To calculate FEV_1, a 1-second interval is taken from the 200-ml point or about 5 mm up from the start. This ensures overcoming the initial friction of the spirometer. The FEV_1 value is also converted to liters by using the spirometer factor.

A number of instruments provide instantaneous measurement of peak flow, vital capacity, and various forced expiratory volume measurements from a single forced expiration. An example is the *pulmonary function indicator* shown in Figure 8.6. This instrument utilizes a heated platinum-wire transducer, protected by stainless-steel screens on both sides, installed in a breathing tube. When exposed to the moving air, the platinum wire is

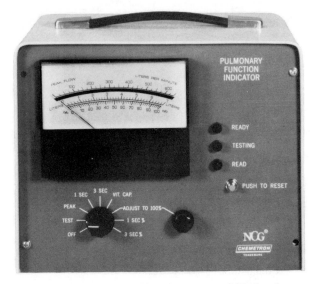

Figure 8.6. Pulmonary function indicator. (Courtesy of National Cylinder Gas Division of Chemetron Corporation, Chicago, Ill.)

cooled. Its resistance changes as a function of the rate of airflow. The transducer element has a small thermal mass and is a good heat conductor, thereby ensuring extremely fast response time. Disposable paper mouthpieces are used on the breathing tube, and the transducer may be sterilized so that one unit can be used for a number of patients. By use of timing and memory circuits, this instrument is able to provide all its output measures from a single forced expiration. Figure 8.7 shows the pulmonary function indicator as used with a patient.

The maximum voluntary ventilation (MVV) may be measured with a Venturi tube device through which the patient breathes rapidly from a rubber balloon filled with a measured amount of air. As the patient breathes, air is removed from the balloon for 15 seconds, and the volume is measured. Rough values for MVV can be obtained from a table provided with the equipment.

Spirometric values are best interpreted by measuring changes because they are so dependent on body size, sex, age, height and degree of disease. The Veterans Administration and the U.S. Army, in a cooperative study, have produced predictive nomograms for spirometric values, based on height and age for both males and females.

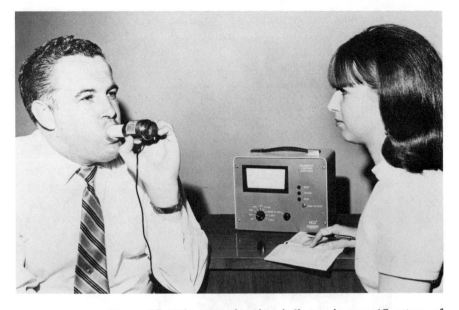

Figure 8.7. Pulmonary function indicator in use. (Courtesy of National Cylinder Gas Division of Chemetron Corporation, Chicago, Ill.)

Residual volume, functional residual capacity, and similar parameters which cannot be obtained directly with a spirometer are measured by using foreign gas mixtures. The *closed-circuit* technique involves rebreathing from a spirometer charged with a known volume and concentration of a measurable gas, such as hydrogen or helium. Helium is usually used. After several minutes of breathing, complete mixing of the spirometer and pulmonary gases is assumed, and the residual volume is calculated by simple proportion of gas volumes and concentrations.

The *open-circuit* or *nitrogen washout* method involves the inspiration of pure oxygen and expiration into an oxygen-purged spirometer. If the patient has been breathing air, the gas remaining in his lungs is 78 percent nitrogen. As he begins to breathe the pure oxygen, it will mix with the gas still in his lungs and a certain amount of nitrogen will "wash out" with each breath. By measuring the amount of nitrogen in each expired breath, a washout curve is obtained from which the volume of air initially in the lungs can readily be calculated. The preferred breathing level for beginning this measurement is the end of normal expiration, which is the functional residual capacity.

A gas analyzer is used for these tests. A description of several types of gas analyzers is presented in Section 8.3.1, which is concerned with gas distribution and diffusion.

Since pressure measurements are required for both compliance and airway resistance determinations, methods of measuring intra-alveolar and pleural pressures are important in pulmonary function laboratories where such determinations are performed. Unfortunately, measurement of these pressures presents some problems, primarily because of the inaccessibility of the chambers in which the measurements are to be made. For this reason, indirect methods have been developed.

Intra-alveolar pressure measurements are normally performed by placing the patient in a *body plethysmograph* of the type shown in Figure 8.8.

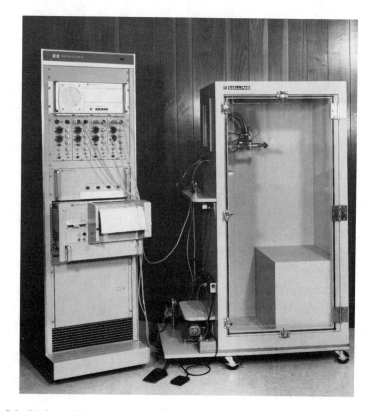

Figure 8.8. Body plethysmograph. (Courtesy of Warren E. Collins, Inc., Braintree, Mass.)

The body plethysmograph is an airtight box in which the patient is seated. As the patient breathes air from within the box through a tube, the pressure or the volume inside the box changes, depending on the type of system—rising during inspiration and falling during expiration when the alveolar gas is compressed. The pressure or volume variations in the box are the inverse of the pressure variations in the lungs, as the gas within the lungs expands and is compressed due to the positive and negative pressures in the lungs. As part of the procedure, the patient's breathing tube is blocked for a few seconds, during which the patient is asked to "pant" while mouth pressure is measured. Since mouth pressure and lung pressure are the same when there is no airflow, these data can be used in calibration of the measurement.

For measurement of intrathoracic pressures, a balloon is placed in the patient's esophagus, which is within the thoracic cage. Since the balloon is exposed to the intrathoracic pressure, its pressure, measured with respect to mouth pressure by using some form of differential pressure transducer, represents the difference between pressures.

A variety of instruments can be used to measure airflow. One of the most widely used is the *pneumotachometer,* often called the *pneumotachograph.* This device utilizes the principle that air flowing through an orifice produces a pressure difference across the orifice that is a function of the velocity of the air. In the more common pneumotachometer, the orifice consists of a set of capillaries or a metal screen. Since the cross section of the orifice is fixed, the pressure difference can be calibrated to represent flow. Two pressure transducers or a differential pressure transducer can be used to measure the pressure difference.

Another method of measuring airflow is the transducer described earlier for the pulmonary function indicator shown in Figures 8.6 and 8.7, in which a heated wire is cooled by the flow of air, and the resistance change due to the cooling is measured as representative of airflow. Because the cooling effect is the same regardless of the direction of airflow, this transducer is insensitive to direction, whereas the pneumotachograph described above indicates not only the amount of flow but also the direction.

Ultrasonic-airflow-measuring devices utilizing the Doppler effect (see Chapters 6 and 9) have also been developed.

Since flow is the first derivative or rate of change of volume, some volume-measuring devices also produce a measurement of flow.

In some applications, the flow or volume of respiration is not required, but a measure of *respiration rate* (number of breaths per minute) is needed. Respiration rate can, of course, be obtained from any instrument that records the volume changes during the respiratory cycle. There are, however, other instruments that are difficult to calibrate for volume changes but that well serve the purpose of measuring respiration rate. Such instruments are much simpler and easier to use than the spirometer or other devices intended for volume measurements. These instruments include a mercury plethysmograph of the type described in Chapters 6 and 9 and the impedance pneumograph in which impedance changes due to respiration can be measured across the chest.

8.3. GAS EXCHANGE AND DISTRIBUTION

Once air is in the lungs, oxygen and carbon dioxide must be exchanged between the air and the blood in the lungs and between the blood and the cells in the body tissue. In addition, the gases must be transported between the lungs and the tissue by the blood. The physiological processes involved in this overall task were presented briefly in Section 8.1. A number of tests have been devised to determine the effectiveness with which these processes are carried out. Some of these tests and the instrumentation required for their performance are described and discussed in this section. The tests connected with the exchange of gases are treated first, after which measurements pertaining to the transport of oxygen and CO_2 in the blood are covered.

8.3.1. MEASUREMENTS OF GASEOUS EXCHANGE AND DIFFUSION. The mixing of gases within the lungs, the ventilation of the alveoli, and the exchange of oxygen and carbon dioxide between the air and blood in the lungs all take place through a process called *diffusion*. Diffusion is the movement of gas molecules from a point of higher pressure to a point of lower pressure to equalize the pressure difference. This process can occur when the gas is unequally distributed in a chamber or wherever a pressure difference exists in the gas on two sides of a membrane permeable to that gas.

Measurements required for determining the amount of diffusion involve the partial pressures of oxygen and carbon dioxide, P_{O_2} and P_{CO_2} respectively. There are many methods by which these measurements can be obtained, including some chemical analysis methods, diffusing capacity measurements using a carbon monoxide (CO) analyzer, paramagnetic oxygen analysis, a sonic-gas analysis, and various methods of carbon dioxide analysis. Each of these methods is discussed below.

CHEMICAL ANALYSIS METHODS. The original gas analyzers developed by Haldane, and modified by Scholander, were of the chemical type. The Scholander device is shown in Figure 8.9. A 0.5–ml (approx.) gas sample is introduced into a reaction chamber by use of a transfer pipette at the upper end of the reaction chamber capillary. An indicator droplet in this capillary allows the sample to be balanced against a trapped volume of air in the thermobarometer. Absorbing fluids for CO_2 and O_2 can be transferred in from side arms without causing any change in the total volume of the system. The micrometer is adjusted so as to put

Figure 8.9. Scholander gas analyzer. (Courtesy of Mark Company, Division of M.B. Claff & Sons, Inc., Brockton, Mass.)

mercury into the system in place of the gases being absorbed. The volume of the absorbed gases is read from the micrometer barrel calibration.

DIFFUSING CAPACITY USING CO INFRARED ANALYZER. To determine the efficiency of perfusion of the lungs by blood and the diffusion of gases, the most important tests are those that measure O_2, CO_2, pH, and bicarbonate in arterial blood. In trying to measure the diffusion rate of oxygen from the alveoli into the blood, it is usually assumed that all alveoli have an equal concentration of oxygen. Actually, this condition does not exist because of the unequal distribution of ventilation in the lung; hence the terms *diffusing capacity* or *transfer factor* (rather than *diffusion*) are used to describe the transfer of oxygen from the alveoli into the pulmonary capillary blood.

Carbon monoxide (CO) resembles oxygen in its solubility and molecular weight and also combines with hemoglobin reversibly. Its affinity for hemoglobin is about 200 to 300 times that of oxygen, however. Carbon monoxide can thus be used as a tracer gas in measuring the diffusing capacity of the lung. It passes from the alveolar gas into the alveolar walls, then into the plasma, from which it enters the red blood cells, where it combines with hemoglobin.

A relationship may be obtained that is a function of both the diffusing capacity of the alveolar membrane and the rate at which CO combines with hemoglobin in the alveolar capillaries. This relationship may be expressed as follows:

$$\frac{1}{TF} = \frac{1}{D_m} + \frac{1}{\theta V_c} \qquad \text{mm Hg/ml/min}$$

where

TF = the diffusing capacity for the lung for CO
D_m = the diffusing capacity for the alveolar membrane
V_c = the volume of blood in the capillaries
θ = the reaction rate of CO with oxyhemoglobin

TF, or the diffusing capacity for the whole lung, in normal adults ranges from 20 to 38 ml/min/mm Hg. It varies with depth of inspiration, increases during exercise, and decreases with anemia or low hemoglobin.

The principal methods of measuring diffusing capacity involve the inhalation of low concentrations of carbon monoxide. The concentration is less than 0.25 percent and usually ranges from 0.05 to 0.1 percent. The concentration of CO in the alveoli and the rate of its uptake into the blood per minute are measured by either the steady-state method or the single-breath method, both of which are described below. In either method, uptake of carbon monoxide is calculated by measuring the concentration and the volume of the air-CO mixture. Since the concentration of CO fluctuates throughout the respiratory cycle, end-tidal expired air is collected and the CO in the air is measured.

In the single-breath method, the last 75 to 100 ml of the expired air is collected so that enough end-tidal air is available for the measurement. This is the CO in the alveolar gas. In the steady-state method, the patient rebreathes the gas until equilibrium is reached.

The small amount of CO in the blood is negligible, for it combines with the hemoglobin in the red blood cells and exerts no significant back pressure. By estimating the P_{CO} in the blood by the rebreathing method, the diffusing capacity can be calculated as

$$TF \text{ or diffusing capacity} = \frac{\text{ml CO taken up/min}}{P_{CO} \text{ in alveoli mm/Hg}}$$

For this measurement, as well as for all methods requiring carbon monoxide determination, a carbon monoxide analyzer or a gas chromatograph is used. The commonly-used carbon monoxide analyzer utilizes an infrared energy source, a beam chopper, sample and reference cells, plus a detector and amplifier. A milliammeter or a digital meter may be used for display. Two infrared beams are generated, one directed through the sample and the other through the reference. The CO gas mixture flowing through the sample cell absorbs more infrared energy than does the reference gas. The two infrared beams are each measured by a differential infrared detector. The output signal is proportional to the amount of monitored gas in the sample cell. The signal is amplified and presented to the output display meter or to a recorder.

PARAMAGNETIC OXYGEN ANALYZER. Faraday first demonstrated that oxygen has the unique property among gases of concentrating a magnetic field in the same manner as iron. Oxygen is attracted into a magnetic field and in so doing attempts to displace the less-magnetic gases from the field. A device to show this action consists of a nitrogen-filled dumbbell of glass suspended on a quartz fiber in an asymmetrical magnetic field. Oxygen around the dumbbell results in torsion, as both of the nitrogen-filled ends are forced out of the field. A mirror on the fiber reflects a light beam on a scale, indicating the O_2 concentration. The simplest clinical instrument of this type reads from 0 to 760 mm Hg P_{O_2}, and also from 0 to 100 percent O_2, and requires only a flashlight cell to provide the indicator light. Increased accuracy is obtained by thermostatic control of the field, by longer optical paths, and by using electrostatic fields to repel the dumbbell back to a null balance point. In this latest method, the electrical balancing potential is linearly related to P_{O_2}, and a very high accuracy can be obtained. The accuracy of the reading depends on the background gases, most of which are diamagnetic.

SONIC GAS ANALYZER. The velocity of sound in a gas depends on the specific heat ratio y, the density of the gas d, and the barometric pressure p, according to the formula

$$V = \sqrt{yp/d}$$

Various instruments have made use of this principle by determining the delay of pulses or the phase shift of high-frequency audio oscillations. A commercial instrument intended for respiratory gas analysis employs an ultrasonic frequency (150 kc) and a long tube (100 cm). With this device, it is possible to obtain excellent sensitivities (0.004 percent for O_2 and 0.0008 percent for CO_2). The inflowing gas is dried and a reading is made with and without absorption of CO_2. The ultra-high-frequency oscillator output is fed through a calibrated variable delay network to the vertical deflection plates of a cathode-ray tube. The sound, which traverses the gas system, is picked up by a crystal microphone, amplified, and applied to the horizontal plates of the cathode-ray tube. The delay is read by adjusting the calibrated delay network to keep the two signals in phase (indicated by a straight diagonal line on the oscilloscope screen).

This method has certain disadvantages, however. The change of phase observed when O_2 is altered from 0 to 21 percent is somewhat over seven cycles, thus making it necessary for the user to watch the oscilloscope and count cycles. The computation of phase shift from the calibrated delay plus the total cycle count is apt to be confusing. Several minutes are required to rinse out the gas to equilibrium. The instrument is extremely sensitive to helium or hydrogen contamination, which precludes the use of compressed gases containing O_2. The present instrument cannot analyze concentrations of CO_2 above about 8 percent, since this gas absorbs sound and too little reaches the receiving end of the tube. An additional complication is that the computation of O_2 involves a correction for the amount of CO_2 that was removed before anlysis. The volume of sample required is about 100 ml for analysis of both O_2 and CO_2. The sensitivity of this instrument, however, tends to outweigh the complexity and expense, at least for some applications.

CARBON DIOXIDE ANALYZER. A simple CO_2 analyzer for student use is based on the fact that the rate of flow of gas into a CO_2 absorber, A, is a hyperbolic function of the CO_2 concentration, C_{CO_2}, if the outflow rate B is held constant, according to the formula

$$A = B \left(\frac{1}{1 - C_{CO_2}} \right)$$

A constant outflow of about 100 ml per minute is obtained by a negative pressure ($\frac{1}{2}$ atm or less) beyond a "critical" orifice diameter of 0.003 to 0.004 inch in a sheet of 0.002-inch brass.

8.3.2. MEASUREMENTS OF GAS DISTRIBUTION. The distribution of oxygen from the lungs to the tissues and carbon dioxide from the tissues to the lungs takes place in the blood. The process by which each gas is transported, however, is quite different. As mentioned earlier, oxygen is carried by the hemoglobin of the red blood cells. On the other hand, carbon dioxide is carried through chemical processes in which CO_2 and water combine to produce carbonic acid, which is dissolved in the blood. The amount of carbonic acid in the blood, in turn, affects the pH of the blood. In assessing the performance of the blood in its ability to transport respiratory gases, then, measurements of the partial pressures of oxygen (P_{O_2}) and carbon dioxide (P_{CO_2}) in the blood, the percent of oxygenation of the hemoglobin, and the pH of the blood are most useful.

Electrodes for measurement of P_{O_2}, P_{CO_2}, and pH are described in detail in Chapter 4. These electrodes, together with amplification and read-outs, provide a fairly simple method for this type of analysis. Measurements both in vitro and in vivo are possible with these electrodes. A blood gas analyzer that utilizes such electrodes and provides a digital output of the pH, P_{CO_2} and P_{O_2} readings is shown in Figure 8.10. This device utilizes

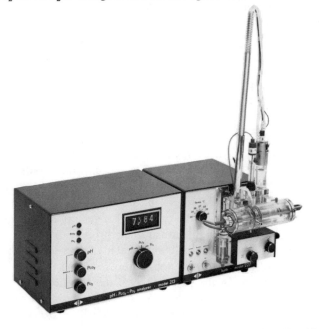

Figure 8.10. Digital blood gas analyzer (IL213). (Courtesy of Instrumentation Laboratory, Inc., Lexington, Mass.)

direct-coupled field-effect transistor amplifiers to provide the high impedance input for the electrodes and the high gain required for such small signals. A water-bath temperature controller, accurate to 0.05°C, and a vacuum pump or roller assembly for inserting samples in the chamber are all part of the system.

Another in vitro method for analyzing both P_{O_2} and P_{CO_2} is known as the *Van Slyke apparatus*. In this device, a measured quantity of blood is used and the O_2 and CO_2 are extracted by vacuum. The quantity of these two gases is measured manometrically, after which the CO_2 is absorbed. The quantity is measured again, the oxygen is absorbed, and the remaining gas, which is nitrogen, is measured. The amount of O_2 and CO_2 may be calculated from these measurements as a percentage of the total gas.

Another method involving the measurement of pH as part of the blood gas determination is called the *Astrup technique* and utilizes a semilogarithmic paper with a special nomogram. In this method a pH determination is made on a heparinized microsample of blood. Two other pH determinations are made on the same sample after it has been equilibrated with two known CO_2 tensions, obtained from cylinders accompanying the apparatus. These three points are plotted on special graph paper and connected by a straight line. The slope of the line is an index of the buffering capacity of the blood, which is calculated using this nomogram.

When hemoglobin is oxygenated, its light-absorption properties change as a function of the percentage of oxygen saturation. At a wavelength of 6500 Å (angstrom units) the difference in absorption between oxygenated and nonoxygenated blood is greatest, whereas at 8050 Å the absorption is the same. Thus by measuring the absorption of a sample of blood at both wavelengths on a special photometer, the percentage of oxygenation can be determined.

A similar principle can be used to measure the percentage of oxygenation of the blood in vivo. Here an instrument called an *ear oximeter* is used. The ear oximeter is composed of an ear clip that holds a light source on one side of the earlobe and two sensors on the opposite side, so that the light passing through the earlobe is picked up by both of the sensors. As the blood in the capillaries of the earlobe changes color, these changes are reflected in the amount of light transmitted through the ear at each of the two aforementioned wavelengths. Since each of the sensors receives and filters transmitted light so that its maximum response is at one of the two wavelengths, variations in the percentage of oxygenation can be measured. This method should only be used to measure differences in oxyhemoglobin saturation rather than exact oxygen blood level or exact percentage of oxygenation.

· 9 ·

MEASUREMENT OF
PHYSICAL VARIABLES

Although most physiological measurements are so specialized that they require instrumentation specifically designed for that purpose, a number of measurements can be made with essentially the same types of instruments that are used for similar measurements in other fields. About the only specialization for the medical application of these devices is in the sensitivity and range of the instruments and, in some cases, in the physical size and configuration. This group includes the measurement of temperature, position, displacement, velocity, force, and acceleration. This chapter deals with each of these physical variables as they are encountered in the field of biomedical instrumentation and the methods and instruments commonly used for their measurement.

9.1. TEMPERATURE MEASUREMENT

Body temperature is one of the oldest known indicators of the general well-being of a person. Techniques and instruments for the measurement of temperature have been commonplace in the home for years

195

and throughout all kinds of industry, as well as in the hospital. Except for the narrow range required for physiological temperature measurements and the size and shape of the sensing element, instrumentation for measurement of temperature in the human body differs very little from that found in various industrial applications.

Two basic types of temperature measurements can be obtained from the human body: systemic and skin surface measurements. Both provide valuable diagnostic information, although the systemic temperature measurement is much more commonly used.

Systemic temperature is the temperature of the internal regions of the body. This temperature is maintained through a carefully controlled balance between the heat generated by the active tissues of the body, mainly the muscles and the liver, and the heat lost by the body to the environment. Measurement of systemic temperature is accomplished by temperature-sensing devices placed in the mouth, under the armpits, or in the rectum. The normal oral (mouth) temperature of a healthy person is about 98.6°F (37°C). The underarm temperature is about one degree lower, whereas the rectal temperature is about one degree higher than the oral reading. The systemic body temperature can be measured most accurately at the tympanic membrane in the ear, which is believed to approximate the temperature at the "inaccessible" temperature control center in the brain. For some still unknown reason, the body temperature, even in a healthy person, does not remain constant over a 24-hour period but is often 1 to 1½ degrees lower in the early morning than in late afternoon. Although strenuous muscular exercise may cause a temporary rise in body temperature from 1° to 4°F, the systemic temperature is not affected by the ambient temperature, even if the latter drops to as low as 0°F or rises to 100°F. This balance is upset only when the metabolism of the body cannot produce heat as rapidly as it is lost or when the body cannot rid itself of heat fast enough.

The temperature control center for the body is located deep within the brain (in the forepart of the hypothalamus) (see Chapter 10). Here the temperature of the blood is monitored and its control functions are coordinated. In warm, ambient temperatures, cooling of the body is aided by production of perspiration due to secretion of the sweat glands and by increased circulation of the blood near the surface. In this manner, the body acts as a radiator. If the external temperature becomes too low, the body conserves heat by reducing blood flow near the surface to the minimum required for maintenance of the cells. At the same time, metabolism is increased. If these measures are insufficient, additional heat is produced by increasing the tone of skeletal muscles and sometimes by involuntary contraction of skeletal muscles (shivering) and of the arrector muscles in the skin (gooseflesh).

In addition to the central "thermostat" for the body, temperature sensors at the surface of the skin permit some degree of local control in the event a certain part of the body is exposed to local heat or cold. Cooling or heating is accomplished by control of the surface blood flow in the region affected.

The only deviation from normal temperature control is a rise in temperature called "fever," experienced with certain types of infection. The onset of fever is caused primarily by a deliberate shutdown of the mechanisms for heat elimination. The body temperature increases as though the "thermostat" in the brain were suddenly turned "up," thus causing additional metabolism because the increased temperature accelerates the chemical reactions of the body. At the beginning of a fever the skin is often pale and dry and shivering usually takes place, for the blood that normally keeps the surface areas warm is shut off, and the skin and muscles react to the coolness. At the conclusion of the fever, as the body temperature is lowered to normal, increased sweating ("breaking of the fever") is often noted as the means by which the additional body heat is eliminated.

Surface or skin temperature is also a result of a balance, but here the balance is between the heat supplied by blood circulation in a local area and the cooling of that area by conduction, radiation, convection, and evaporation. Thus skin temperature is a function of the surface circulation, environmental temperature, and air circulation around the area from which the measurement is to be taken. In order to obtain a meaningful skin temperature measurement, it is usually necessary to have the subject remain with no clothing covering the region of measurement in a fairly cool ambient temperature (around 70°F). Care must be taken, however, to avoid chilling and the reactions relative to chilling. If a surface measurement is to include the reaction to the cooling of a local region, it should be recognized that the cooling of the skin increases surface circulation, which in turn causes some local warming of adjacent areas. Heat transferred into the site of measurement from adjacent areas of the body must also be accounted for.

9.1.1. MEASUREMENT OF SYSTEMIC BODY TEMPERATURE. Since the internal or systemic body temperature is a good indicator of the health of a person, measurement of this temperature is considered one of the vital signs of medicine. For this reason, temperature measurement constitutes one of the more important physiological measurements. Although a high degree of accuracy is not always important, methods of temperature measurement must be reliable and easy to perform. In the case of continuous monitoring, the temperature measurement must not cause discomfort to the patient.

Where continuous recording of temperature is not required, the *mer-*

cury thermometer is still the standard method of measurement. Since these devices are inexpensive, easy to use, and sufficiently accurate, they will undoubtedly remain in common use for many years to come. Even so, electronic thermometers, such as that shown in Figure 9.1, are available as

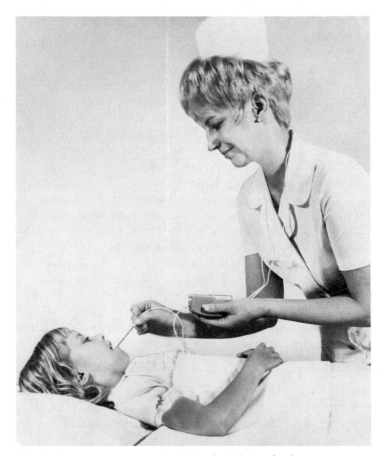

Figure 9.1. Oral temperature measurement using electronic thermometer. (Courtesy of Diagnostic, Inc., Indianapolis, Ind.)

replacement for mercury thermometers. With disposable tips, these instruments require much less time for a reading and are much easier to read than the conventional thermometer. Where continuous recording of the temperature is necessary, or where greater accuracy is needed than can be obtained with the mercury thermometer or its electronic counterpart, more sophisticated measuring instruments must be used.

Two types of electronic temperature-sensing devices are found in biomedical applications. They are the *thermocouple,* a junction of two dissimilar metals that produces an output voltage nearly proportional to the temperature, and the *thermistor,* a semiconductor element whose resistance varies with temperature. Both types are available for medical temperature measurements, although thermistors are used more frequently than thermocouples. This preference is primarily because of the greater sensitivity of the thermistor in the temperature range of interest and the requirement for a reference junction for the thermocouple.

To obtain a voltage proportional to variations in temperature in a thermocouple, a second junction maintained at a reference temperature is required. In practice, the circuit is opened at the second junction for measurement of the potential. The output voltage measured at this point is roughly proportional to the difference between the temperature at the measuring junction and that of the reference junction. This voltage, called the *contact potential,* ranges from a very few microvolts to a few hundred microvolts per degree Centigrade, depending on the two metals used. Generally the output voltage of a thermocouple is measured directly by using a meter or measured indirectly by comparing the measured voltage with a precisely known voltage obtained by using a potentiometer. Care must be taken to minimize current through the thermocouple circuit, for the current not only causes heating at the junctions but also an additional error due to the *Peltier effect,* wherein one junction is warmed and the other is cooled. (The two junctions at which the leads join the two dissimilar metals constitute a single junction.)

Thermistors are variable resistance devices formed into disks, beads, rods, or other desired shapes. They are manufactured from mixtures of oxides (sometimes sulfates or silicates) of various elements, such as nickel, copper, magnesium, manganese, cobalt, titanium, and aluminum. After the mixture is compressed into shape, it is sintered at a high temperature into a solid mass. The result is a resistor with a large temperature coefficient. Where most metals show an increase of resistances by about 0.3 to 0.5 percent per °C temperature rise, thermistors decrease their resistance by 4 to 6 percent per °C rise.

Unfortunately, the relationship between resistance change and temperature change is nonlinear. The resistance R_{t_1} of a thermistor at a given temperature T_1 can be determined by the following equation :

$$R_{t_1} = R_{t_0} e^{\,\beta(1/T_1 - 1/T_0)}$$

where

R_{t_1} = the resistance at temperature T_1
R_{t_0} = the resistance at a reference temperature T_0
e = the base of the natural logarithms (approx. 2.718)

$\beta =$ the temperature coefficient of the material, usually in the range of about 3000 to 4000

$T_1 =$ the temperature at which the measurement is being made (in degrees Kelvin)

$T_0 =$ the reference temperature (in degrees Kelvin)

To overcome the nonlinear characteristics of thermistors, the instrumentation in which the resistance is measured often incorporates special linearizing circuits. Some such circuits employ pairs of matched thermistors as part of the linearizing network.

Semiconductor devices with large positive-temperature coefficients that essentially overcome some of the disadvantages of the conventional thermistor have been developed. A comparison of resistance versus temperature curves for copper, a thermistor, and the posistor® (one of the positive coefficient devices) is given in Figure 9.2.

In addition to nonlinearity, the use of thermistors can result in other problems, such as the danger of error due to self-heating, the possibility of hysteresis, and the changing of characteristics because of aging.

The effect of self-heating can be reduced by limiting the amount of current used in measuring the resistance of the thermistor. If the power dissipation of the thermistor can be kept to about a milliwatt, the error should not be excessive, even when temperature differences as small as 0.01°C are sought.

The most important characteristics to consider in selecting a thermistor probe for a specific biomedical application are

1. The physical configuration of the thermistor probe. This is the interface with the site from which the temperature is to be measured. The configuration includes the size, shape, flexibility, and any special features required for the measurement. Commercial probes are available for almost any biomedical application, particularly for measurement of oral and rectal temperatures. Some of these probes are shown in Figure 9.3.

2. The sensitivity of the device. This is its ability to measure accurately small changes in temperature, but it can also be interpreted as the resistance change produced by a given temperature change. Usually overall sensitivity is a function of both the thermistor probe and the circuitry used to measure the resistance, but the limiting factor is the resistance-temperature characteristic of the thermistor (see Figure 9.2).

3. The absolute temperature range over which the thermistor is designed to operate. This is usually no problem with body temperature measurements, for the temperature range to be measured is so limited, but often, if a general-type temperature measuring instrument is used, the range is so wide that the desired resolution is not attainable.

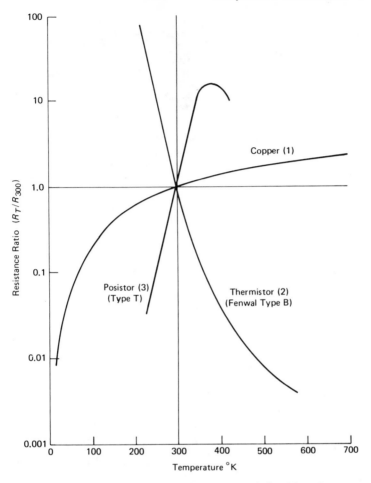

Figure 9.2. Resistance-temperature relationship of copper, thermistor and positor. (From L.A. Geddes and L.E. Baker, *Principles of Applied Biomedical Instrumentation*. John Wiley & Sons, Inc., 1969, by permission.)

4. Resistance range of the probe. Thermistor probes are available with resistances from a few hundred ohms to several megohms. A probe should be selected with a suitable resistance range corresponding to the temperature range of interest to match the impedance of the bridge or other type of circuit used to measure the resistance.

Although the resistance of a thermistor can be measured by use of an ohmmeter, most thermistor thermometers use a Wheatstone bridge or similar circuit to obtain a voltage output proportional to temperature variations. Generally the bridge is balanced at some reference temperature and

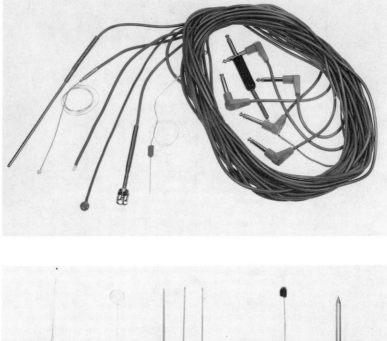

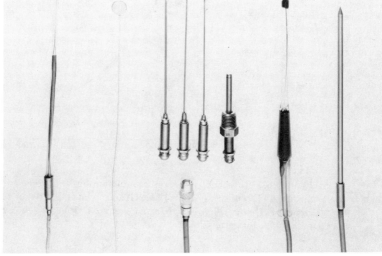

Figure 9.3. Thermistor probes. (Courtesy of Yellow Springs Instruments Company, Yellow Springs, Ohio.)

calibrated to read variations above and below that reference. Either ac or dc excitation can be used for the bridge. If the temperature difference between two measurement sites is desired, thermistors at the two locations are placed in adjacent legs of the bridge.

9.1.2. SKIN TEMPERATURE MEASUREMENTS. Although the systemic temperature remains very constant throughout the body, skin temperatures can vary several degrees from one point to another. The range is usually from 85 to 95°F. Exposure to ambient temperatures, the covering of fat over capillary areas, and local blood circulation patterns are just a few of the many factors that influence the distribution of temperatures over the surface of the body. Often skin temperature measurements can be used to detect or locate defects in the circulatory system by showing differences in the pattern from one side of the body to the other.

Skin temperature measurements from specific locations on the body are frequently made by using small, flat thermistor probes taped to the skin. The simultaneous readings from a number of these probes provide a means of measuring changes in the spatial characteristics of the circulatory pattern over a time interval or with a given stimulus.

Although the effect is insignificant in most cases, the presence of the thermistor on the skin slightly affects the temperature at that location. Other methods of measuring skin temperature that draw less heat from the point of measurement are available. They include the infrared methods of measurement, heat-quenched phosphors, and liquid crystals.

The human skin has been found to be an almost perfect emitter of infrared radiation. That is, it is able to emit infrared energy in proportion to the surface temperature at any location of the body. If a person is allowed to remain in a room at about 70°F without clothing over the area to be measured, a device sensitive to infrared radiation can accurately read the surface temperature. Such a device, called an *infrared thermometer,* is shown in Figure 9.4. Infrared thermometers in the physiological temperature range are available commercially and can be used to locate breast cancer and other unseen sources of heat. They can also be used to detect areas of poor circulation and other sources of coolness and to measure skin temperature changes that reflect the effects of circulatory changes in the body.

An extension of this method of skin temperature measurement is the *Thermograph,*® shown in Figure 9.5(a). This device is an infrared thermometer incorporated into a scanner so that the entire surface of a body, or some portion of the body, is scanned in much the same way that a television camera scans an image, but much slower. While the scanner scans the body, the infrared energy is measured and used to modulate the intensity of a light beam that produces a map of the infrared energy on photographic paper. This presentation is called a *thermogram.* Figure 9.5(b) shows a photograph of two men and a corresponding thermogram. The thermogram shows that each of the two men has an artificial leg. The advantage of this method is that relatively warm and cool areas are immediately evident. By calibrating against known temperature sources, the picture can be read quantitatively.

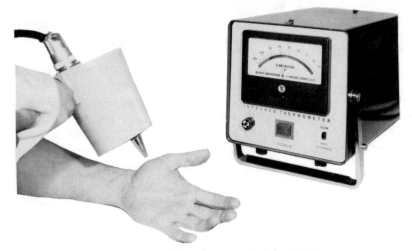

Figure 9.4. Infrared thermometer. Barnes model MT-3 noncontact thermometer provides fast, accurate measurements of skin temperature. (Courtesy of Barnes Engineering Company, Stamford, Conn.)

A similar device, called *Thermovision*,® has a scanner that operates at a rate sufficiently high to permit the image to be shown on an oscilloscope. The raster has about 100 vertical lines per frame, and the horizontal resolution is also about 100 lines, which seems to be adequate for good representation. The intensity of the measured infrared radiation is reproduced by Z-axis modulation (brightness variation) of the oscilloscope beam. One advantage of this system is that certain portions of the gray scale can be enhanced to bring out specific features of the picture. Also, the image can be changed so that warm spots appear dark instead of light, as they usually do. All these enhancement measures can be performed while the subject is being scanned. A Thermovision® system is shown in Figure 9.6.

Related to the infrared techniques is the use of *heat-quenched phosphors* to indicate skin temperature. These are phosphor materials that, when exposed to ultraviolet light, glow with a brightness inversely proportional to the surface temperature. The phosphors in a neutral solution are usually sprayed onto the region of the body to be measured. The glow under ultraviolet light is either photographed or televised by using special filters that remove the ultraviolet light itself and let only the emitted light reach the photographic film or TV vidicon. As with the infrared methods, calibration is possible by using known temperature sources in the picture. It is claimed that a skin temperature increase of 1°C can result in a decrease of 25 percent in the brightness of the picture. Resolution can be further enhanced in the television system by expansion of the gray scale.

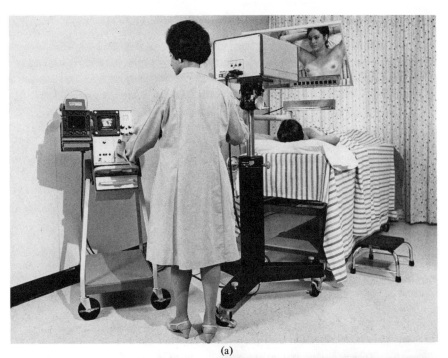

(a)

(b)

Figure 9.5. Thermography: (a) high resolution thermograph; (b) thermogram (see explanation in text). (Courtesy of Barnes Engineering Company, Stamford, Conn.)

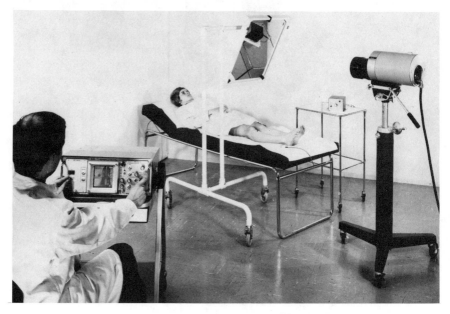

Figure 9.6. Thermovision system. (Courtesy of AGA AKTIE-BOLAG, Lidingö, Sweden.)

The heat-quenched phosphor method is less expensive than thermography, but there are some definite disadvantages. The method is not easy to use. The intensity and uniformity of the ultraviolet lighting are critical. It is difficult to spray the phosphor onto the subject with a completely uniform coating, and so on. Efforts have been made to produce a plastic tape with the phosphor uniformly distributed on the surface, but it is difficult to keep tape of this type from wrinkling. Furthermore, experimental tapes developed so far do not seem to be as sensitive as the phosphors sprayed directly onto the body. Another possible problem arises from the fact that the tape has some tendency to conduct a certain amount of heat away from the point of measurement.

Another method of indicating skin temperature is the use of "liquid crystals," which reflect light at different wavelengths (different colors) at different temperatures. These "crystals" are actually neither crystal nor liquid but organic compounds that can be poured like a liquid, yet reflect light like a crystal. As with the heat-quenched phosphors, they are sprayed onto the surface of the body, but only after the skin has been covered with a black undercoating. Again the uniformity with which the crystals are applied and the light intensity used are critical. Continued experimentation with both the liquid crystals and heat-quenched phosphors might eventually result in some useful quantitative techniques for measuring skin temperature variations.

9.2. DISPLACEMENT, FORCE, VELOCITY, AND ACCELERATION MEASUREMENTS

Requirements involving displacement, motion and associated measurements in biomedical instrumentation are many and varied. They range from the measurement of the distance a person moves his arm (perhaps against a load in order to relate muscle action to body movement) to the minute palpitations of the valves in the heart. The requirements for accuracy also vary from strictly qualitative measurements (detection of the fact that movement has taken place) to very precise quantitative measurements. One specific type of displacement measure used in the biomedical field is the *myogram*—the measure of the amount a muscle contracts when stimulated. The instrument on which myograms are measured is called a *myograph.* Since most methods of measuring displacement can be applied to myography, this type of measurement is used as an example for some of the methods described.

Muscles in the body contract in two different modes. If the muscle contracts (shortens) without greatly changing its applied force or *tone,* the contraction is said to be *isotonic.* If, however, the muscle applies a force, but without appreciably changing its length, its action is *isometric.* In a similar vein, a displacement transducer that is able to be moved through its range with only a negligible amount of force required to move it is an *isotonic transducer,* whereas a stiff transducer that measures force without significant displacement is basically an *isometric transducer.*

9.2.1. POTENTIOMETER TRANSDUCERS. The simplest device for measuring displacement is the *linear* or *rotational potentiometer.* Examples of both are shown in Figure 9.7. Figure 9.7(a) shows schematic diagrams for both linear and rotational potentiometers, while Figure 9.7(b) shows a diagram and Figure 9.7(c) shows a photograph of a commercially available linear potentiometer. The point at which displacement is to be measured is simply attached to a sliding contact that can move along a path of linear electrical resistance, such as a wire or a section of carbon or wirewound resistor. If, in Figure 9.7(a), a potential is applied between terminals A and B of either device, the potential at point C with respect to point A is proportional to the linear or rotational distance from A to C. One possible difficulty with this type of transducer is the amount of friction of the wiper on the wire or resistor. A true *displacement transducer* should not require any force to move the contact, whereas a *force transducer* should require a force proportional to the amount of movement to move the sliding contact. Where a potentiometer is used as a force trans-

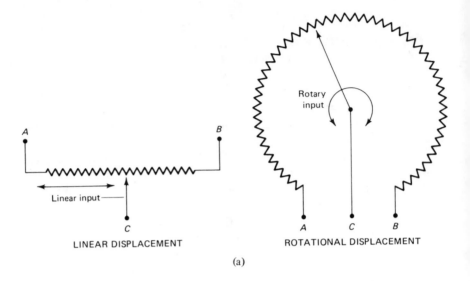

A   B

Linear input

C

LINEAR DISPLACEMENT

Rotary input

A   C   B

ROTATIONAL DISPLACEMENT

(a)

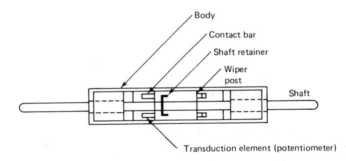

Body

Contact bar

Shaft retainer

Wiper post

Shaft

Transduction element (potentiometer)

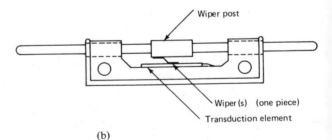

Wiper post

Wiper(s) (one piece)

Transduction element

(b)

Figure 9.7. Potentiometer displacement transducers: (a) schematics of linear and rotary potentiometer transducers; (b) physical diagram of linear transducers; (c) photograph of a linear transducer. (Courtesy of Bourns, Inc., Riverside, Calif.)

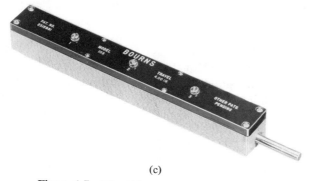

(c)

Figure 9.7. (*Contd.*)

ducer, the object producing the force must be allowed to move a sufficient amount to produce a suitable change in the position of the contact.

Velocity is the rate of change or first derivative of position or displacement:

$$V = \frac{dD}{dt}$$

where
$\qquad V =$ velocity
$\qquad D =$ displacement
and

$\qquad \dfrac{d}{dt} =$ indicates the first derivative or instantaneous rate of change.

Similarly, acceleration, A, is the rate of change or first derivative of velocity. Acceleration is also the second derivative of displacement:

$$A = \frac{dV}{dt} = \frac{d^2D}{dt^2}$$

The inverse of these relationships can be expressed as the integrals:

$$V = \int A\, dt \qquad D = \int V\, dt = \int \int A\, (dt)^2$$

From these relationships it can be seen that if it is possible to measure one of these variables (displacement, velocity, or acceleration), the other two variables can be calculated or obtained by either analog or digital differentiation or integration methods. Generally integration is preferred, because of the adverse effect of noise in differentiation circuitry.

It should also be recognized that force can be determined by measuring displacement against a linear spring. Thus force becomes proportional to displacement.

9.2.2. THE DIFFERENTIAL TRANSFORMER. Where smaller displacements than can be adequately detected by a potentiometer type of transducer must be measured, more sensitive displacement transducers are required. One such device is the *differential transformer* shown in diagram form in Figure 9.8. The differential transformer consists basically of a

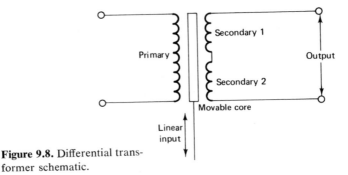

Figure 9.8. Differential transformer schematic.

transformer with a primary and two secondary windings, connected so that the two secondary voltages oppose each other. The transformer has a movable core that is attached to the point at which displacement is to be measured. When energized by an ac source, the primary induces equal voltages into the two secondaries as long as the movable core is exactly in the center, producing a net output of zero volts. However, when the core is moved in the direction of one of the secondaries, that secondary has an increased induced voltage, whereas the voltage in the other secondary is reduced. The ouput is the difference between the voltages induced in the two secondaries, which is proportional to the amount of displacement of the core, and in a polarity corresponding to the direction in which the core is moved. Thus the differential transformer is able to measure extremely small variations in displacement, even as small as one millionth of an inch. One problem with this type of displacement transducer, however, is that it is not linear for very large amounts of displacement.

The core is usually made of a material with high permeability. The excitation frequency must be several times the highest frequency component of the variations of input displacement. Unless a higher frequency is required, 60 Hz is conveniently used. For example, one commercially available differential transformer operates with an excitation voltage of 6.3 volts. Under these conditions the sensitivity is on the order of 10 mV per 0.001-inch displacement.

Like the potentiometer, the differential transformer is a true displacement transducer that can be built with low compliance and thus is suitable for isotonic measurements. A commercially available transducer utilizing a differential transformer is shown in Figure 9.9.

Figure 9.9. Differential transducer transformer transducer. This is used for blood pressure measurement. (Courtesy of Biotronex Laboratory, Inc., Silver Springs, Md.)

9.2.3. STRAIN GAGE TRANSDUCERS. Even more common than differential transformers in the measurement of displacement and its associated physical variables are transducers that employ one or more strain gages. These have been treated in Chapter 6 in the context of blood pressure measurements but a more general explanation is presented here. A *strain gage* is a length of conductive material, such as a fine wire or a piece of semiconductor material, that is stretched or compressed in proportion to displacement. When stretched, the wire or other device elongates and, at the same time, is reduced in its cross-sectional area, thereby increasing its electrical resistance. Similarly, when the gage is compressed, the resistance is lowered. Because a force is required to stretch or compress the device, the strain gage is basically a force (isometric) transducer.

As an example of a device to measure displacement, suppose, as shown in Figure 9.10, that a strain gage is placed on each side of a flexible

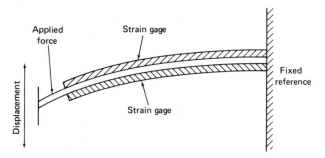

Figure 9.10. Flexible arm (cantilevered beam) displacement-force transducer utilizing two strain gages.

arm, such as a segment of spring metal. The strain gages are cemented to the arm so that when the arm is bent one of the strain gages is stretched and the other is compressed, both in proportion to the amount of bending. If one end of the flexible arm is fixed to a reference point (cantilevered) and the other end is attached to the point at which displacement is to be measured, the arm will be bent by the displacement and the resistances of the two strain gages will vary accordingly. Since the flexible arm acts as a spring, the force required to produce the measured displacement is proportional to the displacement. If *displacement* is the desired measurement, the arm is made extremely flexible so that the force required to bend it is negligible compared to the energy causing the displacement. On the other hand, if *force* is to be measured, the arm is made stiff so that movement is negligible, and the small amount of displacement is proportional to the applied force.

The sensitivity of the strain gage is determined by the *gage factor*. As mentioned in Section 6.2.2, this term denotes the percentage of resistance change per unit change in length. Thus the gage factor can be expressed as

$$G = \frac{\Delta R/R}{\Delta L/L}$$

where

$\Delta R/R =$ the ratio of change in resistance
$\Delta L/L =$ the corresponding ratio of change in length

The average gage factor for metals is about 2.0, whereas the gage factor for silicon (a semiconductor material) is about 120. To increase sensitivity, several lengths of wire are usually stretched or compressed simultaneously and connected in series, as shown in Figure 9.11, to form a typical wire strain gage.

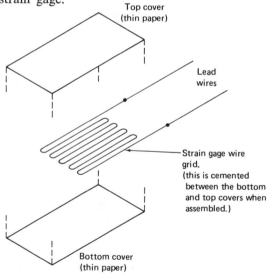

Figure 9.11. Typical bonded strain gage configuration.

Wire strain gages are of two basic types, bonded and unbonded. In the *bonded strain gage* the wire is bonded to a thin piece of plastic or paper that is cemented to the material to which the stress is applied. This is the type of strain gage shown in Figures 9.10 and 9.11. The strength of the strain gage itself is not significant, for the material to which the strain gage is cemented actually bears the applied load. A force-displacement transducer employing a bonded wire strain gage is shown in Figure 9.12; a pressure transducer utilizing semiconductor strain gages is shown in Figure 9.13.

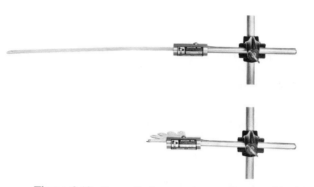

Figure 9.12. Force-displacement transducer with bonded strain gage. (Courtesy of Biocom, Inc., Culver City, Calif.)

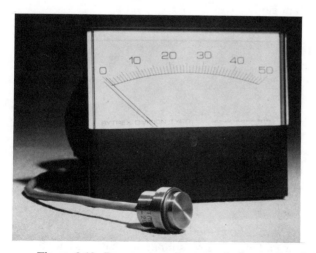

Figure 9.13. Pressure transducer employing semiconductor bonded strain gages with readout device. (Courtesy of Tyco Instrument Division, Tyco Laboratories, Inc., Watertown, Mass.)

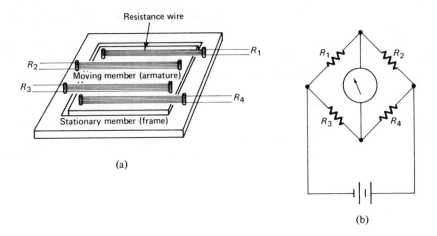

Resistance wire

R_2

Moving member (armature)

R_3

R_1

R_4

Stationary member (frame)

(a)

R_1 R_2

R_3 R_4

(b)

Figure 9.14. Unbonded strain gage transducer. (From D. Bartholomew, *Electrical Measurement and Instruments.* Allyn & Bacon, Inc., Boston, Mass., by permission.)

In the *unbonded strain gage* the wires of the strain gage are wound under tension between isolating posts, as shown in Figure 9.14. Here four unbonded gages are attached to two otherwise isolated members, called the armature and the frame. Mechanical stops are provided to prevent an overload from breaking the wires. Without applied force, the wires are stretched with the armature centered in the frame. As force is applied, tension on two of the gages is increased, while tension on the other two gages is decreased. Figure 9.15 shows a displacement transducer utilizing an unbonded strain gage.

Figure 9.15. Displacement transducer utilizing unbonded strain gage. (Courtesy of Statham Instruments, Inc., Oxnard, Calif.)

Instead of wire, some modern gages are made of foil on a substrate material by an etching process similar to that used in the fabrication of circuit boards. These foil gages have characteristics similar to bonded wire strain gages.

214

As mentioned earlier, a semiconductor strain gage made of silicon has a gage factor some 60 times that of metal. This means that with silicon strain gages, much smaller transducers can be used to obtain force and displacement measurements. This higher sensitivity also permits stiffer gages with higher frequency responses. With semiconductor strain gages, frequency responses up to 2000 Hz are not difficult to attain. The major disadvantage of the semiconductor strain gage is the variation in resistance of the silicon with temperature. This variation can be partially compensated for by having at least two individual strain gage elements in different legs of a bridge, a technique also employed with wire and foil gages for the same reason. Another disadvantage seems to be a tendency of semiconductor materials to be more brittle than wire strain gages.

Both bonded and unbonded strain gages are found in many forms of transducers designed for biomedical applications. Bonded strain gages attached to a flexible diaphragm that is bent in proportion to the pressure difference on the two sides of the diaphragm form an excellent differential pressure transducer. A weight attached to the armature of an unbonded strain gage serves as an accelerometer.

As with thermistors, strain gages are usually connected into one or more arms of a bridge circuit, with the output of the bridge amplified and recorded. When two or four strain gages are used in a single measurement, they are placed in different arms of the same bridge. This step not only increases sensitivity but also provides temperature compensation. Wire strain gages typically have nominal resistances of around 120 or 600 ohms.

In addition to the conventional strain gages described, several special types of strain gage devices have been developed especially for biomedical applications. One type, called the *mercury strain gage,* consists of a small piece of silicon rubber tubing filled with mercury. As the tubing is stretched, the resistance path of the mercury elongates and becomes smaller in diameter, thereby increasing the resistance. Mercury strain gages are much more compliant than metallic or semiconductor strain gages and thus are very useful in certain physiological applications, such as measuring changes in chest diameter during breathing to obtain respiration rate and diameter changes in a muscle or of the vascular bed of a finger or toe. Because of their very low resistance (one to a few ohms), mercury strain gages require special circuits for their use. The effect of temperature changes, even in the input wires, can become a problem. The mercury strain gage plethysmograph shown in Figure 6.35 (Chapter 6) utilizes this type of transducer.

Similar to the mercury strain gage is a strain gage in which the tubing is filled with an electrolyte instead of mercury. These strain gages have a more convenient resistance range and are quite inexpensive, but they are seldom used because of the difficulty in calibrating and maintaining them.

Another related device is the "rubber resistor," which is made of conductive elastomer. As the elastic is stretched, the resistance changes in about the same fashion as the mercury in the mercury strain gage.

As in the case of potentiometer transducers and differential transformers, strain gage transducers are usually able to measure absolute position on a displacement or force scale once they have been calibrated to that scale. The exception is the strain gage accelerometer, which registers only the amount of acceleration.

9.2.4. OTHER PHYSICAL TRANSDUCERS. Although most commercially available transducers for measurement of force, displacement, and related variables employ potentiometers, differential transformers, or strain gages, some other transducers for these types of measurement are occasionally found in biomedical instrumentation. A few of these devices are described briefly below.

CAPACITANCE TRANSDUCERS. If one plate of a capacitor is connected to a point at which displacement or movement is to be measured and the other plate is fixed as a reference, any change in the relative position of the two plates (distance between the plates) is reflected as a change in the capacity. The absolute value of the capacity can be measured on a capacitance bridge, or the capacitor can be connected with an inductance to form a tuned circuit. When excited by an ac voltage at a frequency just off resonance, the tuned circuit presents an impedance proportional to the value of the capacity, which in turn reflects the displacement. If instead the tuned circuit is incorporated into an oscillator, a frequency modulated signal can be obtained for demodulation via an FM discriminator. By causing the moveable plate to be inserted between two fixed plates, two capacitors are formed, such that as the moveable plate moves toward one fixed plate, thus increasing the value of that capacitor, it moves away from the other, thereby reducing the capacity of the second capacitor. In this way a differential capacitance transducer is formed.

VARIABLE INDUCTANCE TRANSDUCERS. If the inductance of a coil is caused to vary with displacement, a variable inductance transducer is formed. This effect can be accomplished by moving the core or by varying an air gap in the magnetic circuit. Although used less frequently than the capacitance transducer because of mechanical limitations, transducers employing the variable inductance principle do have a few practical applications. Inductance transducers are usually linear over a very limited mechanical range. Figure 9.16 shows an arrangement wherein a metal core is moved into and out of the center of a coil to vary the inductance as the core is displaced. The absolute value of the inductance can be mea-

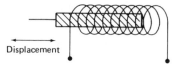

Figure 9.16. Example of variable inductance displacement transducer.

sured by using an ac bridge, or the variable inductance can be part of a tuned circuit, which can be treated in the same manner as described for the capacitance transducer.

INDUCTION-TYPE TRANSDUCERS. The principle that a conductor in a changing magnetic field or a moving conductor in a fixed magnetic field will develop an induced voltage proportional to the velocity of movement is used in the design of an induction-type transducer. Transducers of this type, which respond to velocity rather than displacement, produce ac voltages that can be amplified directly. One example of this type of transducer is the so-called *variable reluctance phonograph cartridge,* in which movement of the stylus changes the magnetic flux through a coil and induces in the coil a signal proportional to the velocity of the stylus.

PIEZOELECTRIC TRANSDUCERS. Certain substances have the characteristic that, when physically deformed, they produce an electric charge proportional to the amount of deformation. This characteristic, called the *piezoelectric effect,* is found naturally in rochelle salt, quartz, ammonium, and dihydrogen phosphate and can be introduced by treatment into barium titantate.

When contacts are applied to the face of the piezoelectric material, an electrical signal can be obtained. The piezoelectric element is electrically equivalent to a voltage generator, delivering a voltage proportional to the applied force, connected in series with a capacitor. The electrical signal at the output, therefore, depends on the impedance of the amplifier to which it is connected. When the piezoelectric element is connected to an amplifier with a very low input impedance, the electrical *current* into this amplifier is proportional to the rate of change of the applied force or deformation of the element. However, when the piezoelectric element is connected to an electrometer amplifier (with an extremely high input impedance), the *voltage* at the input of the amplifier is directly proportional to the applied force or deformation of the element.

When the piezoelectric element is connected to an amplifier with a moderately high input impedance, which is usually the case, the voltage at the input of this amplifier is proportional to the applied force or deformation of the element immediately after the force, or deformation, has changed. However, as stated earlier, the piezoelectric element forms a capacitor that is gradually discharged over the input impedance of the amplifier, causing a decay of the voltage. In the latter mode, like the induction-type transducer described above, piezoelectric transducers respond only to velocity of movement and do not measure position or displacement directly.

Piezoelectric elements are found most frequently in microphones or other transducers that measure sound or some type of pressure or vibration pattern, such as the movement of the heart against the chest wall, tremor, and activity of animals in a cage. The output is a voltage that can be amplified directly and displayed or recorded. Important characteristics of piezoelectric crystal transducers include sensitivity (the voltage output for a given change in stress) and frequency response.

PHOTOELECTRIC DISPLACEMENT TRANSDUCERS. Another method of transforming displacement into an electrical signal is found in the use of a photoelectric system consisting of a light source and a photoresistive cell with a shutter partially blocking the light path. The shutter is movable and is attached to the point at which displacement is to be measured. Movement of that point causes the shutter either to increase or decrease the amount of light reaching the photocell. This, in turn, changes the resistance of the photocell, which can be measured by using a bridge circuit. Even this type of transducer is capable of measuring rather small changes in displacement. A diagram and photograph of this device are shown in Figure 9.17.

ULTRASOUND DISPLACEMENT AND MOVEMENT TRANSDUCERS. When ultrasonic energy is transmitted into the body through the skin, echoes are produced wherever the energy passes through a change of medium. Furthermore, if any reflecting surface is in motion, the frequency of the reflected ultrasound is shifted from that transmitted due to the Doppler effect. The change in frequency indicates not only the velocity with which the reflecting object is moving but also the direction of movement relative to the ultrasonic source.

The use of ultrasound to measure distance, movement, and sometimes the size of an organ or other distinguishable part of the body has application in both medical research and clinical medicine. Where location or distance is to be measured, a pulsed ultrasound signal is transmitted, and the time from transmission of each pulse until receipt of a corresponding echo is measured, usually on an oscilloscope triggered by the transmitted pulse. Even relatively small movements, such as those associated with the operation of the individual valves of the heart, can be measured by this method. For measurements of this type, ultrasound energy at a frequency near 2.25 MHz is transmitted in one-microsecond bursts at a rate between 200 and 1000 pulses per second.

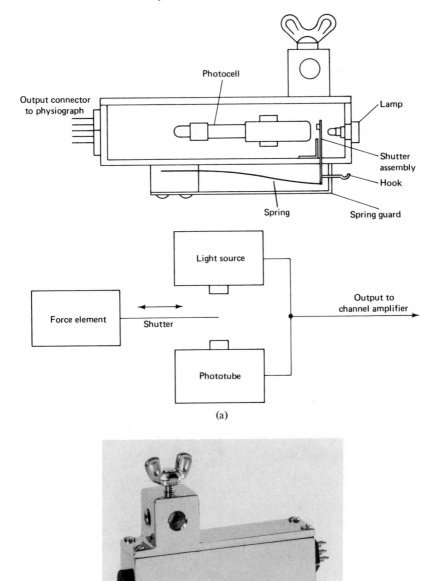

(a)

(b)

Figure 9.17. Photoelectric displacement transducer: (a) block diagram; (b) photograph. (Courtesy of Narco BioSystems, Houston, Tex.)

A well-established application of pulsed ultrasound is in the location of the midline of the brain by simultaneously measuring the distance to the midline from probes at both sides of the head. The midline echoes from the two probes are simultaneously displayed on an oscilloscope, one producing upward deflection of the oscilloscope beam and the other producing a downward deflection, as shown in Figure 9.18. In the normal brain these two deflections line up, indicating equal distance from the midline to each side of the head. Nonalignment of these deflections indicates the possibility of a tumor or some other disorder that might cause the midline of the brain to shift from its normal position. The instrument for this measurement produces ultrasound energy at a frequency of from 1 to 10 MHz. The pulse rate is 1000 per second.

Pulsed ultrasound is also used for detection of lesions in the breast and other parts of the body, ventricular size estimation, and fetal size determination. For measurement of movement, however, instruments employing the Doppler effect are often superior to those using pulsed ultrasound. For

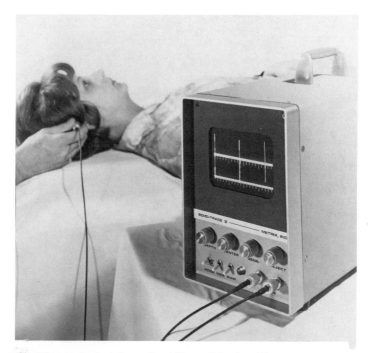

Figure 9.18. Ultrasonic location of midline of brain. (Courtesy of Metrix, Inc., Denver, Colo.)

example, the Doppler principle is used in the ultrasonic blood flow meters described in Chapter 6. Another application of the Doppler principle is the fetal pulse detector shown in Figure 9.19. With this instrument, the

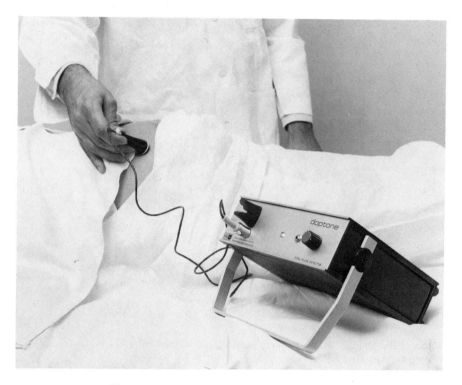

Figure 9.19. Ultrasonic fetal pulse detector. (Courtesy of Gould, Inc., Palo Alto, Calif.)

difference in frequencies between the continuously transmitted 2-MHz ultrasound and reflections from moving organs or flowing blood within the pregnant uterus are presented audibly via a built-in loudspeaker. Several distinctive sounds, including the fetal pulse, can be heard to provide an indication of any disorder or possible complication. The progress of labor can also be monitored using ultrasonic methods.

DIGITAL ENCODERS. In the preceding devices displacement or force variations (and sometimes velocity or acceleration) appear as analog voltages or currents proportional to the variables being measured. Angular displacement, however, is sometimes converted directly into a digital code that can be read into a digital computer or displayed in digital form. The digital code (see Chapter 15) is a set of 1s and 0s that represent, in binary form, the angle of a shaft. The number of 1s and 0s in the set determines the accuracy with which the angular measurement can be made. Most encoders utilize an optical principle in which a pattern similar to one of those shown in Figure 9.20 is scanned along one radius

Figure 9.20. Digital shaft encoder patterns. (Courtesy of Itek, Wayne George Division, Newtown, Mass.)

by a light transmission or reflection mechanism. At any given distance from the center along a specific radius, the pattern is either light (1) or dark (0). Each concentric ring at which there can be a difference in the pattern represents one bit of digital information. The set of bits along the specified radius depends on the angle to which the shaft is turned and constitutes a binary code that indicates the angle. A commercial digital-shaft-angle encoder is shown in Figure 9.21.

A form of linear digital encoder that converts linear movement and displacement into a digital code is also available. In this device an optical scanner moves over two sets of alternate light and dark regions, producing two phased pulsatile outputs. These outputs correspond in frequency to the velocity of the object whose displacement or movement is being measured. The relative phase of the two outputs indicates the direction of movement, whereas the number of alternations of the pulsatile outputs indicates the amount of movement. An electronic counter that receives

Figure 9.21. Digital shaft angle encoder. This is a 256-count incremental encoder with integral electronics providing a square wave output. (Courtesy of Renco Corporation, Santa Barbara, Calif.)

both sets of pulses indicates the displacement of the object being measured. If the counter has a digital output, that output provides a coded representation of the displacement.

In this section many methods for measuring displacement (and, of course, force, velocity, and acceleration) have been presented. It should be realized, however, that this number represents only a small sampling of the possible measurement methods available to the biomedical instrumentation engineer or technician. In the area of physical measurement, as perhaps in no other, the engineer can directly apply measurement methods developed for other fields. Caution must be exercised, however, to ensure that the measuring device does not interfere with the mechanism being measured and that the characteristics of the transducer are applicable and suitable within the range of the measured variables in the biological system. A final caution is to beware of unexpected reactions to the measurement by the system being measured. For example, a muscle pulling the extra load of a force transducer may not act in the same way as it does under its normal load. If the engineer is aware of these problems, he should be able to adapt, from his available sources of ideas, many new and useful ways of measuring physical variables from the human body.

· 10 ·

THE NERVOUS
SYSTEM

The task of controlling the various functions of the body and coordinating them into an integrated living organism is not simple. Consequently, the nervous system, which is responsible for this task, is the most complex of all the systems in the body. It is also one of the most interesting. Composed of the brain, numerous sensing devices, and a high-speed communication network that links all parts of the body, the nervous system not only influences all the other systems but is also responsible for the behavior of the organism. In this broad sense, *behavior* includes the ability to learn, remember, acquire a personality, and interact with its society and the environment. It is through the nervous system that the organism achieves autonomy and acquires the various traits that characterize it as an individual.

A complete study of the nervous system, with all of its ramifications, would be far beyond the scope of this book. However, an overall view can be given that provides the reader with a physiological background for measurements within the nervous system, as well as some understanding of the effect of the nervous system on measurements from other systems of the body. To make this presentation more useful in the study of biomedical instrumentation, many of the concepts and theories are greatly simplified. This simplification is not intended to detract from the reader's understanding of the concepts and theories, but it should facilitate visualization of an extremely complex system and provide a better perspective for further detailed study, if required. The simplification requires, however, that caution must be used in attempting to extrapolate or generalize from the information presented.

10.1. THE ANATOMY OF THE NERVOUS SYSTEM

The basic unit of the nervous system is the *neuron*. A neuron is a single cell with a *cell body,* sometimes called the *soma,* one or more "input" fibers, called *dendrites,* and a long transmitting fiber, called the *axon.* Often the axon branches near its ending into two or more terminals. Examples of three different types of neurons are shown in Figure 10.1.

The portion of the axon immediately adjacent to the cell body is called the *axon hillock.* This is the point at which action potentials are usually generated. Branches that leave the main axon are often called *collaterals.* Certain types of neurons have axons or dendrites coated with a fatty insulating substance called myelin. The coating is called a *myelin sheath* and the fiber is said to be *myelinated.* In some cases, the myelin sheath is interrupted at rather regular intervals by the *nodes of Ranvier,* which help speed the transmission of information along the nerves. Outside of the central nervous system, the myelin sheath is surrounded by another insulating layer, sometimes called the *neurilemma.* This layer, thinner than the myelin sheath and continuous over the nodes of Ranvier, is made up of thin cells, called *Schwann cells.*

As can be seen from Figure 10.1, some neurons have long dendrites whereas others have short ones. Axons of various length can also be found throughout the nervous system. In appearance, it is difficult to tell a dendrite from an axon. The main difference is in the function of the fiber and the direction in which it carries information with respect to the cell body.

Both axons and dendrites are called *nerve fibers,* and a bundle of individual nerve fibers is called a *nerve.* Nerves that carry sensory information from the various parts of the body *to the brain* are called *afferent nerves,* whereas those that carry signals *from the brain* to operate various muscles are called *efferent nerves.*

The *brain* is an enlarged collection of cell bodies and fibers located inside the skull, where it is well protected from light as well as from physical, chemical, or temperature shock. At its lower end, the brain connects

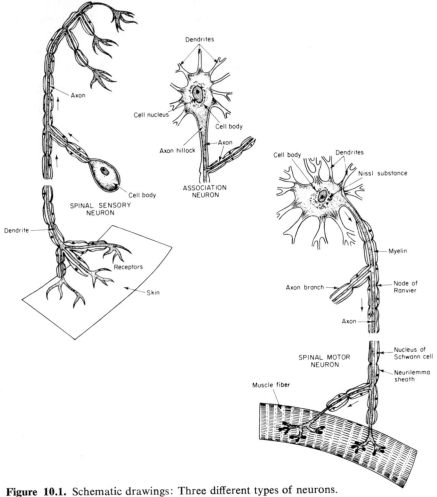

Figure 10.1. Schematic drawings: Three different types of neurons. (From W.F. Evans, *Anatomy and Physiology, The Basic Principles.* Englewood Cliffs, N.J., Prentice-Hall, Inc., 1971, by permission.)

with the *spinal cord,* which also consists of many cell bodies and fiber bundles. Together the brain and spinal cord comprise one of the main divisions of the nervous system, the *central nervous system* (CNS). In addition to a large number of neurons of many varieties, the central nervous system also contains a number of large fatty cell bodies called *glial cells.* About half the brain is composed of glial cells. At one time it was believed that the main function of glial cells was structural and that they physically supported the neurons in the brain. Later it was postulated, however, that the glial cells play a vital role in ridding the brain of foreign substances and seem to have some function in connection with memory.

Cell bodies and small fibers in fresh brain are gray in color and are often called *gray matter,* whereas the myelin coating of larger fibers has a white appearance, so that a collection of these fibers is sometimes called *white matter.* Collections of neuronal cell bodies within the central nervous system are called *nuclei,* while similar collections outside the central nervous system are called *ganglia.*

The central nervous system is generally considered to be *bilaterally symmetrical,* which means that most structures are anatomically duplicated on both sides. Even so, some functions of the central nervous system in humans seem to be located nonsymmetrically. Several of the functions of the central nervous system are *crossed over,* so that neural structures on the left side of the brain are functionally related to the right side of the body and vice versa.

Nerve fibers outside the central nervous system are called *peripheral nerves.* This name applies even to fibers from neurons whose cell bodies are contained within the central nervous system. Throughout most of their length, many peripheral nerves are mixed, in that they contain both afferent and efferent fibers. Afferent peripheral nerves that bring sensory information into the central nervous system are called *sensory nerves,* whereas efferent nerves that control the motor functions of muscles are called *motor nerves.* Peripheral nerves leave the spinal cord at different levels, and the nerves that *innervate* a given level of body structures come from a given level of the spinal cord.

The interconnections between neurons are called *synapses.* The word "synapse" can be used as both a noun and a verb. Thus the connection is called a *synapse,* and the act of connecting is called *synapsing.* All synapses occur at or near cell bodies. As explained in Section 10.2, mammalian neurons that synapse do not touch each other but do come into close proximity, so that the axon (output) of one nerve can activate the dendrite or cell body (input) of another by producing a chemical that stimulates the membrane of dendrite or cell body. In some cases, the chemical is produced by one axon, near another axon, to inhibit the second axon from activating a neuron with which it can normally communicate. This action is explained more fully below. Because of the chemical method of transmission across a synapse from axon to dendrite or cell body, the communication can take place in one direction only.

The peripheral nervous system actually consists of several subsystems. The system of afferent nerves that carry sensory information from the sensors on the skin to the brain is called the *somatic sensory nervous sys-*

tem. Visual pathways carry sensory information from the eyes to the brain, whereas the *auditory nervous system* carries information from the auditory sensors in the ears to the brain.

Another major division of the peripheral nervous system is the *autonomic nervous system,* which is involved with emotional responses and controls smooth muscle in various parts of the body, heart muscle, and the secretion of a number of glands. The autonomic nervous system is composed of two main subsystems that appear to be somewhat antagonistic to each other, although not completely. These are the *sympathetic nervous system,* which speeds up the heart, causes secretion of some glands, and inhibits other body functions, and the *parasympathetic nervous system,* which tends to slow the heart and controls contraction and secretion of the stomach. In general, the sympathetic nervous system tends to mobilize the body for emergencies, whereas the parasympathetic nervous system tends to conserve and store bodily resources.

A very general look at the anatomy of the brain should be helpful in understanding the functions of the nervous system. Figure 10.2 shows a side view of the brain and spinal cord, and Figure 10.3 is a cutaway showing some of the major structures.

The part of the brain that connects to the spinal cord and extends up into the center of the brain is called the *brainstem.* The essential parts of the brainstem are the *medulla* (sometimes called the *medulla oblongata*), which is the lowest section of the brainstem itself, the *pons* located just above the medulla and protruding somewhat in front of the brainstem, and the upper part of the brainstem called the *midbrain.* Above and slightly forward of the midbrain are the *thalamus* and *hypothalamus.* Behind the brainstem is the *cerebellum.* Almost completely surrounding the midbrain, thalamus, and hypothalamus are the structures of the *cerebrum.* The outer surface of the cerebrum is called the *cerebral cortex.* The *corpus callosum* is the interconnection between the left and right hemispheres of the brain. Structurally the two hemispheres appear to be identical, but as indicated earlier, they seem to differ functionally in man. Just forward of the hypothalamus is the *hypophysis* or *pituitary gland,* which produces hormones that control a number of important hormonal functions of the body. Not shown in the figures, but surrounding the thalamus, is the *reticular activating system.* The specific functions of each of these major portions of the brain, as far as they have been discovered to date, are discussed in Section 10.3.

10.2. NEURONAL COMMUNICATION

As discussed in detail in Chapter 3, neurons are among the special group of cells that are capable of being excited and that, when

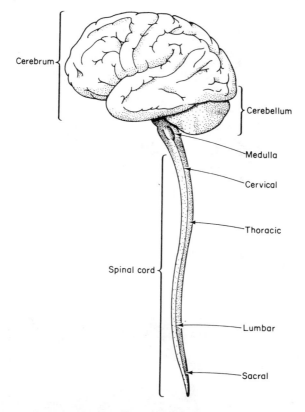

Figure 10.2. The brain and spinal cord. (From W.F. Evans, *Anatomy and Physiology, The Basic Principles,* Englewood Cliffs, N.J., Prentice-Hall, Inc., 1971, by permission.)

excited, generate action potentials. In neurons, these action potentials are of very short duration and are often called *neuronal spikes* or *spike discharges.* Information is usually transmitted in the form of *spike discharge patterns.* These patterns, which are simply the sequences of spikes that are transmitted down a particular neuronal pathway, are shown in Figures 10.4 and 10.5. The form of a given neuronal pattern depends on the firing patterns of other neurons that communicate with the neuron generating the pattern and the refractory period of that neuron (see Chapter 3). When an action potential is initiated in the neuron, usually at the cell body or axon hillock, it is propagated down the axon to the axon terminals where it can be transmitted to other neurons.

Given sufficient excitation energy, most neurons can be triggered at almost any point along the dendrites, cell body, or axon and generate action potentials that can move in both directions from the point of initiation. This process does not normally happen, however, because in their

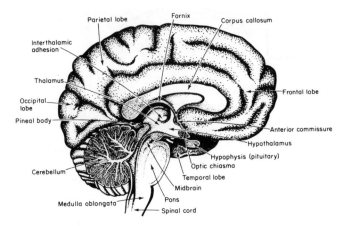

Figure 10.3. Cutaway section of the human brain. (From W.F. Evans, *Anatomy and Physiology, The Basic Principles,* Englewood Cliffs, N.J., Prentice-Hall, Inc., 1971, by permission.)

Figure 10.4. Spike discharge pattern from a single neuron in the red nucleus of a cat. The red nucleus is involved with motor functions. (Courtesy of Neuropsychology Research Laboratory, Veterans Administration Hospital, Sepulveda, Calif.)

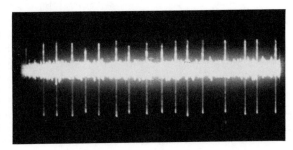

natural function, neurons synapse only in a certain way; that is, the axon of one neuron excites the dendrites or cell body of another. The result is a one-way communication path only. If an action potential should somehow be artificially generated in the axon and caused to travel up the neuron to the dendrites, the spike cannot be transmitted the wrong way across the gap to the axon of another neuron. Thus the one-way transmission between neurons determines the direction of communication.

It was believed for many years that transmission through a synapse was electrical and that an action potential was generated at the input of a neuron due to ionic currents or fields set up by the action potentials in the adjacent axons of other neurons. More recent research, however, has disclosed that in mammals, and in most synapses of other organisms, the transmission times across synapses are too slow for electrical transmission. This

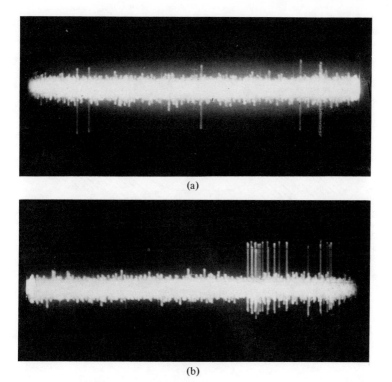

(a)

(b)

Figure 10.5. Spike discharge patterns from a single thalamic cell in a cat: (a) Random quiet pattern; (b) Burst pattern. (Courtesy of Neuropsychology Research Laboratory, Veterans Administration Hospital, Sepulveda, Calif.)

has led to the presently accepted chemical theory, which states that the arrival of an action potential at an axon terminal releases a chemical— probably *acetylcholine* in most cases—that excites the adjacent membrane of the receiving neuron. Because of the close proximity of the transmitting axon terminal to the receiving membrane, the time of transmission is still quite short. The possibility that some of the chemical may still be present after the refractory period is eliminated by the presence of *acetylcholine esterase,* another chemical that breaks down the acetylcholine as soon as it is produced, but not before it has been able to initiate its intended action potential in the nearby membrane. This chemical theory of transmission is diagrammed in Figure 10.6.

Actually, the situation is not quite as simple as has been described. There are really two kinds of communication across a synapse, excitatory and inhibitory. The same chemical appears to be used in both. In general, several axons from different neurons are in communication with the "input" of any given neuron. Some act to excite the membrane of the receiver,

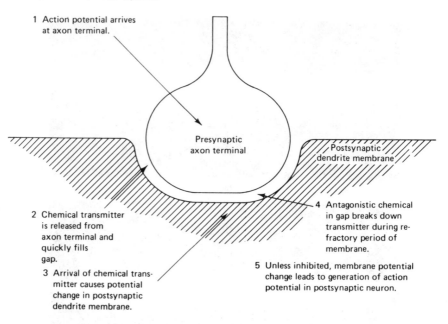

1 Action potential arrives at axon terminal.

Presynaptic axon terminal

Postsynaptic dendrite membrane

2 Chemical transmitter is released from axon terminal and quickly fills gap.

3 Arrival of chemical transmitter causes potential change in postsynaptic dendrite membrane.

4 Antagonistic chemical in gap breaks down transmitter during refractory period of membrane.

5 Unless inhibited, membrane potential change leads to generation of action potential in postsynaptic neuron.

Figure 10.6. Sequence of events during chemical transmission across a synapse.

while others tend to prevent it from being excited. Whether the neuron fires or not depends on the net effect of all of the axons interacting with it.

The effects of the various neurons acting on a receiving neuron are reflected in changes in the graded potential of the receiving neuron. *Graded potentials* are variations around the average value of the resting potential. When this graded potential reaches a certain threshold, the neuron fires and an action potential develops. Regardless of the graded potential before firing, the action potentials of a given neuron are always the same and always travel at the same rate. An excitatory graded potential is called an *excitatory postsynaptic potential* (EPSP), and an inhibitory graded potential is called an *inhibitory postsynaptic potential* (IPSP).

There are several theories as to how inhibitory action takes place. One possibility is that the inhibitory axon somehow causes a graded potential (IPSP) in the receiving neuron which is more negative than the normal resting potential, thus requiring a greater amount of excitation to cause it to fire. Another possibility is that the inhibiting axon acts, not on the receiving neuron but on the excitatory transmitting axon. In this case, the inhibiting axon might set up a premature action potential in the transmitting axon, so that the necessary combination of chemical discharges cannot occur in synchronism as it would without the inhibition. Whatever method is actually used, the end result is that certain action potentials which would otherwise be transmitted through the synapse are prevented from doing so when inhibitory signals are present. Synapses, then, behave much like multiple-input AND and NOR logic gates and, by their widely varied patterns of excitatory and inhibitory "connections," provide a means of switching and interconnecting parts of the nervous system with a complexity far greater than anything yet conceived by man.

10.3. THE ORGANIZATION OF THE BRAIN

Knowledge of the actual function of various parts of the brain is still quite sparse. Experiments in which portions of an animal's brain have been removed (oblated) show that there is a tremendous amount of redundancy in the brain. There is also a great amount of adaptivity in that if a portion of a brain believed responsible for a given function is removed from an infant animal, the animal somehow still seems able to develop that function to some extent. This result has led to the idea of the "law of mass action" in which it is theorized that the impairment caused by damage to some portion of the brain is not so much a function of *what* portions have been damaged but, rather, *how much* was damaged. In other words, when one region of the brain is damaged, another region seems to take over the function of the damaged part. Also, while tests show that a particular region of the brain seems to be related to some specific function, there are also indications of some relationship of that function to other parts of the brain. Thus, when, in the following paragraphs, certain functions are indicated for certain parts of the brain, it must be realized that these parts only seem to play a predominant role in those functions and that other parts of the brain are undoubtedly also involved.

In the brainstem, the *medulla* seems to be associated with control of some of the basic functions responsible for life, such as breathing, heart rate, and kidney functions. For this purpose, the medulla seems to contain a number of timing mechanisms, as well as important neuronal connections.

The *pons* is primarily an interconnecting area. In it are a large number of both ascending and descending fiber tracts, as well as many nuclei. Some of these nuclei seem to play a role in salivation, feeding, and facial expression. In addition, the pons contains relays for the auditory system, spinal motor neurons, and some respiratory nuclei.

The *cerebellum* includes an apparent "low-pass filter circuit" to smooth out what would otherwise be "jerky" muscle motions. The cerebellum also plays a vital role in man's ability to maintain his balance.

The *thalamus* manipulates nearly all sensory information on its way to the cerebrum. It contains main relay points for the visual, auditory, and somatic sensory systems.

The *reticular activation system* (RAS), which surrounds the thalamus, is a nonspecific sensory portion of the brain. It receives excitation from all the sensory inputs and seems to be aroused by any one of them, but it does not seem to distinguish which type of sensory input is active. When aroused, the RAS alerts the cerebral cortex, making it sensitive to incoming information. It is the RAS that keeps a person awake and alert and causes him to pay attention to a sensory input. Most information reaching the RAS is relayed through the thalamus.

The *hypothalamus* is apparently the center for emotions in the brain. It controls the neural regulation of endocrine gland functions via the pituitary gland and contains nuclei responsible for eating, drinking, sexual behavior, sleeping, temperature regulation, and emotional behavior generally. The hypothalamus exercises primary control over the autonomic nervous system, particularly the sympathetic nervous system.

The *basal ganglia* seem to be involved in motor activity and have indirect connections with the motor neurons.

The main subdivision of the *cerebrum* is the *cerebral cortex,* which contains some 9 billion of the 12 billion neurons found in the human brain. The cortex is actually a rather thin layer of neurons at the periphery of the brain, which contains many fissures or inward folds to provide a greater amount of surface area. Some of the deeper fissures, also called *sulci,* are used as landmarks to divide the cortex into certain lobes. Several of the more prominent ones are shown in Figure 10.7, along with the location of the important lobes.

All sensory inputs eventually reach the cortex, where certain regions seem to relate specifically to certain modalities of sensory information. Other regions of the cortex seem to be specifically related to motor functions. For example, all somatic sensory (heat, cold, pressure, touch, etc.) inputs lead to a region of the cortical surface just behind the central sulcus, encompassing the forward part of the *parietal lobe.* Somatic sensory inputs from each part of the body lead to a specific part of this region, with the inputs from the legs and feet nearest the top, the torso next, followed by the arms, hands, fingers, face, tongue, pharynx, and, finally, the intraabdominal regions at the bottom. The amount of surface allotted to each part of the body is in proportion to the number of sensory nerves it contains rather than its actual physical size. A pictorial representation of the layout of these areas, called a *homunculus,* appears as a rather grotesque human figure, upside down, with enlarged fingers, face, lips, and tongue.

Just forward of the central sulcus is the *frontal lobe,* in which are found the primary motor neurons that lead to the various muscles of the

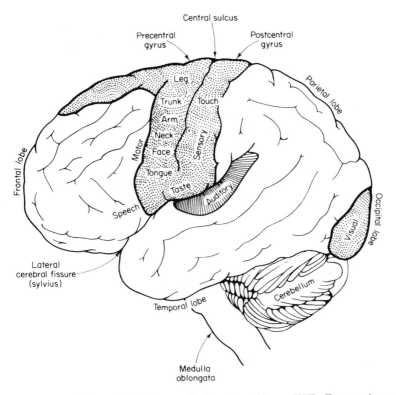

Figure 10.7. The cerebral cortex. (From W.F. Evans, *Anatomy and Physiology, The Basic Principles,* Englewood Cliffs, N.J., Prentice-Hall, Inc., 1971, by permission.)

body. The motor neurons are also distributed on the surface of the cortex in a manner similar to the sensory neurons. The location of the various motor functions can also be represented by a homunculus, also upside down but proportioned according to the degree of muscular control provided for each part of the body.

Figure 10.8 shows both the sensory and motor homunculi, which represent the spatial distribution of the sensory and motor functions on the cortical surface. In each case, the figure shows only one-half of the brain in cross section through the indicated region.

The forward part of the brain, sometimes called the *prefrontal lobe,* contains neurons for some special motor control functions, including the control of eye movements.

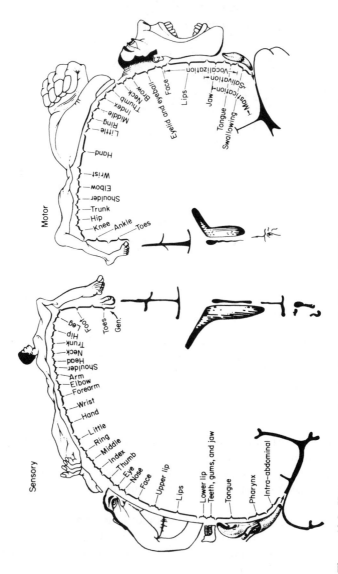

Figure 10.8. Human sensory and motor homunculi. (From W.F. Evans, *Anatomy and Physiology, The Basic Principles*, Englewood Cliffs, N.J., Prentice-Hall, Inc., 1971, by permission.)

The *occipital lobe* is at the very back of the head, over the cerebellum. The occipital lobe contains the visual cortex, in which the patterns obtained from the retina are mapped in a geographic representation.

Auditory sensory input can be traced to the *temporal lobes* of the cortex, located just above the ears. Neurons responding to different frequencies of sound input are spread across the region, with the higher frequencies located toward the front and low frequencies to the rear.

Smell and taste do not have specific locations in the cerebral cortex, although an olfactory bulb near the center of the brain is involved in the perception of smell.

The cerebral cortex has many areas that are neither sensory nor motor. In man, this accounts for the largest portion of the cortex. These areas, called *association areas,* are believed by many scientists to be involved with integrating or associating the various inputs to produce the appropriate output responses and transmit them to the motor neurons for control of the body.

10.4. NEURONAL RECEPTORS

Certain special types of neurons are sensitive to energy in some form other than the usual chemical discharge from axons of other neurons. Those in the retina of the eye, for example, are sensitive to light, while others, such as the pressure sensors at the surface of the body, are sensitive to pressure or touch. Actually, any of these sensors can respond to any type of stimulation if the energy level is sufficiently high, but the response is greatest to the form of energy for which the sensor is intended. In each case, the energy sensed from the environment produces patterns of action potentials that are transmitted to the appropriate region of the cerebral cortex. Coding is accomplished either by having a characteristic of the sensed energy determine which neurons are activated or by altering the pattern in which spikes are produced in a given neuron. In some cases, both methods are employed.

The *somatic sensory system* consists of receptors located on the skin that respond to pain, pressure, light, touch, heat, or coolness—each receptor responding to one of these modalities. The intensity of the sensory input is coded by the frequency of spike discharges on a given neuron, plus the number of neurons involved. There is considerable feedback to provide extremely accurate localization of the source of certain types of inputs, especially fine touch.

In the *visual system,* light sensors, both rods and cones, are located at the retina of the eye. Some processing of the sensed information occurs at the retina, from which it is transmitted via the optic nerve, through the thalamus, to the occipital cortex. The crossover in the visual system is interesting. Instead of all the information sensed by the left eye going to the right brain, as one might expect, information from the left half of each retina (which views the right visual field) is transmitted to the left side of the brain. Thus the demarcation is according to the field of vision and not according to which eye generates the signals. The rods, which are more sensitive to dim light, are not sensitive to color, whereas the less-sensitive cones carry the color information. There is still a certain amount of speculation as to the exact manner in which color information is coded, but it is fairly well agreed that the color information is sensed as some combination of primary colors, each of which is carried by its own set of sensors and neurons.

In the *auditory system,* the frequency of sound seems to be coded in two different ways. After passing through the acoustical system into the inner ear, the sound excites the basilar membrane, a rather stiff membrane coiled in a fluid-filled chamber. The sound vibrations are actually carried to the membrane by the fluid. At lower frequencies, the entire membrane seems to vibrate as a unit, and the sound frequency is coded into a spike discharge frequency by the hair-cell sensory neurons located along the membrane. Above a certain crossover point (about 4000 Hz), however, the situation is different, and the frequencies seem to distribute themselves along the basilar membrane. Thus, for higher frequencies, the frequency is coded according to which sensors are activated, whereas for lower frequencies, the coding is by spike discharge frequency. Auditory information from both ears is transmitted to the temporal lobes on both sides of the cerebral cortex. Timing devices are provided so that if a sound strikes both ears a fraction of a millisecond apart, the ear receiving it first inhibits the response from the other ear. This gives the hearing a sense of direction.

This same directional characteristic also applies to smell and taste. That is, an odor reaching one nostril a fraction of a millisecond before it reaches the other causes inhibition that provides a sense of direction for the odor. Coding for taste and smell is not well understood, although intensity is somehow coded into firing rates of neurons as well as the number of neurons activated.

10.5. THE SOMATIC NERVOUS SYSTEM AND SPINAL REFLEXES

The somatic sensory nervous system carries sensory information from all parts of the body to corresponding sites in the cerebral cortex, whereas motor neurons carry control information to the muscles of the body. The sensory and motor neurons are not necessarily single uninterrupted channels that go all the way from the cortex to the big toe, for example. They may have a number of synapses along the way to permit the action of inhibition as well as excitation. There are, of course, exceptions in which some of the motor control functions are carried out by extremely long axons. In this system countless feedback loops control the action of the muscles. The muscles themselves contain stretch and position receptors that permit precise control over their operation.

Many of the routine muscular movements of the body are not controlled by the brain at all but occur as reflxes of the spinal cord. The spinal cord has many nuclei of neurons that give almost automatic response to input stimuli. Actually, only the more complicated responses are controlled by the brain. In a simplified form, this process could be comparable to a large central computer (the brain) connected to a number of small satellite computers in the spinal cord. Each of the small computers handles the data processing and controls the functions of the system within which it operates. Whenever one of the small computers is faced with a situation beyond its limited capability, the data are sent to the central computer for processing. Thus the spinal reflexes seem to handle all responses except those beyond their capability.

10.6. THE AUTONOMIC NERVOUS SYSTEM

The autonomic nervous system differs from the somatic and motor nervous systems in that its control is essentially involuntary. It was once thought that the autonomic system is completely involuntary, but recent experimentation indicates that it is possible for a person to control portions of this system to some extent.

The major divisions of the autonomic nervous system are the *sympathetic* and *parasympathetic systems*. The sympathetic nervous system receives its primary control from the hypothalamus and is essentially a function of emotional response. It is the sympathetic nervous system that is responsible for the "fight-or-flight" reaction to danger and for such responses as fear and anger. When one or more of the sensory inputs to the brain indicate danger, the body is immediately mobilized for action. The heart rate, respiration, red blood cell production, and blood pressure all increase. Normal functions of the body, such as salivation, digestion, and sexual functions, are all inhibited to conserve energy to meet the situation. Blood flow patterns in the body are altered to favor those functions required for the emergency, and adrenalin, which is the chemical that apparently activates synapses in the sympathetic nervous system, is released throughout the body to maintain the emergency status. Other indications of activation of the sympathetic nervous system are dilation of the pupils of the eyes and perspiration at the palms of the hands, which lowers the skin resistance.

The sympathetic nervous system is designed for "global" action, with short neurons leaving the spine at all levels to innervate the motor systems affected by these nerves. In contrast, the parasympathetic system is responsible for more specific action. Although not completely antagonistic to the sympathetic system, the parasympathetic nervous system causes dilation of the arteries, inhibition or slowing of the heart, contractions and secretions of the stomach, constriction of the pupils of the eyes, and so on. Where the sympathetic system is primarily involved in mobilizing the body to meet emergencies, the parasympathetic system is concerned with the vegetating functions of the body, such as digestion, sexual activity, and waste elimination.

10.7. MEASUREMENTS FROM THE NERVOUS SYSTEM

Direct measurements of the electrical activity of the nervous system are few. However, the effects of the nervous system on other systems of the body are manifested in most physiological measurements. It is possible, in many cases, to stimulate sensor neurons with their specific type of stimulus and measure the responses in various nerves or, in some cases, in individual neurons either in the peripheral or central nervous system. It is also possible to stimulate individual neurons or nerves electrically and to measure either the muscle movement that results from the stimulation or the neuronal spikes that occur in various parts of the system due to the stimulation.

When measuring responses to electrical stimulation, care must be taken to see that the stimulation does not create a wider response than that which would occur if the neuron were stimulated naturally. For example, if electrical stimulation is used, it is very easy to activate other neurons in the vicinity of the intended neuron inadvertently, thus causing responses that are not really related to the desired response.

10.7.1. NEURONAL FIRING MEASUREMENTS. Several methods of measuring the neuronal spikes associated with nerve firings have been developed. They differ basically in the vantage point from which the measurement is taken. A *gross* nerve firing measurement is obtained when a relatively large (greater than 0.1 mm diameter) electrode is placed in the vicinity of a nerve or a large number of neurons. The result is a summation of the action potentials from all the neurons in the vicinity of the electrode. For a more localized measurement, the action potentials of a single neuron can be observed either *extracellularly,* with a microelectrode located just outside the cell membrane, or *intracellularly,* with a microelectrode actually penetrating the cell. Figure 10.9 shows an example of a gross neuronal measurement; Figure 10.10 is an example of an extracellular measurement of a single neuron; and Figure 10.11 shows an intracellular measurement of a single neuron.

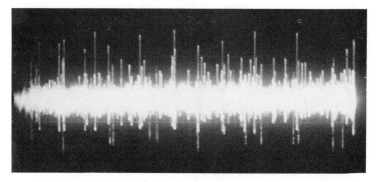

Figure 10.9. Gross measurement of multiple unit neuronal discharge. (Full width covers time span of 500 msec. Maximum peak-to-peak amplitude is approximately 145 microvolts.) (Courtesy of Neuropsychology Research Laboratory, Veterans Administration Hospital, Sepulveda, Calif.)

Because of the difficulty of penetrating an individual cell without damaging it and holding an electrode in that position for any length of time, the use of intracellular measurements is limited to certain specialized cell preparations, usually involving only the largest type of cells. Yet, although action potential spikes can be measured readily with extracellular electrodes, the actual value of resting and action potentials and the measurement of graded potentials require the use of intracellular techniques. Any form of single neuron measurement is much more difficult to obtain than gross measurements. In practice, the microelectrode is inserted into the general area and then moved about slightly until a firing pattern indicative of a single neuron can be observed. Even though this is done, identification of the neuron from which the measurement originates is difficult.

Electrodes and microelectrodes used in the measurement of gross and single neuronal firings are described in detail in Chapter 4. Single neuron measurements require microelectrodes with tips of about 10 microns in diameter for extracellular measurements and as small as 1 micron for intracellular measurements. A fine needle or wire electrode is used for gross neuronal measurements. When the measurement is made between a single electrode and a "distant" indifferent electrode, the measurement is defined as *unipolar.* When the measurement is obtained between two electrodes

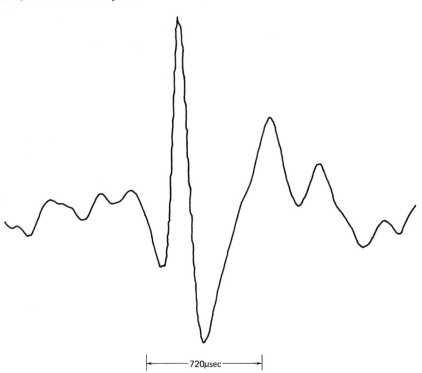

Figure 10.10. Extracellular measurement of unit discharge from red nucelus of a cat. Peak-to-peak height is approximately 180 microvolts.

Figure 10.11. Intracellular measurement of antidromic spike from abduceus nucleus of a cat. (Part of motor control system for the eye.) Spike height is about 61 millivolts. Each horizontal division equals 0.5 millisecond. (Courtesy of Brain Research Institute, UCLA.)

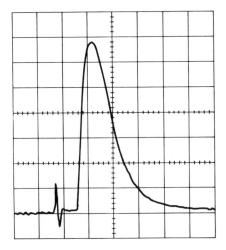

spaced close together along a single axon or a nerve, the measurement is called *bipolar*.

Neuronal firing measurements range from a few hundred microvolts for extracellular single-neuron measurements to around 100 mV for intracellular measurements. For most of these measurements, especially those less than 1 mV, differential amplification is required to reduce the effect of electrical interference. The amplifier must have a very high input impedance to avoid loading the high impedance of the microelectrodes and the electrode interface. Because of the short duration of neuronal spikes, the amplifier must have a frequency response from below one hertz to several thousand hertz.

Ordinary pen recorders are generally unsuitable for recording or display of neuronal firings because of the high upper-frequency requirement. As a rule, an oscilloscope with a camera for photographing the spike patterns or a high-speed light-galvanometer or an electrostatic recorder is used for these measurements.

Another measurement involving neuronal firings is that of nerve conduction time or velocity. Here a given nerve is stimulated while potentials are measured from another nerve or from a muscle actuated by the stimulated nerve. The time difference between the stimulus and the resultant firing is measured on an oscilloscope. Some commercial electromyograph (EMG) instruments, such as those described in Section 10.7.3, have provisions for performing nerve conduction velocity measurement.

10.7.2. ELECTROENCEPHALOGRAM (EEG) MEASUREMENTS. Electroencephalography was introduced in Chapter 3 as the measurement of the electrical activity of the brain. Since clinical EEG measurements are obtained from electrodes placed on the surface of the scalp, these waveforms represent a very gross type of summation of potentials that originate from an extremely large number of neurons in the vicinity of the electrodes.

Originally it was thought that the EEG potentials represent a summation of the action potentials of the neurons in the brain. Later theories, however, indicate that the electrical patterns obtained from the scalp are actually the result of the graded potentials on the dendrites of neurons in the cerebral cortex and other parts of the brain, as they are influenced by the firing of other neurons that impinge on these dendrites. There are still many unanswered questions regarding the neurological source of the observed EEG patterns.

EEG potentials have random-appearing waveforms with peak-to-peak amplitudes ranging from less than 10 μV to over 100 μV. Required bandwidth for adequately handling the EEG signal is from below 1 Hz to over 100 Hz.

Electrodes for measurement of the EEG are described in Chapter 4. For clinical measurements, surface or subdermal needle electrodes are used. The ground reference electrode is often a metal clip on the earlobe. As discussed in Chapter 4, a suitable electrolyte paste or jelly is used in conjunction with the electrodes to enhance coupling of the ionic potentials to the input of the measuring device. To reduce interference and minimize the effect of electrode movement, the resistance of the path through the scalp between electrodes must be kept as low as possible. Generally this resistance ranges from a few thousand ohms to nearly 100 kilohms, depending on the type of electrodes used.

Placement of electrodes on the scalp is commonly dictated by the requirements of the measurement to be made. In clinical practice, a standard pattern, called the *10–20 Electrode Placement System,* is generally used. This system, devised by a committee of the International Federation of Societies for Electroencephalography, is so named because electrode spacing is based on intervals of 10 and 20 percent of the distance between specified points on the scalp. The 10–20 EEG electrode configuration is illustrated in Figure 10.12.

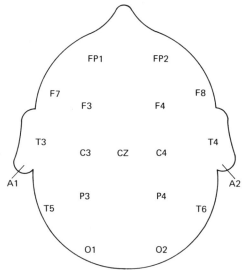

Figure 10.12. 10–20 EEG electrode configuration.

In addition to the electrodes, the measurement of the electroencephalogram requires a readout or recording device and sufficient amplification to drive the readout device from the microvolt-level signals obtained from the electrodes. Most clinical electroencephalographs provide the capability of simultaneously recording EEG signals from several regions of the brain.

For each signal, a complete channel of instrumentation is required. Thus electroencephalographs having as many as 16 channels are available. A clinical instrument with eight channels and a portable unit are shown in Figure 10.13.

(b)

Figure 10.13. Electroencephalographs: (a) Grass model 6. (Courtesy of Grass Instrument Company, Quincy, Mass.); (b) Beckman portable model. (Courtesy of Beckman Instruments, Schiller, Park, Ill.)

Because of the low-level input signals, the electroencephalograph must have high-quality differential amplifiers with good common-mode rejection. The differential preamplifier is generally followed by a power amplifier to drive the pen mechanism for each channel. In nearly all clinical instruments, the amplifiers are ac coupled with low-frequency cutoff below 1 Hz and a bandwidth extending to somewhere between 50 and 100 Hz. Stable dc amplifiers can be used, but possible variations in the dc electrode potentials are often bothersome. Most modern electroencephalographs include adjustable upper- and lower-frequency limits to allow the operator to select a bandwidth suitable for the conditions of the measurement. In addition, some instruments include a fixed 60-Hz rejection filter to reduce powerline interference.

In order to reduce the effect of electrode resistance changes, the input impedance of the EEG amplifier should be as high as possible. For this reason, most modern electroencephalographs have input impedances greater than 10 megohms.

Perhaps the most distinguishing feature of an electroencephalograph is the rather elaborate lead selector panel, which, in most cases, permits any two electrodes to be connected to any channel of the instrument. Either a bank of rotary switches or a panel of push buttons is used. The switch panel also permits one of several calibration signals to be applied to any desired channel for calibration of the entire instrument. The calibration signal is usually an offset of a known number of microvolts, which, because of capacitive coupling, results in a step followed by an exponential return to baseline.

The readout in a clinical electroencephalograph is a multichannel pen recorder with a pen for each channel. The standard chart speed is 30 mm per second, but most electroencephalographs also provide a speed of 60 mm per second for improved detail of higher-frequency signals. Some have a third speed of 15 mm per second to conserve paper during setup time. An oscilloscope readout for the EEG is also possible, but it does not provide a permanent record. In some cases, particularly in research applications, the oscilloscope is used in conjunction with the pen recorder to edit the signal until a particular feature or characteristic of the waveform is observed. In this way, only the portions of interest are recorded. Many electroencephalographs also have provisions for interfacing with an analog tape recorder to permit recording and playback of the EEG signal.

In some research applications, the EEG signals are separated into their conventional frequency bands by means of band-pass filters and the output signals of the individual filters are recorded separately (see Chapter 3). In some cases, they are displayed as feedback to the subject whose EEG is being measured (see Chapter 11). In other situations, the entire EEG signal is digitized for computer analysis and through Fourier analysis is converted into a frequency spectrum. It is also possible, as explained in more detail in Chapter 15, to separate the raw EEG into one-octave bands and, through a process of selective digitizing, provide data that are easily converted into a pseudospectral plot.

A special form of electroencephalography is the recording of evoked potentials from various parts of the nervous system. In this technique the EEG response to some form of sensory stimulus, such as a flash of a light or an audible click, is measured. To distinguish the response to the stimulus from ongoing EEG activity, the EEG signals are time-locked to the stimulus pulses and averaged, so that the evoked response is reinforced with each presentation of the stimulus, while any activity not synchronized to the stimulus is averaged out. Figure 10.14 shows a raw EEG record containing an evoked response from a single presentation of the stimulus and the effect of averaging 8 and 64 presentations respectively.

The averaging of evoked response signals can be accomplished by several methods. In most cases, either a general-purpose digital computer

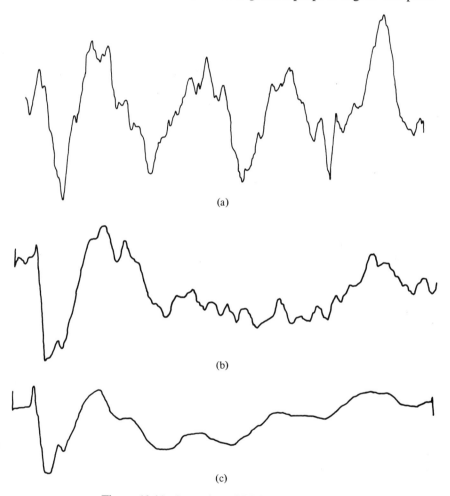

(a)

(b)

(c)

Figure 10.14. Averaging of EEG evoked potentials: (a) raw EEG of single response; (b) average of 8 responses; (c) average of 64 responses. (Courtesy of Dr. Norman S. Namerow, The Center for Health Sciences, Department of Neurology, UCLA, whose research was supported by M.S. Grant #516-C-3.)

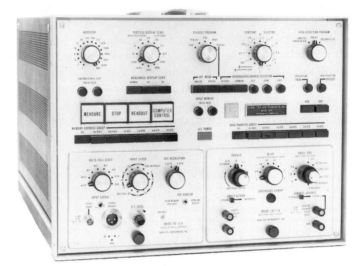

Figure 10.15. Special purpose computer for averaging evoked EEG signals. (Courtesy of Nicolet Instrument Company, formerly Fabri-Tek Instrument, Inc., Madison, Wis.)

or a special-purpose computer of the type shown in Figure 10.15 is used. Analog methods, such as photographing multiple traces from an oscilloscope synchronized to the stimulus, can also be used to enhance the evoked waveform, but the results are not as definitive as with the digital devices.

10.7.3 ELECTROMYOGRAPHIC (EMG) MEASUREMENTS. Like neurons, skeletal muscle fibers generate action potentials when excited by motor neurons via the motor end plates. They do not, however, transmit the action potentials to any other muscle fibers or to any neurons. The action potential of an individual muscle fiber is of about the same magnitude as that of a neuron (see Chapter 3) and is not necessarily related to the strength of contraction of the fiber. The measurement of these action potentials, either directly from the muscle or from the surface of the body, constitutes the electromyogram, as discussed in Chapter 3.

Although action potentials from individual muscle fibers can be recorded under special conditions, it is the electrical activity of the entire muscle that is of primary interest. In this case, the signal is a summation of all the action potentials within the range of the electrodes, each weighted by its distance from the electrodes. Since the overall strength of muscular contraction depends on the number of fibers energized and the time of contraction, there is a correlation between the overall amount of EMG activity for the whole muscle and the strength of muscular contraction. In fact, under certain conditions of isometric contraction, the voltage-time integral of the EMG signal has a linear relationship with the isometric voluntary tension in a muscle. There are also characteristic EMG patterns associated with special conditions, such as fatigue and tremor.

The EMG potentials from a muscle or group of muscles produce a noiselike waveform that varies in amplitude with the amount of muscular activity. Peak amplitudes vary from 50 μV to about 1 mV, depending on the location of the measuring electrodes with respect to the muscle and the activity of the muscle. A frequency response from about 10 Hz to well over 3000 Hz is required for faithful reproduction.

Surface, needle, and fine-wire electrodes are all used for different types of EMG measurement. Surface electrodes are generally used where gross indications are suitable, but where localized measurement of specific muscles is required, needle or wire electrodes that penetrate the skin and contact the muscle to be measured are needed. As in the measurement of neuronal firing measurements, both unipolar and bipolar measurements of EMG are used.

The amplifier for EMG measurements, like that for ECG and EEG, must have high gain, high input impedance and a differential input with good common-mode rejection. However, the EMG amplifier must accommodate the higher frequency band. In many commercial electromyographs, the upper-frequency response can be varied by use of switchable low-pass filters.

Unlike ECG or EEG equipment, the typical electromyograph has an oscilloscope readout instead of a graphic pen recorder. The reason is the higher frequency response required. Sometimes a storage cathode-ray tube is provided for retention of data, or an oscilloscope camera is used to obtain a permanent visual record of data from the oscilloscope screen. A typical commercial electromyograph is shown in Figure 10.16.

Most electromyographs include an audio amplifier and loudspeaker in addition to the oscilloscope display to permit the operator to hear the "crackling" sounds of the EMG. This audio presentation is especially helpful in the placement of needle or wire electrodes into a muscle. A trained operator is able to tell from the sound not only that his electrodes are making good contact with a muscle but also which of several adjacent muscles he has contacted.

Another feature often found in modern electromyographs is a built-in stimulator for nerve conduction time or nerve velocity measurements. By stimulating a given nerve location and measuring the EMG downstream, a latency can be determined from the time difference displayed on the oscilloscope.

Figure 10.16. Electromyograph. (Courtesy of Hewlett-Packard Company, Waltham, Mass.)

The EMG signal can be quantified in several ways. The simplest method is measurement of the amplitude alone. In this case, the maximum amplitude achieved for a given type of muscle activity is recorded. Unfortunately, the amplitude is only a rough indication of the amount of muscle activity and is dependent on the location of the measuring electrodes with respect to the muscle.

Another method of quantifying EMG is a count of the number of spikes, or, in some cases, zero crossings, that occur over a given time interval. A modification of this method is a count of the number of times a given amplitude threshold is exceeded. Although these counts vary with the amount of muscle activity, they do not provide an accurate means of quantification, for the measured waveform is a summation of a large number of action potentials that cannot be distinguished individually.

The most meaningful method of quantifying the EMG utilizes the time integral of the EMG waveform. With this technique, the integrated value of the EMG over a given time interval, such as 0.1 second, is measured and recorded or plotted. As indicated above, this time integral has a linear relationship to the tension of a muscle under certain conditions of isometric contraction, as well as a relationship to the activity of a muscle under isotonic contraction. As with the amplitude measurement, the integrated EMG is greatly affected by electrode placement, but with a given electrode location, these values provide a good indication of muscle activity.

In another technique that is sometimes used in research, the EMG signal is rectified and filtered to produce a voltage that follows the envelope or contour of the EMG. This envelope, which is related to the activity of the muscle, has a much lower frequency content and can be recorded on a pen recorder, frequently in conjunction with some measurement of the movement of a limb or the force of the muscle activity.

· 11 ·

INSTRUMENTATION
FOR SENSORY
MEASUREMENTS
AND THE
STUDY OF BEHAVIOR

The most obvious difference between inanimate and animate objects is that the latter move, respond to their environment, and show changes in their body functions. These properties of animate objects, in a general sense, are called *behavior*. In animals and men the behavior is controlled by the nervous system. The specialized field of medicine in which the nervous system is studied and its diseases are treated is called *neurology*. The *behavior* of organisms, on the other hand, is studied within the various fields of *psychology*. The *experimental psychologist* studies the behavior of animals and men by observing them in experimental situations. The way in which physical stimuli are perceived by men is studied in a specialty called *psychophysics*. The interaction between environmental stimuli and physiological functions of the body is studied in the field of *psychophysiology*. *Clinical psychologists,* as well as *psychiatrists* (who have medical training), deal with the study and treatment of abnormal (pathological) behavior. Behavior is considered abnormal if it interferes substantially with the well-being of the individual and with his interaction with society.

For the treatment of disorders involving the various senses, especially those related to communication, a number of specialized fields have evolved. The *audiologist* determines deficiencies in the acuity of hearing, which often can be improved by the prescription of hearing aids. The *speech pathologist* treats disorders of speech, which may be due to damage to the structures involved in the formation of sounds or may have a neurological cause. The *ophthalmologist* is a physician who specializes in disorders of the eye, whereas the *optometrist* has no medical training and treats only those visual disorders that can be corrected by the prescription of eyeglasses. For the measurement of the acuity of the senses, as well as for the study of behavior, large numbers of instruments have been developed, which can be highly specialized.

The results of behavioral studies seldom show a simple cause-effect relationship but are usually in the form of statistical evidence. This peculiarity requires large numbers of experiments in order to obtain results that are statistically significant. As a result, especially in animal experiments, automated systems are frequently used to control the experiment automatically and record the results. The diversity of the field, on the other hand, has resulted in commercially available instruments that are often in the form of modules and building blocks which can be assembled by the experimenter into specialized systems to suit the requirements of a particular experiment.

One obvious way to study behavior is to measure the electrical signals in the brain and the nervous system that control the behavior, as discussed in Chapter 10. However, because the voltages recorded on an electroencephalograph are the result of many processes that occur simultaneously in the brain, only events that involve larger areas of the brain, such as epileptic seizures, can be readily identified on the EEG recording. For this reason, mental disorders generally cannot be diagnosed from the electroencephalogram, although the EEG is usually used to rule out certain organic disorders of the brain (e.g., tumors), which can show symptoms similar to those of nonorganic types of mental illness. The instrumentation used to measure the EEG is described in Chapter 10.

11.1. PSYCHOPHYSIOLOGICAL MEASUREMENTS

As stated in Chapter 10, many body functions, including blood pressure, heart rate, perspiration, and salivation, are controlled by the autonomic nervous system. This part of the nervous system normally cannot be controlled voluntarily but is influenced by external stimuli and emotional states of the individual. By observing and recording these body functions, insight into emotional changes that cannot be measured directly can be obtained. A practical application of this principle is the *polygraph* (colloquially called the "lie detector"), a device for simultaneously recording several body functions that are likely to show changes when questions asked by the interrogator cause anxiety in the tested person.

For the measurement of blood pressure, heart rate, and respiration rate in psychophysiological studies, the same instruments are used as are utilized for medical applications (see Sections 6.1 and 6.2, and Chapter 8 respectively). For measuring variations in perspiration, a special technique has been developed. In response to an external stimulus, such as a sharp point, the resistance of the skin shows a characteristic decrease, called the *galvanic skin response* (GSR). The baseline value of the skin resistance, in this context, is called the *basal skin resistance* (BSR). The GSR is believed to be caused by the activity of the sweat glands. It does not depend on the overt appearance of perspiration, however, and the actual mechanism of the response is not completely understood. The GSR is measured most readily at the palms of the hands, where the body has the highest concentration of sweat glands. An active electrode, positioned at the center of the palm, can be used together with a neutral electrode, either at the wrist or at the back of the hand. In some devices clips are simply attached to two fingers. Frequently, in order to increase the stability of the measurement, nonpolarizing electrodes, such as silver-silver chloride surface electrodes (see Chapter 4), are used with an electrode jelly that has about the same salinity as the perspiration. In order to minimize the polarization at the electrodes, the current density is kept below 10 μA per cm^2.

Figure 11.1 shows a block diagram of a device that allows the simultaneous measurement, or recording, of both the BSR and the GSR. Here a current generator sends a constant dc current through the electrodes. The voltage drop across the basal skin resistance, typically on the order of several kilohms to several hundred kilohms, is measured with an amplifier and a meter that can be calibrated directly in BSR values. A second meter, coupled through an *RC* network with a time constant of about 3 to 5 seconds, measures the GSR as a change of the skin resistance of from several hundred ohms to several kilohms. The output of this amplifier can be recorded on a suitable graphic recorder. A measurement of the absolute magnitude of the GSR is not very meaningful. The change of the magnitude of the GSR, depending on the experimental conditions and its *latency* (the time delay between stimulus and response), can be used to study emotional changes.

Instead of the change of the skin resistance, the change of the so-called *skin potential* has been used occasionally. This is actually a potential difference of between 50 and 70 mV that can be measured between nonpolarizing electrodes on the palm and the forearm and that also shows a response to emotional changes.

Although the activity of the autonomic nervous system cannot be con-

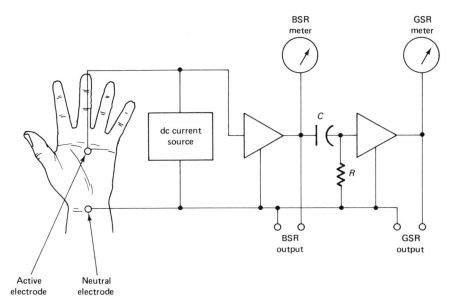

Figure 11.1. Block diagram of a device to measure and record basal skin resistance (BSR) and the galvanic skin response (GSR).

trolled directly, it can be influenced in an indirect way by two mechanisms known as *conditioning* and *feedback*.

Certain physiological responses are normally elicited by certain external stimuli. The view of food, for instance, stimulates the production of saliva and causes "one's mouth to water." As discovered by Pavlov in his famous experiments with dogs, a previously neutral stimulus can be made to elicit the same response as the view of food if it is presented several times just before the natural stimulus. This process of making the autonomic nervous system respond to previously neutral stimuli is called *Pavlovian* (or *classical*) *conditioning*.

Experiments of this type require the continuous recording of one or more of the autonomic responses. Pavlov, for example, measured the flow rate of saliva. Sometimes the autonomic responses can be influenced by simply informing the subject when a change in the response occurs. This, again, requires that the response be measured and that certain characteristics of it be signaled to the subject in a suitable way. This principle is called *biological feedback*. Although this technique had been known for some time, it received renewed interest during the early 1970s for possible therapeutic uses in controlling variables like heart rate, blood pressure and the occurrence of certain patterns in the electroencephalogram.

11.2. INSTRUMENTS FOR TESTING MOTOR RESPONSES

Motor responses, or responses of the skeletal muscles, are under voluntary control but often require a learning process for the proper interaction between several muscles in order to perform the response correctly. Numerous devices have been described in the literature, or are available commercially, to measure motor responses and to study the influence of factors like fatigue, stress or the effects of drugs. Some of these devices are very simple. Several so-called manual dexterity tests, for instance, consist of a number of small parts that the subject has to assemble in a certain way, while the time required for completion of the task is measured. In related instruments called *steadiness testers* (Figure 11.2) a metal stylus must be moved through channels of various shapes without touching the metal walls. An error closes the contact between wall and stylus and advances an electromechanical counter. The *pursuit rotor,* shown in Figure 11.3, uses a similar principle. A light spot moves with adjustable speed along a circular, or star-shaped, pattern on the top surface of the tester. The subject has the task of pursuing the spot with a hook-shaped probe that contains a photoelectric sensor. An indicator and timer automatically measure the percentage of time during which the subject is "on target" during a certain test interval.

The performance of certain muscles or muscle groups can be measured with various *dynamometers,* which measure the force that is exerted either mechanically or with an electric transducer.

11.3. INSTRUMENTATION FOR SENSORY MEASUREMENTS

The human senses provide the information inputs required by man to orient himself in his environment and to protect himself from danger. Many methods and instruments have been developed to measure the performance of the sense organs, study their functioning, and detect impairments. Some of the senses do not require very sophisticated equipment. The temperature senses, for instance, can be studied with several metal objects, or water containers, which are maintained at certain temperatures. Some of the original work on touch perception, early in this century, was performed by stimulating the skin with bristles of horsehair that had been calibrated to exert a known pressure. The same method is still in use today except that nylon has replaced the horsehair. More com-

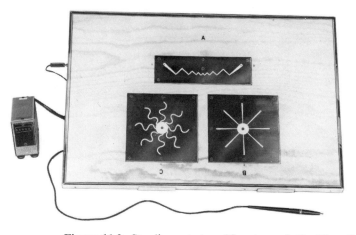

Figure 11.2. Steadiness tester. (Courtesy of Stoelting Company, Chicago, Ill.)

plicated devices are necessary for studying optical perception. An example would be a measurement in which a spot of controllable brightness and size is viewed against a background whose brightness can also be varied. Variations in the size and brightness of the spot and the brightness of the background are all independently controlled. Another special device for studies of visual perception is the *tachistoscope*. Here a display of an illuminated card is presented to the viewer by means of a semitransparent mirror. A second display is then presented for an adjustable short time interval, which may be followed by either a repeat of the original card or by a third display. The change of displays is achieved by switching the illumination or by means of electromechanical shutters. By varying the presentation time for the second display and by using displays of various complexity, the perception and recognition of objects can be studied. The purpose of the presentation of the third display is to mask optical after-images, which might prolong the actual presentation time of the second display.

Acuity of hearing can be measured with the help of an instrument called an *audiometer*. Here the sound intensity in an earphone is gradually increased until the sound is perceived by the subject. The hearing in the other ear during this measurement is often masked by presenting a neutral stimulus (white noise) to this ear. Normally the threshold of hearing is determined at a number of frequencies. This process is automated in the *Békésy audiometer* (named after George von Békésy, its inventor), shown in Figure 11.4. In order to perform a measurement, the subject first presses a control button, thus starting a reversible motor, which drives a volume control potentiometer and increases the amplitude of the stimulus signal

Figure 11.3. Pursuit rotor. (Courtesy of Pentagon Devices Corporation, Syosset, N.Y.)

until it is perceived by the subject. The subject then releases the button, opening the switch, and the motor reverses. By alternately closing and opening the switch, the subject maintains the volume at a level at which the tone can just be heard. A pen, connected to the volume control mechanism, draws a line on a moving paper. At the same time the paper drive mechanism, which is linked to the instrument's frequency control, slowly changes the frequency of the tone. Within about 15 minutes a recording, called an *audiogram,* is obtained. The audiogram often is calibrated, not in absolute values of the perception threshold but in relative values referred to the acuity of normal subjects (which is stored in the instrument in a mechanical cam, not shown in Figure 11.4). The resultant curve corresponds directly to the hearing loss as a function of frequency. Figure 11.5 shows a somewhat simplified version of the original Békésy audiometer, which changes the frequency of the stimulus in steps instead of continuously.

Hearing acuity in infants or uncooperative subjects can be tested with the help of a conditioning method. A light electrical shock can cause a change in the galvanic skin resistance. An audible tone, when paired with the shock, can be made to elicit the same response. Once this conditioning has been completed, the skin reflex can be used to determine whether the subject can hear the same tone presented at a lower volume. This technique, however, is not always completely reliable. A better method is to measure the evoked EEG response when a tone with a certain intensity is

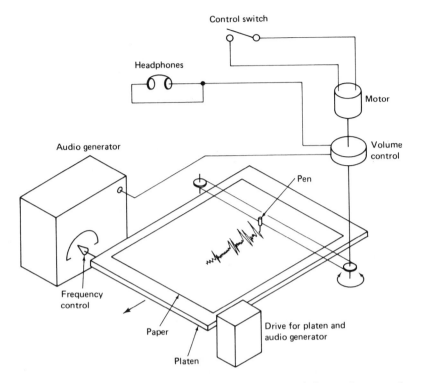

Figure 11.4. Békésy audiometer diagram.

presented. This requires the repeated presentation of the tone and an averaging technique to extract the evoked response from the ongoing activity (see Section 10.7.2).

11.4. INSTRUMENTATION FOR THE EXPERIMENTAL ANALYSIS OF BEHAVIOR

In order to describe and analyze behavior accurately, data must be recorded in terms other than the subjective report of an observer. Especially for a mathematical analysis, numerical values must be assigned to some aspects of behavior. For behavior involving motor responses and motor skills, special testing devices have been developed to obtain a numerical rating—for example, the pursuit rotor just described. Other tests require the completion of some manual or mental task in which the time required for completion is measured. Sometimes the number of errors is also used to compare the performance of individuals.

Many basic behavioral experiments are performed with animals (rats,

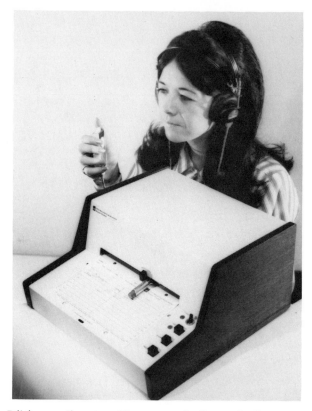

Figure 11.5. Békésy audiometer. (Courtesy of Grason-Stadler. Subsidiary of General Radio Company, Concord, Mass.)

pigeons, monkeys) as subjects. These experiments are made in a neutral environment provided by a soundproof enclosure often called a "Skinner box" (after B. F. Skinner, who pioneered the method), in which the animal is isolated from uncontrolled environmental stimuli. Each experiment must be designed in such a way that the behavior is well defined and can be measured automatically. For example, such events as pressing a bar or pecking on a key, or the presence of an animal in one part of the cage or jumping over a barrier could be measured. In specially instrumented cages, the activity of animals can be quantified.

Behavior emitted by organisms to interact with and modify their environment is called *instrumental* or *operant behavior*. Such behavior, which is controlled by the central nervous system rather than by the autonomic nervous system, can also be conditioned but in a way that differs from classical conditioning. Operant behavior that is positively reinforced (rewarded) tends to occur more frequently in the future; behavior that is negatively reinforced decreases in frequency. In animal experiments, positive reinforcement is usually administered in the form of food or water given to animals that had been deprived of these commodities. This reinforcement can be administered easily by automatic dispensing devices. Negative reinforcement is in the form of harmless, but painful, electric shocks administered through isolated grid bars that serve as the floor of the cage. With suitable reinforcement, the animal can be conditioned to "emit certain behavior," like the pressing of a bar, in response to a certain stimulus. From changes in the behavior that can occur under the influences of drugs, or when the stimulus is modified, valuable insight into the mechanisms of behavior can be obtained.

Figure 11.6 shows a setup as it might be used for the simpler types of

Figure 11.6. Skinner box. (Courtesy of BRS-Foringer, Beltsville, Md.)

such experiments, using rats as subjects. The Skinner box is equipped with a response bar and a stimulus light. Positive reinforcement is administered by an automatic dispenser for food pellets. An electric–shock generator is connected to the grid floor of the cage through a scrambler switch that makes it impossible for the animal to escape the shock by clinging to bars that are of the same electrical potential. An automatic programmer turns on the stimulus at certain time intervals and controls the reinforcements according to the animal's response, following a prescribed schedule (called the *contingency*). In many experiments, these schedules can be very complex. Elaborate modular control systems, either with relays or based on solid-state logic, are therefore available for programming stimulus contingencies and measuring response parameters. Simple behavior is often recorded on a *cumulative-event recorder*. In this device a paper strip is moved with a constant speed (4 in./hour). Each time the bar is pressed by the animal, a solenoid or stepping motor is energized and moves a pen a small distance over the paper perpendicular to the direction of paper movement. The pen is reset, either when it has traveled the full width of the paper or by a timing motor after a certain time interval (for example, every 10 minutes). The position of the pen at any time represents the total number of events (bar presses) that have occurred since the last resetting of the pen. Reinforcement is indicated by a diagonal movement of the pen. This recording method, despite its simplicity, is very informative. The slope of the curve corresponds to the response rate. When reset after fixed time intervals, the pen excursion directly represents a form of time histogram.

Insight into behavior mechanisms obtained in animal experiments has been extrapolated to human behavior. Part of human behavior can be explained as having been conditioned by reinforcements administered by society and the environment. In a form of treatment called *behavior*

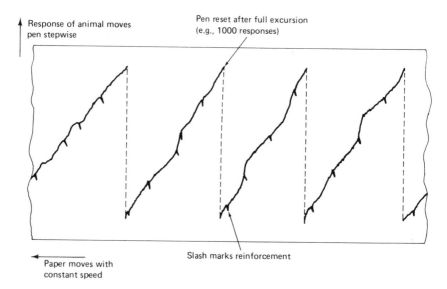

Figure 11.7. Graph from a cumulative event recorder.

therapy, behavioral and emotional problems are treated according to the principles of operant conditioning, sometimes using special equipment.

Perhaps the best-known example of a behavior-therapy method using electronic equipment is the treatment of bed wetting with the so-called Mowrer sheet (named after the psychologists who first used it). This method uses a moisture sensor placed under the bed sheet, which activates an acoustical alarm and turns on the light when the presence of moisture is first detected.

· **12** ·

BIOTELEMETRY

There are many instances in which it is necessary to monitor physiological events from a distance. Typical applications include

1. Radio-frequency transmissions for monitoring astronauts in space.
2. Patient monitoring where freedom of movement is desired, such as in obtaining an exercise electrocardiogram. In this instance, the requirement of trailing wires is both cumbersome and dangerous.
3. Patient monitoring in an ambulance and in other locations away from the hospital.
4. Collection of medical data from a home or office.
5. Research on unrestrained, unanesthetized animals in their natural habitat.
6. Use of telephone links for transmission of electrocardiograms or other medical data.
7. Special internal techniques, such as tracing acidity or pressure through the gastrointestinal tract.
8. Isolation of an electrically susceptible patient (see Chapter 16) from

power-line-operated ECG equipment to protect him from accidental shock.

These applications have indicated the need for systems that can adapt existing methods of measuring physiological variables to a method of transmission of resulting data. This is the branch of biomedical instrumentation known as biomedical telemetry or biotelemetry.

12.1. INTRODUCTION TO BIOTELEMETRY

Literally, *biotelemetry* is the measurement of biological parameters over a distance. The means of transmitting the data from the point of generation to the point of reception can take many forms. Perhaps the simplest application of the principle of biotelemetry is the stethoscope, whereby heartbeats are amplified acoustically and transmitted through a hollow tube system to be picked up by the ear of the physician for interpretation (see Chapter 6).

Historically, Einthoven, the originator of the electrocardiogram as a means of analysis of the electrical activity of the heart, transmitted electrocardiograms from a hospital to his laboratory many miles away as early as 1903. The rather crude immersion electrodes (see Figure 4.4, Chapter 4) were connected to a remote galvanometer directly by telephone lines. The telephone lines in this instance were merely used as conductors for the current produced by the biopotentials.

The use of wires in the transmission of the biodata by Einthoven suited his purpose; however, a major advantage of modern telemetry is the elimination of the use of wires. Certain applications of biotelemetry utilize telephone systems, but essentially these are situations in which "hard-wire" connections are extended by the telephone lines. However, this chapter is concerned primarily with the use of telemetry by which the biological data are put in suitable form to be radiated by an electromagnetic field (radio transmission). This involves some type of modulation of a radio-frequency carrier and is often referred to as *radio telemetry*.

The purpose of this chapter is merely to outline the elements of the subject and to present an example of its application. For a comprehensive treatment, the reader is referred to the book by R. Stuart Mackay listed in the bibliography.

12.2. PHYSIOLOGICAL PARAMETERS ADAPTABLE TO BIOTELEMETRY

Although there had been examples of biotelemetry in the 1940s, they did not receive much attention until the advent of the NASA space programs. For example, in the 1963 report of the Mercury program, the following types of data were obtained by telemetry:

1. Temperature by rectal or oral thermistor
2. Respiration by impedance pneumograph
3. Electrocardiograms by surface electrodes
4. Indirect blood pressure by contact microphone and cuff

As the field progressed, it became apparent that literally any quantity that could be measured was adaptable to biotelemetry. Just as with hard-wire systems, measurements can be applied to two categories:

1. Bioelectrical variables, such as ECG, EMG, and EEG
2. Physiological variables that require transducers, such as blood pressure, gastrointestinal pressures, blood flow, and temperatures

With the first category, a signal is obtained directly in electrical form, whereas the second category requires a type of excitation, for the physiological parameters are eventually measured as variations of resistance, inductance, or capacitance. The differential signals obtained from these variations can be calibrated to represent pressure, flow, temperature, and so on, since some physical relationships exist.

In a typical system, the appropriate analog signal (voltage, current, etc.) is converted into a form or code capable of being transmitted. After being transmitted, the signal is decoded at the receiving end and converted back into its original form. The necessary amount of amplification must also be included. Sometimes it is desirable to store the data for future use. Before discussing these aspects, however, a discussion of the applications for these systems is necessary.

Currently the most widespread use of biotelemetry for bioelectric potentials is in the transmission of the electrocardiogram. Instrumentation at the transmitting end is simple because only electrodes and amplification are needed to prepare the signal for transmission.

One example of ECG telemetry is the transmission of electrocardiograms from an ambulance or site of an emergency to a hospital, where a cardiologist can immediately interpret the ECG, instruct the trained rescue team in their emergency resuscitation procedures, and arrange for any special treatment that may be necessary upon arrival of the patient at the hospital. In this application, the telemetry to the hospital is supplemented by two-way voice communication.

The use of telemetry for ECG signals is not confined to emergency applications. It is used for exercise cardiograms in the hospitals so that the patient can run up and down steps, unencumbered by wires. Also, there have been cases in which individuals with heart conditions wear ECG telemetry units at home and on the job and relay ECG data periodically to the hospital for checking. Other applications include the monitoring of athletes running a race in an effort to improve their performance. ECG telemetry units are also common in human performance laboratories on some college campuses.

The actual equipment worn by the subject is quite comfortable and usually does not impede movement. In addition to the electrodes that are taped into place, the patient or subject wears a belt around the waist with a pocket for the transmitter. A typical transmitter is about the size of a package of king-size cigarettes. The wire antenna can be either incorporated into the belt or hung loosely. Clothing generally has convenient openings to allow for lead wires from the electrodes to come through to the transmitter. Power for the transmitter is from a battery, usually a mercury cell, with a useful life of about 30 hours.

Cardiovascular research performed with experimental animals necessitates some changes in technique. First, the electrodes used are often of the needle type, especially for long-term studies. Second, the animal is likely to interfere with the equipment. For this reason, miniature transmitters have been designed that can be surgically implanted subcutaneously. However, doing so is not always necessary. Many researchers have designed special jackets or harnesses for animals that have been quite successful. Some of the aspects of this particular problem are discussed later.

Telemetry is also being used for transmission of the electroencephalogram. Most applications have been involved with experimental animals for research purposes. One example is in the space biology program in the Brain Research Institute at the University of California, Los Angeles, where chimpanzees have had the necessary EEG electrodes implanted in the brain. The leads from these electrodes are brought to a small transmitter installed on the animal's head, and the EEG is transmitted. Other groups have developed special helmets with surface electrodes for this application. Similar helmets have been used for the collection of EEGs of football players during a game.

Telemetry of EEG signals has also been used in studies of mentally disturbed children. The child wears a specially designed "football helmet" or "spaceman's helmet" with built-in electrodes so that the EEG can be monitored without traumatic difficulties during play. In one clinic the children are left to play with other children in a normal nursery school environment. They are monitored continuously while data are recorded.

One advantage of monitoring by telemetry is to circumvent a problem that often hampers medical diagnosis. Patients frequently experience pains, aches, or other symptoms that give trouble for days, only to have them disappear just before and during a medical examination. Many insidious symptoms behave in this way. With telemetry and long-term monitoring, the cause of these symptoms may be detected when they occur or, if recorded on magnetic tape, can be analyzed later.

One problem often encountered in long-term monitoring by telemetry is that of handling the large amount of data generated. If the time to detect symptoms is very long, it becomes quite a task to record all the information. In many applications, data can be recorded on tape for later playback. A number of types of tape recorders can play back information at a higher speed than that at which data are recorded. Thus one hour's worth of data can be played back in one minute. These rapid playback techniques can be used effectively only if the observer is looking for something specific. That is, a certain voltage amplitude or a certain frequency can be sensed by a discriminator circuit and used to activate a signal, either a light or sound. The observer can then stop the machine and record the vital segment of the data on paper. He does not have to record the whole sequence, only that part of most interest.

The third type of bioelectric signal that can be telemetered is the electromyogram. This device is particularly useful for studies of muscle damage and partial paralysis problems. It is also useful in human performance studies.

Telemetry can also be used in transmitting stimulus signals to a patient or subject. For example, it is well known that an electrical impulse can trigger the firing of nerves (see Chapter 10). It has been demonstrated that if an electrode is surgically implanted and connected to dead nerve endings, an electrical impulse can sometimes cause the nerves to function as they once did. If a miniature receiver is implanted subcutaneously, the electrical signal can be generated remotely. This point brings up the possibility of using telemetry techniques therapeutically. One example is the use of telemetry in the treatment of "dropfoot," which is one of the most common disabilities resulting from stroke. This condition is essentially an inability of the patient to lift his foot, which results in a shuffling, toe-dragging gait.

A method for correcting "dropfoot" by transmitting a signal to an implanted electronic stimulator has been used successfully at Rancho Los Amigos Hospital in Los Angeles. An external transmitter worn by the patient delivers a pulse-modulated carrier signal of 450 kHz to an implanted receiver that demodulates the signal and delivers the resulting signal (a pulse train with a pulse duration of 300 μsec and a frequency that can be varied between 20 and 50 pulses per second) to the peroneal nerve. This nerve, when stimulated, causes muscles in the lower forepart of the leg to contract, thus raising the foot. Stimulation is automatically cycled during gait by a heel switch that turns the transmitter on and off so as to approximate the normal phasic activity of these muscles during gait.

By using suitable transducers, telemetry can be employed for the measurement of a wide variety of physiological variables. In some cases, the transducer circuit is designed as a separate "plug-in" module to fit into the transmitter, thus allowing one transmitter design to be used for different types of measurements. Also, many variables can be measured and transmitted simultaneously by multiplexing techniques.

The transducers and associated circuits are essentially the same as those discussed in earlier chapters. Sometimes they must be modified as to shape, size, and electrical characteristics, but the basic principles of transduction are identical with their hard–wire system counterparts. Not all types of transducers lend themselves to telemetry, however, and usually, in a typical application, a study of adaptable types is necessary. This point is shown later in a case study.

One important application of telemetry is in the field of blood pressure and heart rate research in unanesthetized animals. The transducers are surgically implanted with leads brought out through the animal's skin. A male plug is attached postoperatively and later connected to the female socket contained in the transmitter unit.

Blood flow has also been studied extensively by telemetry. Both Doppler type and electromagnetic type transducers can be employed.

The use of thermistors to measure temperature is also easily adaptable to telemetry. In addition to constant monitoring of skin temperature or systemic body temperature, the thermistor system has found use in obstetrics and gynecology. Long term studies of natural birth control by monitoring vaginal temperature have incorporated telemetry units.

A final application, discussed below in more detail, is the use of "radio pills" to monitor stomach pressure or pH. In this application, a pill that contains a sensor plus a miniature transmitter is swallowed and the data are picked up by a receiver and recorded.

It is interesting to note that biotelemetry studies have been performed on dogs, cats, rabbits, monkeys, baboons, chimpanzees, deer, turtles, snakes, alligators, caimans, giraffes, dolphins, llamas, horses, seals, and elks, as well as on humans.

12.3. THE COMPONENTS OF A BIOTELEMETRY SYSTEM

With the many commercial biotelemetry systems available today, it would be impossible to discuss all the ramifications of each. This section is designed to give the reader an insight into the typical simple system. More complicated systems can be built on this base. In putting together a telemetry system, it should be realized that although parts of it are unique for medical purposes, most of the electronic circuits for oscillators, amplifiers, power supplies, and so on are usually adaptions of circuits in regular use in radio communications.

One of the earliest biotelemetry units was the *endoradiosonde,* developed by Mackay and Jacobson and described in various papers by these two investigators since 1957. The pressure-sensing endoradiosonde is a "radio pill" less than 1 cc in volume so that it can be swallowed by the patient. As it travels through the gastrointestinal tract, it measures the various pressures it encounters. Similar devices have also been built to sense temperature, pH, enzyme activity, and oxygen tension values by the use of different sensors or transducers. Pressure is sensed by a variable inductance, whereas temperature is sensed by a temperature-sensitive transducer.

One version of the circuit is shown in Figure 12.1. Basically, it is a

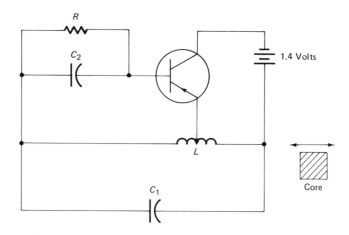

Figure 12.1. Circuit of pressure-sensitive endoradiosonde. (From R. S. Mackay, *Biomedical Telemetry.* New York, John Wiley & Sons, Inc., 1968, by permission.)

transistorized Hartley oscillator having a constant amplitude of oscillation and a variable frequency to communicate information. The ferrite core of the coil is attached to a diaphragm, which causes it to move in and out as a function of pressure and, therefore, varies the value of inductance in the coil. This change in inductance produces a corresponding change in the frequency of oscillations. Inward motion of the ferrite core produces a decrease in frequency. Thus changes in pressure modulate the frequency. An emitter resistor was used in earlier models, and the radio-frequency voltage across it was transmitted by a combined shield and antenna. In later models the oscillator resonator coil also acts as an antenna. The transmitted frequencies, ranging from about 100 kHz to about 100 MHz can be picked up on any simple receiver.

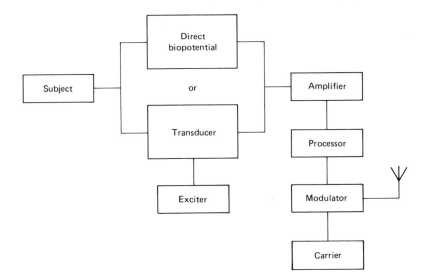

Figure 12.2. Block diagram of a biotelemetry transmitter.

The system described is fairly simple, and most applications do involve more circuitry. The stages of a typical biotelemetry system can be broken down into functional blocks, as shown in Figure 12.2 for the transmitter

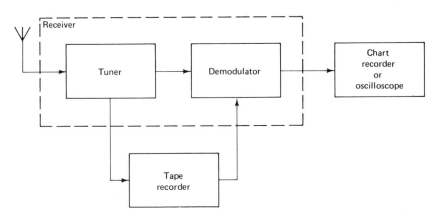

Figure 12.3. Receiver-storage-display units.

and in Figure 12.3 for the receiver. Physiological signals are obtained from the subject by means of appropriate transducers. The signal is then passed through a stage of amplification and processing circuits that include generation of a subcarrier and a modulation stage for transmission.

The receiver (Figure 12.3) consists of a tuner to select the transmitting frequency, a demodulator to separate the signal from the carrier wave, and a means of displaying or recording the signal. The signal can also be stored in the modulated state by the use of a tape recorder, as shown in the block diagram. Some comments on these various stages are provided later.

Since most biotelemetry systems involve the use of radio transmission, a brief discussion of some basic concepts of radio should be helpful to the reader with limited background in this field. A *radio-frequency (RF) carrier* is a high-frequency sinusoidal signal which, when applied to an appropriate transmitting antenna, is propagated in the form of electromagnetic waves. The distance the transmitted signal can be received is called the *range* of the system. Information to be transmitted is impressed upon the carrier by a process known as *modulation*. Various methods of modulation are described below. The circuitry which generates the carrier and modulates it constitutes the *transmitter*. Equipment capable of receiving the transmitted signal and *demodulating* it to recover the information comprise the *receiver*. By tuning the receiver to the frequency of the desired RF carrier, that signal can be selected while others are rejected. The range of the system depends upon a number of factors, including the power and frequency of the transmitter, relative locations of the transmitting and receiving antennas, and the sensitivity of the receiver.

The simplest form of modulation is one in which the transmitter is simply turned on and off to correspond to some code. Such a system does not lend itself to the transmission of physiological data, but is useful for remote control applications. This is called *continuous wave (CW)* transmission.

The two basic systems of modulation are *amplitude modulation (AM)* and *frequency modulation (FM)*. These two methods are illustrated in Figure 12.4.

In an amplitude modulated system, the amplitude of the carrier is caused to vary with the information being transmitted. Standard radio broadcast (AM) stations utilize this method of modulation as does the video (picture) signal for television. Amplitude modulated systems are susceptible to natural and man-made electrical interference, since the interference generally appears as variations in the amplitude of the received signal.

In a frequency modulation (FM) system, the frequency of the carrier is caused to vary with the modulated signal. An FM system is much less susceptible to interference, because variations in the amplitude of the received signal caused by interference can be removed at the receiver before demodulation takes place. Because of this reduced interference, FM transmission is often used for telemetry. FM broadcast stations and television sound also utilize this method of modulation.

In biotelemetry systems, the physiological signal is sometimes used to modulate a low frequency carrier, called a *subcarrier,* often in the audio

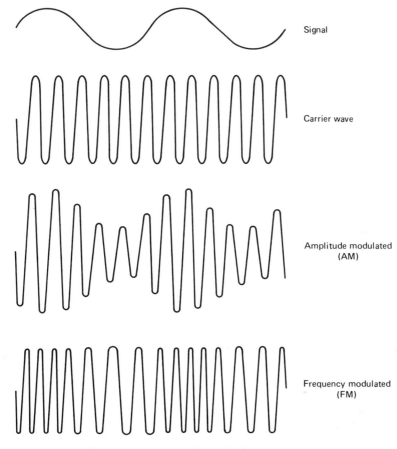

Signal

Carrier wave

Amplitude modulated
(AM)

Frequency modulated
(FM)

Figure 12.4. Types of modulation.

frequency range. The RF carrier of the transmitter is then modulated by
the subcarrier. If several physiological signals are to be transmitted simultaneously, each signal is placed on a subcarrier of a different frequency and
all of the subcarriers are combined to simultaneously modulate the RF
carrier. This process of transmitting many *channels* of data on a single
RF carrier is called *frequency multiplexing,* and is much more efficient and
less expensive than employing a separate transmitter for each channel. At
the receiver, a multiplexed RF carrier is first demodulated to recover each
of the separate subcarriers, which must then be demodulated to retrieve
the original physiological signals. Either frequency or amplitude modulation can be used for impressing data on the subcarriers, and this may or
may not be the same modulation method that is used to place the subcarriers on the RF carrier. In describing this type of system, a designation

is given in which the method of modulating the subcarriers is followed by the method of modulating the RF carrier. For example, a system in which the subcarriers are frequency modulated and the RF carrier is amplitude modulated is designated as FM/AM. An FM/FM designation means that both the subcarriers and the RF carrier are frequency modulated. Both FM/AM and FM/FM systems are used in biotelemetry.

In addition to the modulation schemes described above, some biotelemetry systems make use of a method called *pulse modulation,* in which the transmitter carrier is generated in a series of short bursts or RF *pulses.* If the amplitude of the pulses is used to represent the transmitted information, the method is called *pulse amplitude modulation (PAM),* whereas if the width (duration) of each pulse is varied according to the information, a *pulse width modulation (PWM)* system results. In a related method called *pulse position modulation (PPM),* the timing of a very narrow pulse is varied with respect to a reference pulse. All three of these pulse modulation methods have certain advantages and are used under certain circumstances.

As in amplitude and frequency modulation systems, multiplexing of several channels of physiological data can be accomplished in a pulse modulation system. However, instead of frequency multiplexing, *time multiplexing* is used. In a time multiplexing scheme, each of the physiological signals is sampled briefly and used to control either the amplitude, width or position of one pulse, depending on the type of pulse modulation used. The pulses representing the various channels of data are transmitted sequentially. Thus, in a six-channel system, every sixth data pulse represents a given channel. In order to identify the data pulses an identifiable reference pulse is included in each set. If the sampling rate is several times the highest frequency component of each data signal, no loss of information results from the sampling process.

A system for the monitoring of blood pressure is used to illustrate the FM/FM method of transmission. The transducer used in this case is the flush-diaphragm type of strain gage transducer. Electrically, it can be represented by the bridge circuit of Figure 12.5. Resistors R_1 and R_3 decrease, whereas R_2 and R_4 increase in value as blood pressure increases. Resistor R_b is simply for balancing or zeroing. The transducer is connected in the transmitter circuit as shown in Figure 12.6.

Either direct current or alternating current can be used as excitation for strain gage bridges. When dc is used, the amplifier following the bridge must be a dc amplifier, with its associated problems of stability and drift. When ac is used, the bridge acts as a modulator. A demodulator and filter are required in order to recover the signal.

The exciter unit, which in this example consists of a Colpitts transistor oscillator plus an *RC* coupled common-emitter amplifier stage, excites the

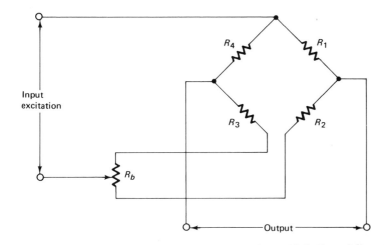

Figure 12.5. Transducer circuit.

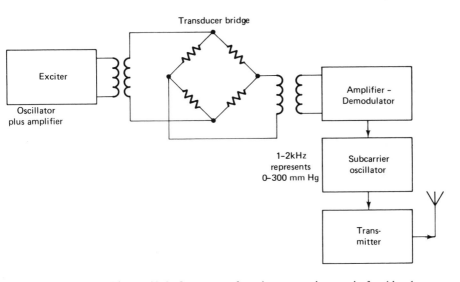

Figure 12.6. One type of exciter-transmitter unit for blood pressure telemetry.

bridge with a constant ac voltage at a frequency of approximately 5 kHz. The exciter unit is coupled to the bridge inductively. The bridge is initially balanced both resistively and capacitively so that any changes in the resistance of the arms of the bridge due to changes in pressure on the transducer will result in changes of the output voltage.

This output voltage is inductively coupled to another common-emitter amplifier stage and *RC* coupled to a further stage of amplification. However, whereas the previous stages are class *A* amplifiers and do not change the waveshape of the input voltage, the latter stage is a class *C* amplifier, which means that the transistor is biased beyond cutoff and the resulting output wave is rectified to obtain a signal representative of the pressure variation.

This rectified wave is put through a resistance-capacitance filter, and the resulting voltage controls the frequency of a unijunction (double base) transistor oscillator. This is the FM subcarrier oscillator that is used to modulate the main carrier.

The system can be arranged so that there is a fairly linear relationship between the subcarrier oscillator frequency and the physiological parameter to be measured. For example, in the system for blood pressure illustrated in Figure 12.6, a frequency range of 1 to 2 kHz represents the range of 0 to 300 mm Hg pressure. The transducer action can be traced very easily. The subcarrier is used to frequency-modulate the main transmitter carrier. This carrier is transmitted at low power on an FM broadcast band specially designated for biotelemetry.

It should be noted that the same exciter-transmitter circuit could be used with small modifications if the blood pressure transducer were replaced by another type or by a thermistor or any other electrical resistance device. Also, the exciter-bridge combination could be replaced by a direct bio-potential signal input, such as an electrocardiogram signal.

It should be noted that, with the transmission of radio frequency energy, legal problems might be encountered. Many systems use very low power and the signals can be picked up only a few feet away. Such systems are not likely to present problems. However, systems which transmit over longer distances are subject to licensing procedures and the use of certain allocated frequencies or frequency bands. Regulations vary from country to country, and in some European countries they are more strict than in the United States.

The regulations that are of concern to persons operating in the United States are contained in Part 15 of the Federal Communications Commission (FCC) regulations for low-power transmission. In case of doubt, this material should be referred to in order to ensure compliance.

Returning to the system under discussion, the signal transmitted at low power on the FM transmitter is picked up by the receiver, which must be tuned to the correct frequency. The audio subcarrier is removed from the RF carrier and then demodulated to reproduce a signal that can be transformed back to the amplitude and frequency of the original data waveform. This signal can then be displayed or recorded on a chart. If it is desirable to store the data on tape for later use, the original data waveform or the modulated subcarrier signal is put on the tape. In the latter case, when playback is desired, the subcarrier signal is passed through the FM subcarrier demodulator.

It should be mentioned that there are systems that convert an analog signal, such as ECG, into digital form prior to modulation. The digital form is useful when used in conjunction with computers, a topic covered in Chapter 15.

12.4. DESIGN OF A SYSTEM: A CASE STUDY

12.4.1. INTRODUCTION. In order to illustrate some of the principles of biometrics and the design of a biotelemetry system, a detailed description of an actual application is presented. This particular work was accomplished in the cardiology research unit of the Cedars-Sinai Research Center at Cedars of Lebanon Hospital in Los Angeles.

The purpose of the system was to monitor continuously the blood pressure and heart rate of a free-roaming dog.

The system requirements were specified for the highest biometrical standards possible, consistent with reasonable cost. These requirements include (a) good sensitivity, (b) linearity, (c) minimum hysteresis or tracking effects, (d) adequate frequency response for blood pressure waves, (e) high signal-to-noise ratio, and (f) maximum accuracy. Also, the system was required to provide for the display, recording, and storage of data, in a form for ready analysis and processing.

In order to achieve these objectives, the first decision to be made concerned the overall type of system. Since the dog was to be free-roaming, a "wireless" system was desirable, and hence a method of telemetry had to be chosen. It was decided to explore the possibility of using a high-sensitivity transducer to sense blood pressure, and to design a system to translate and amplify the resulting data for transmission and reception by telemetry. Concurrently, techniques for implantation of the transducer in dogs were explored, and commercially available equipment was investigated for the display, recording, and processing of the data.

Using engineering design techniques, a block diagram of the system and subsystems was made up, as illustrated in Figure 12.7. Each of these subtasks was investigated separately and then put into the system as a whole.

12.4.2. CHOICE OF TRANSDUCER. In addition to the major characteristics stated in Section 12.3, other physical, mechanical, and physiological factors had to be considered. A search was conducted to determine the types of transducers that were available. The choice was

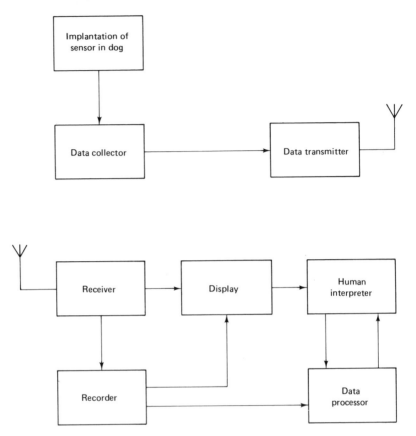

Figure 12.7. System and subsystems.

among (a) the fluid-catheter type, (b) the catheter-tip type, (c) an inductance-frequency changing type, and the (d) flush-diaphragm, direct-implantation type.

The last type was chosen, primarily because of the method of implantation and the fact that, after implantation, the chance of a transducer working loose was less likely. Also, since this was to be a chronic implantation, compatability of the transducer, wiring, and insulation with body fluid and tissues was important. The manufacturer of the transducer reported that the flush-diaphragm type of transducer had been used in chronic implants up to one year in duration with no undue effects on the animal. Table 12.1 shows a comparison of the main types of transducers considered for this application, in terms of the factors involved. The reader is referred back to Chapter 6 for details of the operation of each type of transducer.

TABLE 12.1. COMPARISON OF TRANSDUCERS

Problem Area	Fluid Transducer	Catheter-Tip Type	Flush-Diaphragm Type
Blood temperature	Independent of this.	Very dependent on this.	Partially dependent.
Velocity of blood flow	Pressure error with front hole catheter, but not with side hole type.	Pressure error.	No error.
Mechanical	Catheters can whip about. Domes are bulky for carrying on a dog. Catheter tubes could break.	Too fragile and too bulky for carrying on a dog.	Good; can be locked into aorta by suture. Wiring has good mechanical strength and resistance to bending.
Physiological	Long-term effects of catheter unknown.	Possible clotting at end of catheter.	Transducer and wire are compatible with body fluids.
Zeroing	Dependent on hydrostatic level.	Can be set electrically in wired situation, but not in telemetry situation.	Can be set electrically in wired situation, but not with telemetry.

12.4.3. THE TELEMETRY SYSTEM. Since the transducer chosen was a simple resistance-bridge type, the transmitter unit could easily be selected. Manufacturers of transmitter units can usually supply circuits to make their systems compatible with a particular transducer. Such was the case with the transmitter selected.

The system was put together essentially as described in Section 12.3 and as illustrated in Figure 12.6.

The reason for choosing an FM-FM system was that variations of the blood pressure by themselves are of a very low frequency, on the order of a few hertz. Since these signals would have to be recorded on magnetic tape, it was preferable to have them in the form of a subcarrier at a frequency range that would give a good response on the tape. Therefore the low audio-frequency subcarrier was chosen. Also, it was useful to be able to monitor the signals through a loudspeaker when the investigator's eyes were focused on other things. An ac voltage was used as excitation for the bridge. Power to the system was supplied by an 8.5-volt mercury battery. Since the transmitter unit was miniaturized, the battery occupied a major amount of the space.

The receiver was a commercial FM broadcast type. After the double demodulation, the original variations in blood pressure were displayed with a time base on a chart recorder and an oscilloscope. For storage purposes, the subcarrier was demodulated from the main carrier and fed into an analog instrumentation tape recorder. This subcarrier varied from 1 to 2 kHz. It was stored in this state and, when needed, was fed back into the receiver for demodulation from the subcarrier and fed into the display device as before.

The signal from the transmitter was strong enough to be picked up about 300 feet away (outdoors). In the hospital, signals were received from almost 200 feet away in different rooms.

An important part of the complete system was the display equipment in which the signal was displayed or recorded, or both. In this particular system, the display device was a cathode-ray oscilloscope. A storage oscilloscope was employed with a dual trace vertical amplifier and an appropriate time base generator. The storage feature was very useful, for the basic frequencies of the blood pressure signals are very low. All early calibrations were performed using the oscilloscope. The oscilloscope itself was calibrated both for amplitude and frequency of the sweep, and its square graticule was used for measurements. The limitations on the use of the cathode-ray oscilloscope were that the screen was not large enough to afford sufficient accuracy of reading. Also, in order to record the data, it was necessary to use a camera.

In order to obtain more permanent records, it was decided to use a chart recorder. Although this recorder was equipped with its own amplifiers, the telemetry receiver was able to produce a sufficiently strong signal to be fed directly into the galvanometer of the recorder.

All the component parts of the system were considered in light of the requirements for sensitivity, linearity, hysteresis, and accuracy. These factors had to be tested, as was the signal-to-noise ratio. It was required that the system have a sufficient frequency response to be able to reproduce faithfully blood pressure waves up to a rate of about 180 per minute. As discussed in Chapter 1, the dicrotic notch includes frequencies up to the seventh harmonic. At this heart rate, with a fundamental frequency of 3 Hz, the seventh harmonic would be 21 Hz. The frequency response of all component parts was more than adequate for this purpose.

Prior to any surgical work or implantation of the transducer, it was necessary to test the system in vitro and perform preliminary calibrations. A so-called artificial aorta was designed and built for the in vitro calibrations. The transducer was sealed into the wall of a lucite cylinder through which water flowed, so that it could sense radial water pressure in a manner analogous to blood pressure in the aorta. The temperature and flow of the water were controlled so that 0 to 240-mm Hg pressure variations could be obtained over temperatures around that of the blood (34 to 41°C). Pressure was measured with a mercury manometer and temperature with a mercury thermometer. Calibrations were achieved by recording the electrical output of the receiver on an optical recorder.

Figure 12.8 shows the artificial aorta and Figure 12.9 the complete

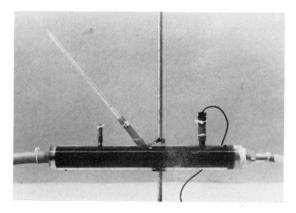

Figure 12.8. The artificial aorta.

experimental setup. The receiver is shown on the left; the pressure manometer is between the receiver and the artificial aorta; the transmitter is in the foreground; and an oscilloscope, used for the output in initially setting up the system, is on the extreme right.

Figure 12.9. The artificial aorta in an experiment.

12.4.4. SURGICAL AND PROTECTION PROCEDURES. Ten problems or requirements were involved in the implantation of the system, with some subrequirements.

1. Choice of dog—this included the psychological attributes of the dog, his temperament, his response to commands, his body build, his health, and his physiological characteristics.
2. Choice of site of implantation—the descending aorta was chosen, but other sites were considered.
3. Choice of surgical techniques in order to arrive at the site of implantation—where to enter the chest, and at what angle to work for a clear view.
4. Method of implantation—position of the transducer in the aorta, consideration of structural strength of the aorta and liability to tearing, methods of clamping to occlude blood during implant, methods of suturing, and so on.
5. Sterilization of the transducer and cable.
6. Consideration of passage of cable through the body to a point of exit, to include protection against mechanical breaks in the cable or insulation and protection against body fluids. This also included consideration of location of point of exit.
7. Protection of the dog's postoperative health.
8. Monitoring of the signal directly after the operation to evaluate overall system performance.
9. Protection of equipment from the dog.
10. The problems of working with a dog as part of a biomachine system.

In arranging the program of experimentation it was necessary to categorize the dogs to be used as acute or chronic. The acute dogs were for the early experiments in which surgical techniques had to be developed and methods of measurements devised. The dogs chosen were hounds weighing about 35 to 45 pounds. They required little preparation prior to the implantation operation.

On the other hand, the chronic dogs were prepared very carefully. The veterinarian at the hospital required 3 weeks to prepare a dog to be chronic. Preferably the dog selected had to be of mild temperament, friendly, and easy to work with. He was given any necessary medication, plus special rations in order to ensure his good health prior to the operation.

The descending aorta was chosen as the site of implantation. This site was picked because the blood pressure there is indicative of the pressure supplied to the systemic circulation and thus would reflect all changes. Also, the aorta is the largest artery; the typical aorta of a large dog has a diameter of 12 to 15 mm. The transducer head has a depth of 1.2 mm, and this would probably interfere with flow in smaller arteries, for even the major arteries have a diameter on the order of only 4 mm. The disadvantage of using the aorta is that it involves open-chest surgery, a complicated and dangerous procedure, whereas other arteries could have been reached with less surgical complication.

Another site suggested was in the wall of the left ventricle, but it involved even greater surgical dexterity and risk.

All operations were performed in the animal surgery suite in the Research Building of the Cedars of Lebanon Hospital. The operating facilities were new and were equipped with full surgical instrumentation. Qualified medical technicians were available to assist. Everything was done under completely sterile conditions.

Prior to each operation, all electrical equipment for the telemetry system was set up to operate with either the oscilloscope or the chart recorder as the output device. The dogs were brought from the vivarium, shaved, and anesthetized with Diabutal® (sodium pentobarbital). They were placed on the operating table and sterilized.

Each transducer was tested in the in vitro setup and calibrated and cold-sterilized for 24 hours prior to the operation. Different surgical techniques were used to ascertain a reliable method. During the actual implantation of the transducer into the aorta, the blood supply had to be occluded. Blalock cardiovascular clamps were used for total occlusion, which was shown to be satisfactory. However, since total occlusion of the blood can be dangerous if prolonged, a technique using Derra clamps, in which the blood flow was only partially occluded, was also tried. This technique was considered as an alternate method.

The early implantations were made with the transducer being placed directly into the aorta wall (see Figure 12.10). However, since the trans-

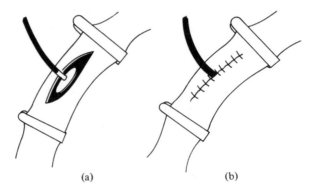

(a)   (b)

Figure 12.10. Direct entry into the aorta. (a) Transducer being lowered into incision. (b) Transducer being sutured in.

ducer worked loose, or was pulled out in some of the dogs, an alternate method was devised. Instead of relying on the wall of the aorta to support the transducer, an intercostal artery was used and then tied off.

Figure 12.11 is an illustration of how the cable from the transducer

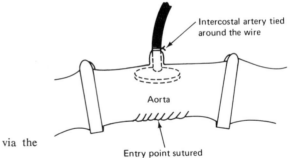

Intercostal artery tied around the wire

Aorta

Figure 12.11. Entry via the intercostal artery.

Entry point sutured

was brought out through an artery. The problem with this method was that the slit was made on the opposite side and the cable threaded through from there. The wire was enclosed in electrical shrink tubing that was filed to a point, sealed at the end, and then sterilized. The result provided a needle point for threading. Figure 12.12 schematically shows the anatomical relationship of the transducer and cable in position.

Signals were obtained while the dogs were still on the operating table. The maximum range of blood pressure between dogs immediately after implantation was from 170/142 to 188/162 mm Hg. The range of heart rates was from 162 to 204 beats per minute.

Another problem was the protection of the wire (preventing the dog from pulling it out). This problem was solved by two complementary procedures. First, a special jacket was designed to protect the dog and the

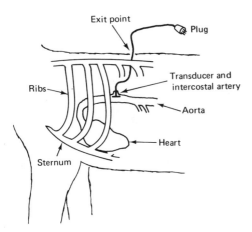

Figure 12.12. Position of transducer and wire.

equipment. It was made out of strong nylon mesh, reinforced at points where the dog could bite. Two pockets were attached, one per side. One pocket was to house the transmitter, the other held a small chemical pump that could be used later to infuse norepinephrine into the bloodstream. The pump would feed directly into the right atrium via a catheter. Since norepinephrine dissipates rapidly in the bloodstream, this device would ensure constant flow.

Figure 12.13 shows a dog with a developmental jacket, and Figure

Figure 12.13. An early jacket.

12.14 shows one with the final jacket design. The latter had the feature of a zip fastener for easy removal, plus lacing on each side so that the jacket could be fitted to different sizes of dogs.

12.4.5. IN VIVO CALIBRATIONS AND USE OF THE SYSTEM. In vivo calibrations were made as a check on the in vitro readings from the artificial aorta when a dog had recovered from surgery, usually about

Figure 12.14. The final
jacket design.

10 days later. They were done by catheterizing the dog through the femoral
artery and measuring blood pressure with a wired fluid-catheter strain-gage
system. This signal was compared with the telemetry signal on a recorder.
A third system, a catheter-tip pressure transducer, was also used for comparison purposes. During the catheterization the dog was inspected by
fluoroscopy to locate the catheter and the transducer as nearly as possible
to the same spot in the aorta. Figure 12.15 shows this comparison. The
uppermost curve is from telemetry. Below it is the curve from the catheter–

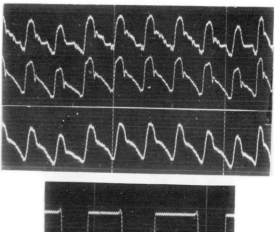

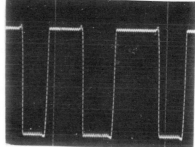

Figure 12.15. In vivo calibration. The top wave is
from telemetry, below it is
the wave from a catheter tip
transducer. The third wave
is taken from a fluid type
transducer. The lowest recording is a 100 mm Hg
calibration signal from
which the fluid transducer
scale is obtained.

tip transducer; the lowest curve is from the fluid system. A calibration signal of 100 mm Hg is shown below the other three curves. The fluid system was calibrated from this latter signal. It should be noted that the telemetry system had a faster response than the fluid strain-gage system and was not subject to oscillations produced in the catheter of the fluid system.

During in vivo calibrations, the flow patterns around the transducer were observed to be quite smooth. This observation was achieved by having a radiopaque dye flow through the bloodstream and then observing it by fluoroscopy on a television screen. Motion pictures were also obtained. Still X rays were taken to show the position of the transducer.

With the system calibrated, and available for studies, postural and exercise effects on blood pressure and heart rate could then be measured.

An antenna system was installed in the vivarium for long-term monitoring. The dog had freedom to run about in an area of about 400 square feet. The signal was picked up very clearly from all parts of this area. The receiving and recording equipment was located in a laboratory two floors above.

Although the results are not of importance to this book, Figure 12.16

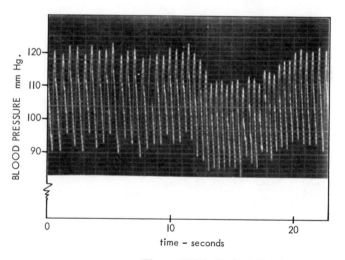

Figure 12.16. Typical blood pressure recording.

shows a typical record to indicate the type of sensitivity that can be obtained. This figure shows the effect of a dog standing, sitting, and then standing again. The change in blood pressure is well displayed. Thus it can be seen that a telemetry system is an excellent tool for physiological studies where mobility of the subject is of prime importance.

12.5. MULTICHANNEL SYSTEMS

Since the late 1950s many manufacturers have produced biotelemetry units, and each has its own methods. Some are simple systems, as described in earlier sections, but others have become more complicated. The field has advanced in two directions, one being in the use of multichannel systems and the other in the use of implantable units. The former are discussed here whereas implantable units are discussed in the next section. It should be noted that all the units described have been designed with miniaturization in mind.

It is often desirable to monitor a number of physiological data channels simultaneously, which means that the biotelemetry system becomes much more complex. It then becomes necessary to use a separate subcarrier frequency for each channel of data in order to avoid any interference between the channels. Perhaps the best way to show this situation is by example.

A typical unit is illustrated in Figure 12.17. The size of the trans-

Figure 12.17. Model 370 PWM multichannel Biolink telemetry system. (Courtesy of BIOCOM, Inc., Culver City, Calif.)

mitter ($\frac{3}{4}$ in. $\times$ $2\frac{1}{2}$ in. $\times$ $6\frac{1}{4}$ in.) can be seen in relation to the man's hand. This system can be used for up to six channels of data on one main carrier through time-multiplexing by pulse width modulation. Plug-in signal conditioners provide flexibility for quantitative multiparameter data acquisition, as well as expansion of the system. The photograph also shows some of these units. It is possible to measure such variables as the ECG, EEG, EMG, EOG, respiration, blood pressure, temperature, galvanic skin response, force, vibration, and the vector ECG with this unit.

An average range for a unit of this type is about 100 yards in free space. The battery provides about 50 hours of continuous use at this range. The system operated on the standard FM broadcast band of 88 to 108 MHz. However, much greater ranges are now possible with modified designs, and the manufacturer states that ranges up to 100 miles can be provided.

In addition to the transmitter and plug-in modules, this system includes a tester-calibrator to ensure artifact-free, quantitative data, a receiver-demodulator, and all necessary electrodes. The circuits used in this system have a high-modulation sensitivity and are quite simple in design. There is minimal cross-talk between channels, and this type of system is low on power consumption. The transmitted signal is a composite of a positive synchronizing pulse and negative signal pulses. The data to be telemetered cause the signal pulses to move back and forth in time (t_1, t_2, t_3, t_4). This method therefore uses pulse position modulation (PPM), since the position of one pulse with respect to another carries the data.

Referring to the block diagram of the transmitting system in Figure 12.18, it can be seen that the sync generator begins the action. Its pulse turns the first channel one-shot multivibrator *ON*. How long it remains on depends on the level of the data being fed into it at that instant of time. Its return to the *OFF* position triggers the next channel, and so on down the line. The resultant square waves are thus width-modulated by the data.

After reception, the composite signal must be separated and reformed to be properly demodulated. The sync-signal separator and amplifiers perform this function, as shown in Figure 12.19. Each channel consists of a flip-flop and an integrating network. The signal pulses are fed through a suitable diode network to all channels. The sync pulse is fed to the first channel only.

In operation, the sync pulse turns the first flip-flop *ON*. The first signal pulse comes in and would turn any *ON* flip-flops *OFF*. Since Channel 1 is the only unit that is *ON,* it is turned *OFF*. When it returns to the *OFF* position, it automatically triggers the Channel 2 flip-flop *ON*. Subsequent signal pulses are used to turn off (or gate) each corresponding flip-flop after it has been turned on. This situation is shown in Figure 12.20. The resulting square wave out of each flip-flop varies in width corresponding to the original square wave in the transmitter. Simple integration yields the original data.

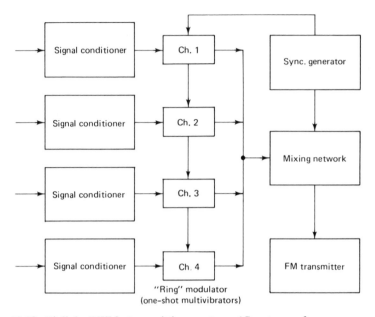

Figure 12.18. Biolink PWM transmitting system. (Courtesy of BIOCOM, Inc., Culver City, Calif.).

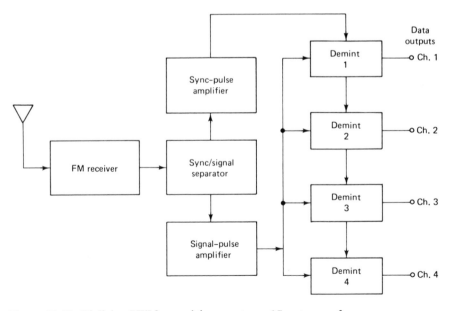

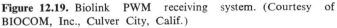

Figure 12.19. Biolink PWM receiving system. (Courtesy of BIOCOM, Inc., Culver City, Calif.)

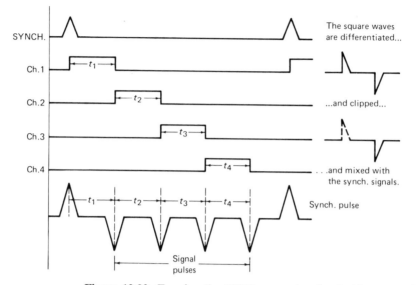

Figure 12.20. Forming the PWM composite signal. (Courtesy of BIOCOM, Inc., Culver City, Calif.)

12.6. IMPLANTABLE UNITS

It was mentioned previously that sometimes it is desirable to implant the telemetry transmitter or receiver subcutaneously. The implanted transmitter is especially useful in animal studies, where the equipment must be protected from the animal. The implanted receiver has been used with patients for stimulation of nerves, as described in Section 12.1.

Although the protective aspect is an advantage, many disadvantages often outweigh this factor, and careful thought should be given before embarking on an implantation. The surgery involved is not too complicated, but there is always risk whenever surgical techniques are used. Also, once a unit is implanted, it is no longer available for servicing, and the life of the unit depends on how long the battery can supply the necessary current.

This section is primarily concerned with completely implanted systems, but there are occasions when a partial implant is feasible. A good example is a system used for the monitoring of the electroencephalogram where the electrodes have been implanted into the brain and the telemetry unit is mounted within and on top of the skull. This type of unit needs a protective helmet.

The use of implantable units also restricts the distance of transmission of the signal. Because the body fluids and the skin greatly attenuate the signal and because the unit must be small to be implanted, and therefore has little power, the range of signal is quite restricted, often to just a few feet. This disadvantage has been overcome by picking up the signal with a nearby antenna and retransmitting it. However, most applications involve monitoring over relatively short distances, and retransmission is not necessary.

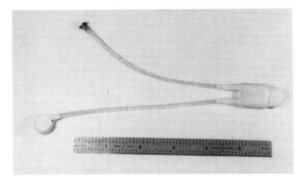

Figure 12.21. Implantable telemetry unit. (Courtesy of Konigsberg Instruments, Inc., Pasadena, Calif.)

Another problem has been the encapsulation of the unit. The outer case and any wiring must be impervious to body fluids and moisture. However, with the plastic potting compounds and plastic materials available today, this condition is easily satisfied.

Implantable telemetry can be used in conjunction with biopotential systems or with transducers. To illustrate, a typical system using various electrodes and transducers in conjunction with a telemetry transmitter is described below. The blood pressure version is illustrated in Figure 12.21.

Three separate units have been developed: one each for temperature, blood pressure, and biopotentials. Each operates on an excitation potential of 1.4 volts and has a transmission distance of about 10 feet. The temperature unit is capable of measuring over a range of 30 to 45°C and has a frequency response of from 0 to 0.5 Hz. The average current drain is 8 μA, and the operating life is approximately 6250 hours. For blood pressure telemetry, the frequency response must be much greater. Therefore this unit has a frequency response of from 0 to 70 Hz, with a pressure range up to 300 mm Hg. The current drain averages 290 μA. With the standard cell rated at 500 milliampere-hours, an operating life of 1700 hours is attainable. A larger cell with twice this capacity is also available. The biopotential unit, with a current drain of 20 μA and a life of about 2500 hours can be used for ECG or EEG. The range is from 0 to 2 mV peak to peak. The frequency response of this system is from 0.1 to 70 Hz.

· 13 ·

INSTRUMENTATION
FOR THE CLINICAL
LABORATORY

Every living organism has within itself a complete and often very complicated chemical factory. In higher animals, food and water enter the system through the mouth, which is the beginning of the digestive tract. In the stomach the food is chemically broken down into basic components by the digestive juices. From there it is transported into the intestine where the nutrients and the excess water are extracted. The extracted nutrients are then further broken down in numerous steps. Some are stored for later use, whereas others are used for the building of new body cells or are metabolized to obtain energy. All life functions, such as the contraction of muscles or the transmission of information through the nervous system, require energy for their operation. This energy is obtained from the nutrients by a series of oxidation processes which consume oxygen and leave carbon dioxide as a waste product. The exchange of oxygen and carbon dioxide with the air takes place in the lungs (see Chapter 8). Many of the chemical processes are performed in the liver, which is an organ specialized for this purpose. Certain soluble waste products are eliminated through the kidneys and the urinary tract. To make all this activity possible, the organism requires an efficient mechanism to transport the various chemical substances between the locations where they are introduced into the organism, are modified, or are excreted.

13.1. THE BLOOD

In very primitive animals, especially in those living in an ocean environment, like the sea anemone, the exchange of nutrients and metabolic wastes between cells and the environment takes place directly through the cell membrane. This simple method is insufficient, however, for larger animals, particularly those that live on land. For these animals, including man, nature has provided a special transport system to exchange chemical products between the specialized cells of the various organs—namely, the blood circulation. The circulatory system of an adult male human contains about 5 quarts of blood. Blood consists of a fluid, called the *plasma,* in which are suspended three different types of *formed elements* or *blood cells.* One cubic millimeter of blood (about $\frac{1}{6}$ drop) contains approximately the following numbers of cells:

Red blood cells (RBC) or erythrocytes	4.5–5.5 million
White blood cells (WBC) or leucocytes	6,000–10,000
Blood platelets or thrombocytes	200,000–800,000

Red blood cells are round disks, indented in the center, with a diameter of about 8 microns. A red blood cell has no cell nucleus, but it has a membrane and is filled with an iron-containing protein, *hemoglobin.* Red blood cells transport oxygen by chemically binding the oxygen molecules to the hemoglobin. Depending on the oxygen content, the hemoglobin changes its color, which accounts for the difference in color between oxygen-rich arterial blood (bright red) and oxygen-depleted venous blood (dark red).

White blood cells are of several different types, with an average diameter of about 10 microns. Each contains a nucleus and, like the amoeba, has the ability to change its shape. White blood cells attack intruding bacteria, incorporate them, and then digest them.

Blood platelets are masses of protoplasm 2 to 4 microns in diameter. They are colorless and have no nucleus. Blood platelets are involved in the mechanism of blood clotting.

By spinning blood in a centrifuge, the blood cells can be sedimented. The blood plasma with the blood cells removed is a slightly viscous, yellowish liquid that contains large amounts of dissolved protein. One of the proteins, *fibrinogen,* participates in the process of blood clotting and forms thin fibers called *fibrin.* The plasma from which the fibrinogen has been removed by precipitation is called *blood serum.*

The mechanism of blood clotting serves the purpose of preventing blood loss in case of injury. This mechanism can, on the other hand, cause undesirable or even dangerous blood clots if foreign bodies, like catheters or extracorporeal devices, are introduced into the blood stream. Blood clotting can be inhibited by the injection of *heparin,* a natural anticoagulant extracted from the liver and lungs of cattle.

Many diseases cause characteristic variations in the composition of blood. These variations can be a characteristic change in the number, size, or shape of certain blood cells (in anemia, for instance, the RBC count is reduced). Other diseases cause changes in the chemical composition of the blood serum (or some other body fluid, like the urine). In diabetes mellitus, for instance, the glucose concentration in the blood (and the urine) is characteristically elevated. A count of the blood cells, an inspection of their size and shape, or a chemical analysis of the blood serum can, therefore, provide important information for the diagnosis of such diseases. Similarly, other body fluids, smears, and small samples of live tissue, obtained by a *biopsy,* are studied through the techniques of *bacteriology, serology,* and *histology* in order to obtain clues for the diagnosis of diseases.

The purpose of *bacteriological tests* is to determine the type of bacteria that have invaded the body, in order to diagnose a disease or prescribe the proper treatment. For such a test, a sample containing the bacteria (e.g., a smear from a strep throat) is innoculated to the surface of various growth media (nutrients) in test tubes or flat Petri dishes. These cultures are then incubated at body temperature to accelerate the growth of the bacteria. When the bacteria have grown into colonies, they can be identified by the color and shape of the colony, by their preference for certain growth media, or by a microscopic inspection, which may make use of the fact that certain stains show a selectivity for certain bacteria groups.

Serological tests serve the same purpose as bacteriological tests but are based on the fact that the organism, when invaded by an infectious disease, develops antibodies in the blood, which defend the body against the infection. These antibodies are selective to certain strains of organisms, and their action can be observed in vitro by various methods. In some methods, for example, agglutination (collecting in clumps) becomes visible under a microscope when a test serum containing the antigen of the organism is added. Because the tests are based not on the organism itself but on the antigen developed by the organism, serological tests are not limited to bacteria but can be used for virus infections and infections by other microorganisms.

Histological tests involve the microscopical study of tissue samples, which are sliced into very thin sections by means of a precision slicer called a *microtome.* The tissue slices are often stained with certain chemicals to enhance the features of interest.

Blood counts and chemical blood tests are often ordered routinely on admission of a patient to a hospital and may be repeated daily to monitor the process of an illness. These tests, therefore, must be performed in very large numbers, even in the smaller hospital. The physician in private practice often has samples analyzed by commercial laboratories specializing in this service. Automated methods of performing the tests have found widespread acceptance, and special instruments have been developed for this purpose.

13.2. TESTS ON BLOOD CELLS

When whole blood is centrifuged, the blood cells sediment and form a packed column at the bottom of the test tube. Most of this column consists of the red blood cells, with the other cells forming a thin, *buffy layer* on top of the red cells. The volume of the packed red cells is called the *hematocrit*. It is expressed as a percentage of the total blood volume. If the number of (red) blood cells per cubic millimeter of blood is known, this number and the hematocrit can be used to calculate the *mean cell volume* (MCV). As stated above, the active component in the red blood cells is the hemoglobin, the concentration of which is expressed in grams per 100 ml. From the hemoglobin, the hematocrit and the blood cell count, the *mean cell hemoglobin* (MCH) (in *picograms*) and the *mean cell hemoglobin concentration* (MCHC) (in percent) can be calculated.

The hematocrit can be determined by aspirating a blood sample into a capillary tube and closing one end of the tube with a plastic sealing material. The tube is then spun for 3 to 5 minutes in a special high-speed centrifuge. Because the capillary tube has a uniform diameter, the blood and cell volumes can be compared by measuring the lengths of the columns. This is usually done with a simple nomogram, as shown in Figure 13.1. When lined up with the length of the blood column, the nomogram allows the direct reading of the hematocrit.

The red blood cells have a much higher electrical resistivity than the blood plasma in which they are suspended, and so the resistivity of the blood shows a high correlation with the hematocrit. This factor provides an alternate method of determining the hematocrit that is obviously more adaptable to automation than the centrifugal sedimentation method.

The *hemoglobin concentration* can be determined by lysing the red blood cells (destroying their membranes) to release the hemoglobin and chemically converting the hemoglobin into another colored compound (acid hematin or cyanmethemoglobin). Unlike that of the hemoglobin, the color concentration of these components does not depend on the oxygenation of the blood. Following the reaction, the concentration of the new component can be determined by colorimetry, as described in Section 13.3.

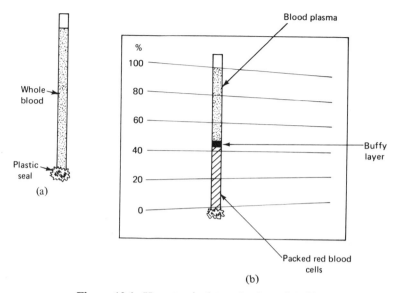

Figure 13.1. Hematocrit determination: (a) blood sample drawn in capillary and sealed with plastic putty; (b) capillary after centrifuging, placed on nomogram to read hematocrit (reading 43%).

Manual blood cell counts are performed by using a microscope. Here the blood is first diluted 1:100 or 1:200 for counting red blood cells (RBC) and 1:10 or 1:20 for white blood cell count (WBC). For counting WBC, a diluent is used that dissolves the RBCs; whereas for counting RBCs, an isotonic diluent preserves these cells. The diluted blood is then brought into a counting chamber 0.1 mm deep, which is divided by marking lines into a number of squares. When magnified about 500 times, the cells in a certain number of squares can be counted. This rather time-consuming method is still used quite frequently when a *differential count* is required for which the WBCs are counted, according to their distribution, into a number of different subgroups. An automated differential blood cell analyzer uses differential staining methods to discriminate between the various types of white blood cells.

Today simple RBC and WBC counts are normally performed by automatic or semiautomatic blood cell counters. The most commonly used devices of this kind are based on the conductivity (Coulter) method, which makes use of the fact that blood cells have a much lower electrical conductivity than the solution in which they are suspended. Such a counter (Figure 13.2) contains a beaker with the diluted blood into which a closed glass tube with a very small orifice (1) is placed. The conductance between the solution in the glass tube and the solution in the beaker is mea-

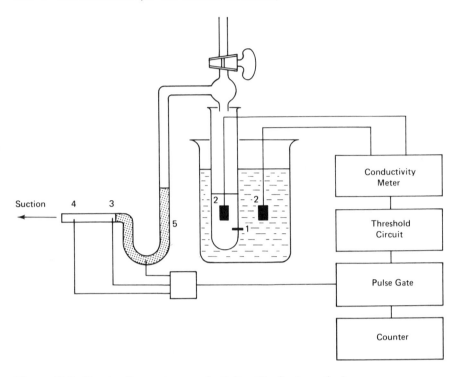

Figure 13.2. Blood cell counter, conductivity (Coulter) method. (Explanation in text.)

sured with two electrodes (2). This conductance is mainly determined by the diameter of the orifice, in which the current density reaches its maximum. The glass tube is connected to a suction pump through a U-tube filled with mercury (5). The negative pressure generated by the pump causes a flow of the solution from the beaker through the orifice into the glass tube. Each time a blood cell is swept through the orifice, it temporarily blocks part of the electrical current path and causes a drop in the conductance measured between the electrodes (2). The result is a pulse at the output of the conductance meter, the amplitude of which is proportional to the volume of the cell. A threshold circuit lets only those pulses pass that exceed a certain amplitude. The pulses that pass this circuit are fed to a pulse counter through a pulse gate. The gate opens when the mercury column reaches a first contact (3) and closes when it reaches the second contact (4), thus counting the number of cells contained in a given volume of the solution passing through the orifice. A count is completed in less than 20 seconds. With counts of up to 100,000, the result is statistically meaningful. Great care, however, must be taken to keep the aperture from clogging. Counters based on this principle are available with varying degrees of automation. The most advanced device of this type (shown in Figure 13.3) accepts a new blood sample every 20 seconds,

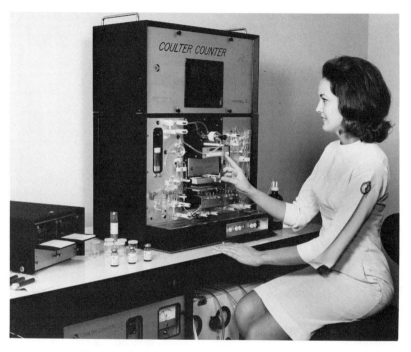

Figure 13.3. Coulter Model S. (Courtesy of Coulter Electronics, Inc., Hialeah, Fla.)

performs the dilutions automatically, and determines not only the WBC and RBC counts but also the hematocrit and the hemoglobin concentration. From these measurements, the mean cell volume, the mean cell hematocrit, and the mean cell hematocrit concentration are calculated and all results are printed out on a preprinted report form.

A second type of blood cell counter uses the principle of the dark-field microscope (Figure 13.4). The diluted blood flows through a thin cuvette (4). The cuvette is illuminated by a cone-shaped light beam obtained from a lamp (1) through a ring aperture (3) and an optical system (2). The cuvette is imaged on the cathode of a phototube (7) by means of a lens (5) and an aperture (6). Normally no light reaches the phototube until a blood cell passes through the cuvette and reflects a flash of light on the phototube. This method has also been used in the design of an automated blood analyzer that automatically determines either four or seven blood variables.

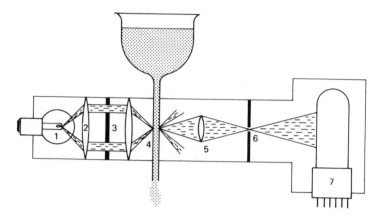

Figure 13.4. Blood cell counter, dark field method. (Explanation in text.)

13.3. CHEMICAL TESTS

Blood serum is a complex fluid that contains numerous substances in solution. The determination of the concentration of these substances is performed by specialized chemical techniques. Although there are usually several different methods by which any particular analysis can be performed, most tests used are based on a chemical color reaction followed by a colorimetric determination of the concentration. This principle makes use of the fact that many chemical compounds in solution appear colored, with the saturation of the color depending on the concentration of the compound. For instance, a solution that appears yellow when being held against a white background actually absorbs the blue component of the white light and lets only the remainder—namely, yellow light—through. The way in which this light absorption can be used to determine the concentration of the substance is shown in Figure 13.5.

In Figure 13.5(a) it is assumed that a solution of concentration C is placed in a cuvette with a length of the light path L. Light of an appropriate color or wavelength is obtained from a lamp through filter F. The light that enters the cuvette has a certain intensity I_0. With part of the light being absorbed in the solution, the light leaving the cuvette has a lower intensity I_1. One way of expressing this relation is to give the *transmittance* T of the solution in the cuvette as the percentage of light that is transmitted:

$$T = \frac{I_1}{I_0} \times 100 \%$$

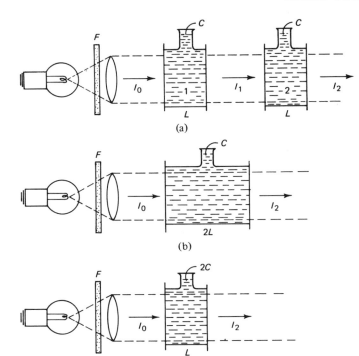

Figure 13.5. Principle of colorimeter analysis. (Explanation in text.)

If a second cuvette with the same solution were brought into the light path behind the first cuvette, only a similar portion of the light entering this cuvette would be transmitted. The light intensity I_2 behind the second cuvette is

$$I_2 = TI_1$$
or
$$I_2 = T^2 I_0$$

The light transmitted through successive cuvettes decreases in the same manner (multiplicatively). For this reason, it is advantageous to express transmittance as a logarithmic measure (in the same way as expressing electronic gains and losses in decibels). This measure is the *absorbance* or optical *density, A*.

$$A = -\log \frac{I_1}{I_0}$$

or
$$A = \log \frac{1}{T}$$

The total absorbance of the two cuvettes in Figure 13.5(a) is, therefore, the sum of the individual absorbances.

The amount of the light absorbed depends only on the number of molecules of the absorbing substance that can interact with the light. If, instead of two cuvettes, each with path length L, one cuvette with path length $2L$, were used [Figure 13.5(b)], the absorbance would be the same. The absorbance is also the same if the cuvette has a path length L but the concentration of the solution were doubled [Figure 13.5(c)]. This relation can be expressed by the equation

$$A = aCL \qquad \text{(Beer's law)}$$

where

$L =$ the path length of the cuvette
$C =$ concentration of the absorbing substance
$a =$ *absorbtivity,* a factor which depends on the absorbing substance and the optical wavelength at which the measurement is performed.

The absorbtivity can be obtained by measuring the absorption of a solution with known concentration, called a *standard.* If A_s is the absorption of the standard, A_u the absorption of an unknown solution, and C_s the concentration of the standard, then the concentration of the unknown is

$$C_u = C_s \frac{A_u}{A_s}$$

Corrections may have to be applied for light losses due to reflections at the cuvette or absorption by the solvent. Figure 13.6 shows the principle of

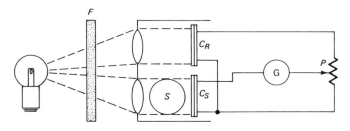

Figure 13.6. Colorimeter (filter-photometer).

a *colorimeter* or *filter-photometer* used for measuring transmittance and absorbance of solutions. A filter F selects a suitable wavelength range from the light of a lamp. This light falls on two photoelectric (selenium) cells: a reference cell C_R and a sample cell C_S. Without a sample, the output of both cells is the same. When a sample is placed in the light path for the sample cell, its output is reduced and the output of C_R has to be divided by a potentiometer P until a galvanometer (G) shows a balance. The potentiometer can be calibrated in transmittance or absorbance units over a range of 1 to 100 percent transmittance, corresponding to 2 to 0 absorbance units.

Other colorimeters, instead of using the potentiometric method, use a meter calibrated directly in transmittance units (a linear scale) and in absorbance. Figure 13.7 shows such a device; the instrument allows measurement at different colors with a built-in filter wheel.

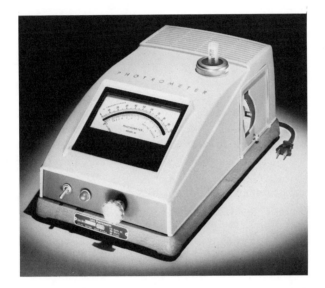

Figure 13.7. Leitz colorimeter. (Courtesy of E. Leitz, Inc., Rockleigh, N.J.)

In order to use the colorimeter to determine the concentration of a substance in a sample, a suitable method for obtaining a colored derivative from the substance is necessary. Thus a chemical reaction that is unique for the substance to be tested and that does not cause interference by other substances which may be present in the sample must be found. The reaction may require several steps of adding reagents and incubating the sample at elevated temperatures until the reaction is completed. Most reactions require that the protein first be removed from the plasma by adding a precipitating reagent and filtering the sample.

TABLE 13.1. THE MOST COMMONLY USED CHEMICAL BLOOD TESTS

Test	Normal Ranges	Unit
1. Blood urea nitrogen (BUN)	8–16	mg N/100 ml
2. Glucose	70–90	mg/100 ml
3. Phosphate (inorg.)	3–4.5	mg/100 ml
4. Sodium	135–145	mEq/liter
5. Potassium	3.5–5	mEq/liter
6. Chloride	95–105	mEq/liter
7. CO_2 (total)	24–32	mEq/liter
8. Calcium	9–11.5	mg/100 ml
9. Creatinine	0.6–1.1	mg/100 ml
10. Uric acid	3–6	mg/100 ml
11. Protein (total)	6–8	g/100 ml
12. Albumin	4–6	g/100 ml
13. Cholesterol	160–200	mg/100 ml
14. Bilirubin (total)	0.2–1	mg/100 ml

The most commonly required tests for blood samples are listed in Table 13.1. This table also shows the units in which the test results are expressed * and the normal range of concentration for each test. Most of these tests can be performed by color reaction even though, in most cases, several different methods have been described that can often be used alternately.

For the measurement of sodium and potassium, however, a different property is utilized, one that causes a normally colorless flame to appear yellow (sodium) or violet (potassium) when their solutions are aspirated into the flame. This characteristic is used in the *flame photometer* (Figure 13.8) to measure the sodium or potassium concentration in samples. The sample is aspirated into a gas flame that burns in a chimney. As a reference, a known amount of a lithium salt is added to the sample, thus causing a red flame. Filters are used to separate the red light produced by the lithium from the yellow or violet light emitted by the sodium or potassium. As in the colorimeter, the output from the sample cell C_S is compared with a fraction of the output from a reference cell C_R. The balance potentiometer P is calibrated directly in units of sodium or potassium concentration.

For the determination of chlorides, a special instrument (*chloridimeter*) is sometimes used that is based on an electrochemical (*coulometric*)

* Depending on the test, the concentration is expressed in either grams or milligrams per 100 milliliters (.1 liter) or in milli-equivalents per liter, which is obtained by dividing the concentration in milligrams per liter by the molecular weight of the substance.

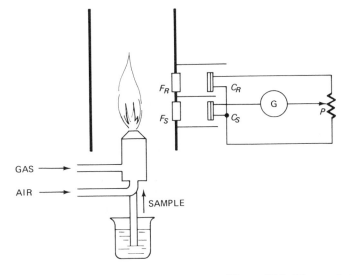

Figure 13.8. Flame photometer.

method. For this test, the chloride is converted into silver chloride with the help of an electrode made of silver wire. By an electroplating process with a constant current, the silver chloride is percipitated. When all the chloride has been used up, the potential across the cell changes abruptly and the change is used to stop an electric timer, which is calibrated directly in chloride concentration.

The simple colorimeter (or filter-photometer) shown in Figures 13.6 and 13.7 has a sophisticated relative, the *spectrophotometer* shown in Figure 13.9. In this device the simple selection filter of the colorimeter is re-

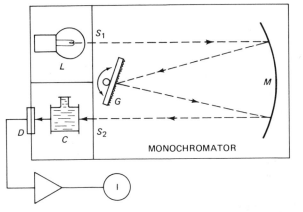

Figure 13.9. Spectrophotometer.

placed by a *monochromator*. A monochromator uses a diffraction grating *G* (or a prism) to disperse light from a lamp that falls through an entrance slit S_1 into its spectral components. An exit slit S_2 selects a narrow band of the spectrum, which is used to measure the absorption of a sample in cuvette *C*. The narrower the exit slit, the narrower the bandwidth of the light, but also the smaller its intensity. A sensitive photodetector *D* (often a photomultiplier) is therefore required, together with an amplifier and a meter *I*, which is calibrated in units of transmittance or absorbance. The wavelength of the light can be changed by rotating the grating. A mirror *M* folds the light path to reduce the size of the instrument.

The spectrophotometer allows the determination of the absorption of samples at various wavelengths. The light output of the lamp, however, as well as the sensitivity of the photodetector and the light absorption of the cuvette and solvent, varies when the wavelength is changed. This situation requires that, for each wavelength setting, the density reading be set to zero, with the sample being replaced by a blank cuvette, usually filled with the same solvent as used for the sample. In *double-beam* spectrophotometers this procedure is done automatically by switching the beam between a sample light path and a reference light path, generally with a mechanical shutter or rotating mirror. By using a computing circuit, the readings from both paths are compared and only the ratio of the absorbances (or the difference of the densities) is indicated.

Certain chemicals, when illuminated by light with a short wavelength in the ultraviolet (UV) range, emit light with a longer wavelength. This phenomenon is called *fluorescence*. Fluorescence can be used to determine the concentration of such chemicals using a *fluorometer*, which, like the photometer, can be either a *filter-fluorometer* or a *spectrofluorometer*, depending on whether filters or monochromators are used to select the excitation and emission wavelengths.

13.4. AUTOMATION OF CHEMICAL TESTS

Even though most chemical tests basically consist of simple steps like pipetting, diluting, and incubating, they are rather time consuming and require skilled and conscientious technicians if errors are to be avoided. Attempts to replace the technicians by an automatic device, however, were not very successful at first. The first automatic analyzer that found wide acceptance and that is still used at most hospitals is the *Autoanalyzer*,® the principle of which is shown in Figure 13.10.

The basic method used in the Autoanalyzer® departs in several respects from that of standard manual methods. The mixing, reaction, and colorimetric determination take place, not in an individual test tube for each

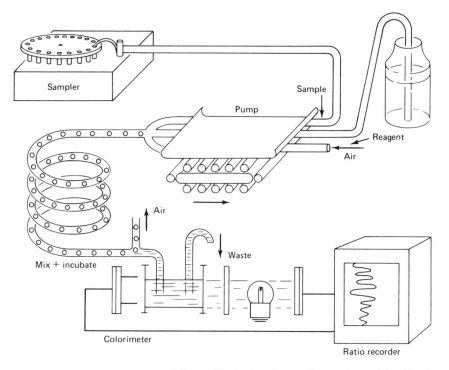

Figure 13.10. Continuous flow analyzer (simplified).

sample but sequentially in a continuous stream. The sampler feeds the samples into the analyzer in time sequence. A proportioning pump, which is basically a simple peristaltic pump working simultaneously on a number of tubes with certain ratios of diameters, is used to meter the sample and the reagent. Mixing is achieved by injecting air bubbles. The mixture is incubated while flowing through heated coils. The air bubbles are removed, and the solution finally flows through the cuvette of a colorimeter or is aspirated into the flame of a flame photometer. An electronic ratio recorder compares the output of the reference and sample photocells. The recording shows the individual samples as peaks of a continuous transmittance or absorbance recording. The samples of a "run" are preceded by a number of standards that cover the useful concentration range of the test. The concentration of the samples is determined from the recording by comparing the peaks of the samples with the peaks of the standards. In this way the effects of errors (e.g., incomplete reaction in the incubator) are eliminated because they affect standards and samples in the same way.

Suitable adaptations of almost all standard tests have been developed for the Autoanalyzer® system. The removal of protein from the plasma is achieved in the continuous-flow method with a *dialyzer* (not shown in Figure 13.10), which consists of two flow channels separated by a cellophane membrane that is impermeable to the large protein molecules, but not to the smaller molecules. The smallest model of the Autoanalyzer® performs a single test at a rate up to 120 samples per hour. Larger models (one of which is shown in Figure 13.11) perform up to 12 different tests on each of 60 samples per hour. The results of these tests are directly provided in

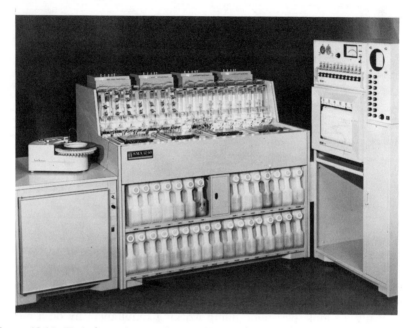

Figure 13.11. Technicon Autoanalyzer SMA 12/60. (Technicon Autoanalyzer and SMA are registered trademarks of Technicon Instruments, Tarrytown, N.Y.)

the form of a "chemical profile," drawn by a recorder on a preprinted chart. By the use of additional equipment, the results may also be provided as a digital output signal for recording on a storage medium, like punch cards or paper tape, or may be usable for direct computer processing.

A major problem with the continuous-flow process is the "carryover" that can occur when a sample with an excessively high concentration is followed by a sample with normal or low concentration. Methods of "carryover" correction are available.

Although the continuous-flow analyzer was the first to find wide acceptance, numerous other analyzers that use discrete samples are now available. Some of these analyzers perform all tests in test tubes mounted on a carousel-type carrier, or a chain belt, with the test tubes being rinsed

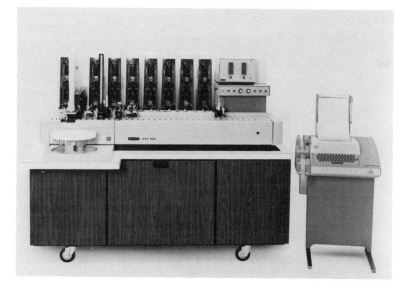

Figure 13.12. Beckman DSA discrete sample analyzer. (Courtesy of Beckman Instruments, Inc., Fullerton, Calif.)

once the analysis has been completed. Another device (Figure 13.12), uses disposable plastic trays as reaction vessels. All automatic analyzers of this type use automatic syringe-type pumps to dispense the sample and add the reagents. After the incubation, the sample is aspirated into a colorimeter cuvette.

One other automatic tester uses a completely different principle. In this device all the reagents for a given test are contained in premeasured quantities in pockets of a plastic pouch. The sample is injected into another section of the pouch and the reagents added by destroying the separations between the pockets. This procedure is accomplished by squeezing the reagent pockets. After the color reaction has been completed, the colorimetric determination is performed in the (transparent) plastic pouch with a special colorimeter.

A basic problem in all automatic chemical analyzers is the positive identification of the samples. In most of the devices, the small sample cups are identified only by their position in the sample tray and the results can be identified only by the sequence in which they are recorded. A machine-readable identification on the test tube is highly preferable to the sequential identification method, especially when automatic data processing methods are to be used for the results.

· 14 ·

X-RAY
AND RADIOISOTOPE
INSTRUMENTATION

In 1895 Conrad Röntgen, a German physicist, discovered a previously unknown type of radiation while experimenting with gas-discharge tubes. He found that this type of radiation could actually penetrate opaque objects and provide an image of their inner structure. Because of these mysterious properties, he called his discovery *X rays.* In many countries X rays are referred to as *Röntgen-rays.* One year after Röntgen's discovery, Henry Becquerel, the French physicist, found a similar type of radiation emanating from samples of uranium ore. Two of his students, Pierre and Marie Curie, traced this radiation to a previously unknown element in the ore, to which they gave the name *radium,* from the Latin word *radius,* the ray. The process by which radium and certain other elements emit radiation is called *radioactive decay,* whereas the property of an element to emit radiation is called *radioactivity.*

Today numerous man-made radioisotopes have joined the X-ray tube and radium as sources of *nuclear* or *ionizing radiation.* The ability of this radiation to penetrate materials that are opaque to visible light is utilized in numerous different techniques for medical diagnosis and research. The ionizing effects of radiation are used for the treatment of certain diseases in *X-ray* and *radium therapy.*

There are three different types of radiation, each with its own distinct properties. More than one type of radiation can emanate from a given sample of radioactive material. The properties of the three types of radiation are defined below.

Alpha rays are positively charged particles that consist of helium nuclei and that travel at the moderate velocity of 5 to 7 percent of the velocity of light. They have a very small penetration depth, which in air is only about 2 inches.

Beta rays are negatively charged electrons. Their velocity can vary over a wide range and can almost reach the velocity of light. Their ability to penetrate the surrounding medium depends on their velocity, but generally it is not very great. Both alpha and beta rays, when traveling through a gaseous atmosphere, interact with the gas molecules, thereby causing ionizing of the gas.

Gamma rays and *X rays* are both electromagnetic waves that have a much shorter wavelength than radio waves or visible light. Their wavelengths can vary between approximately 10^{-6} and 10^{-10} cm, corresponding to a frequency range of between 10^{10} and 10^{14} MHz, with the X rays at the lower and the gamma rays at the higher end of this range. The ability of these rays to penetrate matter depends on their wavelengths, but it is much greater than that of the alpha and beta rays. Gamma rays do not interact with gases directly but can cause ionization of the gas molecules via photoelectrons released when the rays interact with solid matter.

Gamma rays are usually not characterized by their frequency but by their energy, which is proportional to the frequency. This relationship is expressed in the Planck equation

$$E = hf$$

where

$E =$ energy in ergs
$h =$ Planck's constant $= 6.624 \times 10^{27}$ erg sec
$f =$ frequency in hertz

The energy of radiation is usually expressed in electron volts (eV) with 1 eV $= 1.602 \times 10^{-12}$ erg

Figure 14.1 shows the position of gamma rays and X rays within the spectrum of electromagnetic waves.

14.1. GENERATION AND DETECTION OF NUCLEAR RADIATION

X rays are generated when fast-moving electrons are suddenly decelerated by impinging on a target. An X-ray tube is basically a

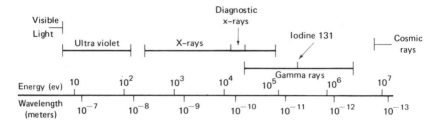

Figure 14.1. Part of the electromagnetic spectrum showing the location of the X rays and gamma rays.

high-vacuum diode with a heated cathode located opposite a target anode (Figure 14.2). This diode is operated in the saturated mode with a fairly

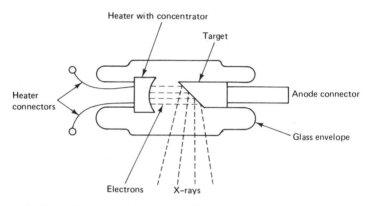

Figure 14.2. X-ray tube, principle of operation.

low cathode temperature so that the current through the tube does not depend on the applied anode voltage.

The intensity of X rays depends on the current through the tube. This current can be varied by varying the heater current, which in turn controls the cathode temperature. The wavelength of the X rays depends on the target material and the velocity of the electrons hitting the target. It can be varied by varying the target voltage of the tube. X-ray equipment for diagnostic purposes uses target voltages in the range of 30 to 100 kV, while the current is in the range of several milliamperes. These voltages are obtained from high-voltage transformers that are often mounted in oil-filled tanks to provide electrical insulation. When ac voltage is used, the X-ray tube conducts only during one half-wave and acts as its own rectifier. Otherwise high-voltage diodes (often in voltage-doubler or multiplier configurations) are used as rectifiers. For therapeutic X-ray equipment, where even higher radiation energies are required, linear or circular particle accelerators have been used to obtain electrons with sufficiently high energy. When the electrons strike the target, only a small part of their energy is converted into X rays; most of it is dissipated as heat. The target, therefore, is usually made of tungsten, which has a high melting point. It may also be water or air cooled, or it may be in the form of a motor-driven rotating cone to improve the dissipation of heat. The electron beam is concentrated to form a small spot on the target. The X rays emerge in all directions from this spot, which therefore can be considered a point source for the radiation.

Radioactive decay is the other source of nuclear radiation, but only a very small number of chemical elements exhibit natural radioactivity. Artificial radioactivity can be induced in other elements by exposing them to neutrons generated with a cyclotron or in an atomic reactor. By introducing an extraneous neutron into the nucleus of the atom, an unstable form of the element is generated that is chemically equivalent to the original form (*isotope*). The unstable atom disintegrates after some time, often through several intermediate forms, until it has assumed the form of another, stable element. At the moment of the disintegration, radiation is emitted, the type and energy of which are characteristic of a particular decay step in the process. The time after which half of the original number of radioisotope atoms have decayed is called the *half-life*. Each radioisotope has a characteristic half-life that can be between a few seconds and thousands of years.

Radioisotopes are chemically identical to their mother element. Chemical compounds in which a radioisotope has been substituted for its mother element are thus treated by the body exactly like the nonradioactive form. With the help of the emitted radiation, however, the path of the substance can be traced and its concentration in various parts of the organism determined. If this procedure is to be done in vivo, the isotope must emit gamma radiation that penetrates the surrounding tissue and that can be measured with an extracorporeal detector. When radioactive material is introduced into the human body for diagnostic purposes, great care must be taken to ensure that the radiation dose that the body receives is at a safe level. For reasons explained below, it is desirable that the radioactivity be as great as possible during the actual measurement. For safety reasons, however, the activity should be reduced as fast as possible as soon as the measurement is completed. In certain measurements, the radioactive material is excreted from the body at a rapid rate and the activity in the body decreases quickly. In most measurements, this "biological decay" of the introduced radioactivity occurs much too slowly. In order to remove the source of radiation after the measurement, isotopes with a short half-life must be used. However, there is a dearth of gamma-emitting isotopes of elements naturally occurring in biological substances that have a half-life of suitable length. The radioisotopes most frequently used for medical purposes are listed in Table 14.1. Iodine 131 is the only gamma-emitting isotope of an element that occurs in substantial quantities in the body. H-3 (tritium) and carbon 14 are beta emitters; hence their concentration in biological samples can be measured only in vitro because the radiation does not penetrate the surrounding tissue.

TABLE 14.1. RADIOISOTOPES

Isotope	Radiation	Half-Life
H-3	beta	12.3 days
C-14	beta	5570 years
Cr-51	gamma	27.8 days
Tc(Technetium)-99m	gamma	6 hours
I-131	gamma	8.07 days
Au-198	gamma	2.7 days

Pierre and Marie Curie discovered that radioactivity can be detected by three different physical effects: (a) the activation it causes in photographic emulsions, (b) the ionization of gases, and (c) the light flashes the radiation causes when striking certain minerals. Most techniques used today are still based on the same principles. Photographic films are the most commonly used method of visualizing the distribution of X rays for diagnostic purposes. For the visualization of radioisotope concentrations in biological samples, a photographic method called *autoradiography* is used. In this technique thin slices of tissue are laid on a photographic plate and left in contact (in a freezer) for extended time periods, sometimes for months. After processing, the film shows an image of the distribution of the isotope in the tissue.

When the gas ions caused by radiation are subjected to the forces of an electric field between two charged capacitor plates, they move toward these plates and cause a current flow. Above a certain voltage, all ion pairs generated reach the plates, and further increases of the voltage cause no additional increase of the current (saturation). The current flow (normally very small) can be used to measure the intensity of the radiation. This device is called an *ionization chamber*.

The number of ion pairs generated depends on the type of radiation. The number is greatest for alpha and lowest for gamma radiation. If the voltage is increased beyond a certain value, the ions are accelerated enough to ionize additional gas molecules (gas amplification, *proportional counter*). If the voltage is increased even further, a point can be reached at which any initial ion pair causes complete ionization of the tube (*Geiger counter*). Further increase of the voltage, therefore, does not increase the current (plateau). The ion generation, however, is self-sustaining and must be terminated, usually by reducing the voltage briefly. The Geiger counter cannot discriminate between the different types of radiation, but it has the advantage of providing large output pulses.

The physical configuration of the various detectors based on the principle of gas ionization can actually be the same. The mode of operation, as shown in Figure 14.3, is determined solely by the operating voltage applied to the device.

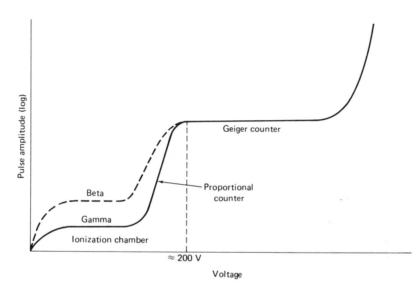

Figure 14.3. Detection of nuclear radiation by the ionization of gas between two capacitor plates. The curve shows the logarithm of the current pulse amplitude as a function of the applied voltage for a constant rate of nuclear disintegrations generating either beta or gamma radiation.

Certain metal salts (e.g., zinc sulfide) show fluorescence when irradiated with X rays or radiation from radioisotopes. When observed under a microscope under favorable circumstances, the minute light flashes (scintillations) caused by individual radiation events can actually be seen. In earlier days these scintillations were used to measure radioactivity by simply counting them. Both scintillation and fluorescence, however, are light events of such low intensity that they can be seen only with eyes that are well adapted to the dark. Only through use of electronic devices for the detection and visualization of low-level light has their usefulness been increased to such an extent that today most isotope instrumentation is based on this principle.

14.2. INSTRUMENTATION FOR DIAGNOSTIC X RAYS

The use of X rays as a diagnostic tool is based on the fact that various components of the body have different densities for the rays. When X rays from a point source penetrate a body section, the internal structure of the body absorbs varying amounts of the radiation. The radiation that leaves the body, therefore, has a spatial intensity variation that is an image of the internal structure of the body. When, as shown in Figure 14.4, this intensity distribution is visualized by a suitable device, a shadow

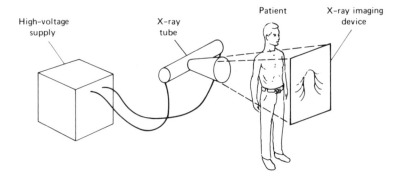

Figure 14.4. Use of X rays to visualize the inner structure of the body.

image is generated that corresponds to the X-ray density of the organs in the body section. Bones and foreign bodies, especially metallic ones, and air-filled cavities show up well on these images because they have a much higher or a much lower density than the surrounding tissue. Most body organs, however, differ very little in density and do not show up well on the X-ray image. For some organs (stomach, colon, etc.), the contrast can be improved by filling the organ with a *contrast medium* (barium sulfate). The ventricles of the brain can be made visible by filling them with air (*pneumo-encephalography*). Blood vessels can also be made visible by injecting a bolus of a special contrast fluid (an iodized organic compound) before the X-ray exposure (*angiography*).

When the image does not change rapidly, photographic film is used extensively to visualize the X-ray distribution. Film has the advantage that it provides a permanent record. By comparing the X rays taken at different times, it is possible to observe the progress of a disease. The X-ray film is normally used in light-tight cassettes that can be handled in normal light. The front of the cassette is thin plastic that is opaque to light but that lets X rays pass readily. The contrast of the X-ray photos can often be increased by bringing an *intensifier screen* in contact with the film emulsion.

The film cassette is normally brought in close contact with the body section to be photographed. The X-ray photo, therefore, is slightly larger than the actual body section. Depending on the part of the body, film cassettes of different sizes are used. The most commonly used (14 in. by 17 in.) accommodates chest and pelvic X rays. By means of electrically operated cassette changers, several X rays can be taken in fast sequence as required for angiography. For this procedure, the injection of contrast material by a power-operated syringe, X-ray exposure, and changing of cassettes can all be synchronized by an electrical timer.

In certain radiological procedures, such as when a catheter is threaded through the blood vessels into the heart, a direct observation of the X-ray image is necessary. It was originally achieved by directly watching the faint image on a fluorescent screen. This direct fluoroscopy is seldom used nowadays, for it requires a high radiation intensity and an exposure for extended time intervals. Where a true time presentation of the image is necessary, the radiation intensity, and therefore the radiation dose, can be decreased substantially by the use of an *image intensifier,* shown in Figure 14.5.

The intensifier tube contains a fluorescent screen, the surface of which is coated with a suitable material to act as a photocathode. The electron image thus obtained is projected onto a phosphor screen at the other end of the tube by means of an electrostatic lens system. The resulting brightness gain is due to the acceleration of the electrons in the lens system and the fact that the output image is smaller than the primary fluorescent image. The gain can reach an overall value of several hundred, not only allowing the reduction of the radiation dose but also resulting in a much brighter image than the one on a normal fluoroscopic screen.

The output screen can be observed by means of mirrors, while angiographic procedures are recorded with a 16-mm movie camera. The flow of the radiopaque dye through the blood vessels is then studied by projecting the film, frame by frame, with a special projector. In modern installations of this kind, the movie camera is replaced by a television camera. The output of the image-intensifier tube is then observed on a television monitor and can be recorded on a videotape recorder.

In order to perform a radiographic examination, the X-ray tube and the film cassette or image intensifier must be brought into the appropriate

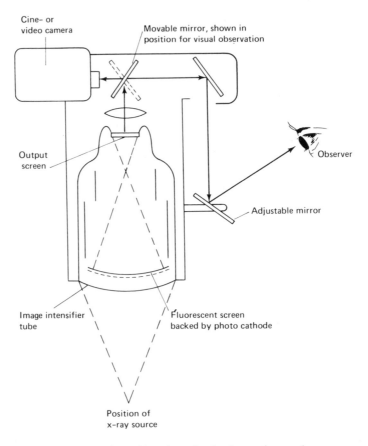

Figure 14.5. X-ray image intensifier for visual observation and recording of the picture with a cine (movie) camera or with a video tape recorder—diagram simplified.

position with respect to the body section to be examined. Elaborate suspension systems for the X-ray tube and image intensifier, plus special positioning tables for the patient, are used for this purpose. Normally they are permanently installed in a special room. Mobile X-ray machines, either for X-ray photography only or including a small image intensifier with TV monitor (Figure 14.6), are available for examinations in the operating room or at the patient's bed, but they do not have the capabilities of the larger stationary units.

Great care must be taken to protect personnel and patients from unnecessary X-ray exposure. The housing for the X-ray tube is shielded with lead, and the emerging X-ray beam is limited to the examination area by adjustable shutters. When taking X-ray photos, the technician retreats

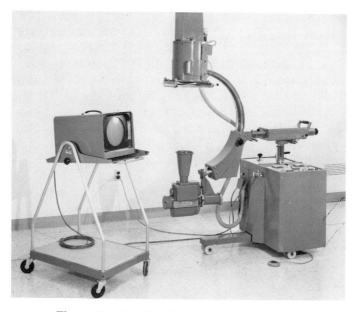

Figure 14.6. Mobile X-ray unit (with image intensifier and television monitor). (Courtesy of Picker Corporation, Cleveland, Ohio.)

to a shielded cubicle during the actual exposure. For fluoroscopic examinations, all personnel surrounding the patients wear aprons made of plastic or rubber containing lead powder.

14.3. INSTRUMENTATION FOR THE MEDICAL USE OF RADIOISOTOPES

The radiation exposure during X-ray examinations occurs only during a very short time interval. In diagnostic methods involving the introduction of radioisotopes into the body, on the other hand, the exposure time is much longer, and therefore the radiation intensity must be kept much smaller in order not to exceed a safe radiation dose. For this reason, the techniques used for radiation detection and visualization with radioisotopes differ greatly from those used for X-rays. Radioisotope techniques are all based on actually counting the number of nuclear disintegrations that occur in a radioactive sample during a certain time interval or on counting the radiation quanta that emerge in a certain direction during this time. Because of the random nature of radioactive decay, any measurement performed in this way is afflicted with an unavoidable statistical error. When the same sample is measured repeatedly, the observed counts are not the same each time but follow a gaussian (normal) distribution. If the mean number of counts observed is n, the standard deviation of this distribution curve will be the square root of n. The concentration of radioactive material in an unknown sample can be determined by comparing the count with that of a known standard. A much greater accuracy is obtained if the number of disintegrations counted for the measurement is high. Higher counts can be obtained either by counting over a longer time interval or by increasing the activity of the sample, both ways being limited in medical applications in which the radioactivity is measured inside the body.

Almost all nuclear radiation detectors used for medical applications utilize the light flashes caused by radiation in a suitable medium. Such *scintillation detectors* (also called *scintillation counters*) for gamma rays use a crystal made from thallium-activated sodium iodide, which is in close contact with the active surface of a photomultiplier tube. Each radiation quantum passing the crystal causes an output pulse at the photomultiplier, the amplitude of which is proportional to the energy of the radiation. This property of the scintillation detector is used to reduce the *background,* (counts due to natural radioactivity) by means of a *pulse height analyzer.* This is an electronic circuit that passes only pulses within a certain amplitude range. The limits of this circuit are adjusted in such a way that only pulses from the radioisotope used can pass, whereas pulses with other energy levels are rejected. Figure 14.7 shows two types of scintillation detectors used for the determination of the concentration of gamma-emitting radioisotopes in medical applications. In the *well counter,* the scintillation crystal has a hole into which a test tube with the sample is inserted. In this configuration almost all radiation from the sample passes the crystal and is counted while a lead shield reduces the background count.

For activity determinations inside the body, a *collimated detector,* also shown in Figure 14.7, is used. In this detector, a lead shield around the scintillation crystal has holes arranged in such a way that only radiation from a source located at one particular point in front of the detector can reach the crystal. Only a very small part of the radiation coming from this source, however, passes the crystal. This detector, therefore, is much less sensitive than the well counter type.

Figure 14.8 shows the other building blocks that constitute a typical instrumentation system for medical radioisotope measurements. The pulses from the photomultiplier tube are amplified and shortened before they pass through the pulse height analyzer. A timer and gate allow the pulses that occur in a set time interval to be counted by means of a *scaler* (decimal counter with readout). A *rate meter* (frequency meter) shows the rate of the pulses. Its reading can be used in aiming the detector toward the

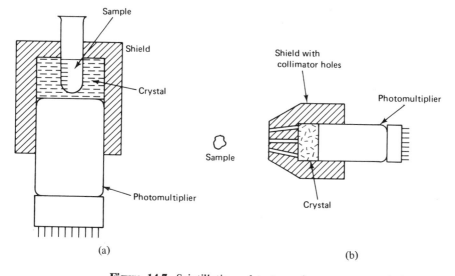

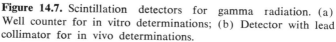

Figure 14.7. Scintillation detectors for gamma radiation. (a) Well counter for in vitro determinations; (b) Detector with lead collimator for in vivo determinations.

location of maximal radioactivity and to set the pulse height analyzer to where it passes all pulses from the particular isotope used.

Instrumentation systems of this kind can be used in a large variety of diagnostic tests. An example of a test that uses an in vitro activity measurement is the determination of the circulating blood volume, which may be administered when internal hemorrhaging is suspected. Here the activity of a premeasured sample of radioiodated human serum albumen (RIHSA) is measured, and subsequently injected to mix with the circulating blood. After a short time has elapsed, a certain volume of blood is withdrawn and its radioactivity determined. From the ratio of the two measured activities and the volume of the two samples, the blood volume can be calculated. Corrections must be applied for any radioactivity present in the blood before the sample injection and any radioactive material remaining in the syringe used for the injection. This procedure has been automated in a special device, shown in Figure 14.9, which consists of three well counters and a computer circuit and which gives a direct readout of the blood volume.

The most frequently administered in vivo radioisotope test is the iodine uptake determination, which is used to measure the activity of the thyroid gland. Two equal samples of radioactive sodium iodide solution are prepared. One of the samples is given to the patient orally, while the other one is kept as a reference. The thyroid gland uses iodine in the production of thyroid hormone, and after some time a certain fraction of the adminis-

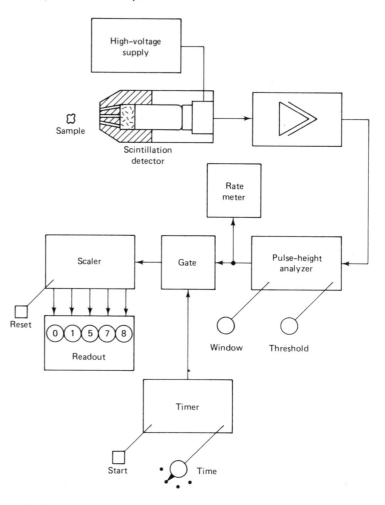

Figure 14.8. Block diagram of an instrumentation system for radioisotope procedures.

tered radioiodide will have accumulated in the thyroid gland. The magnitude of this accumulation, which is indicative of the activity of the gland, can be determined by measuring the radioactivity of the neck area. This activity is compared with that of the reference sample, which is placed in a so-called *phantom*—that is, a plastic model which approximates the human neck in geometry and absorption properties. In this way, errors due to nuclear decay or tissue absorption are eliminated.

Figure 14.10 shows a typical setup for in vivo measurements. This particular device, which consists of two collimated detectors, two rate meters, and a two-channel recorder, is used to measure the kidney function.

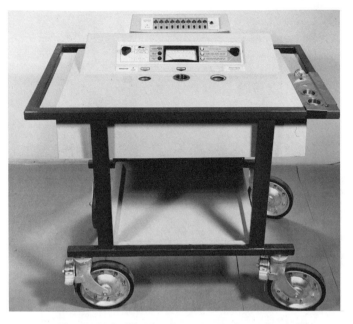

Figure 14.9. Blood volume measuring system. (Courtesy of Ames Company, Division of Miles Laboratories, Inc., Elkhart, Ind.)

The principle of the collimated scintillation detector can also be used to visualize the spatial distribution of radioisotopes in a body organ. In a *radioisotope scanner* (Figure 14.11), the detector is slowly moved over the area to be examined in a zigzag fashion. Attached to the mounting arm of the detector is a recording mechanism that essentially produces a plot of the distribution of the radioactivity. In the simplest case, this recorder is a solenoid-operated printing mechanism that is connected to the output of a binary divider and that produces a dot after a certain number of detector pulses have occurred. The density of dots along a scanning line reflects the amount of radioactivity, and when observed from a distance, the completed scan resembles a half-tone picture. Interesting medical details are often manifested in rather small differences of the activity, which are not readily visible in this simple kind of scan presentation.

Such variations can be made more easily visible by use of contrast enhancement methods, which usually employ a photographic recorder. In this recorder, a flashing light leaves a dot on an X-ray film. While the light source is triggered from the output of a digital divider, its intensity is also modulated by a rate meter circuit and, therefore, also depends on the radioactivity. The rate meter signal is manipulated by amplification and zero suppression so that a small range of variation in radioactivity occupies the

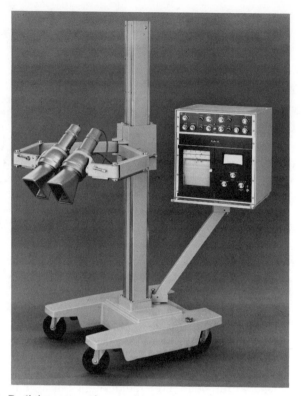

Figure 14.10. Radioisotope equipment for measuring kidney function. (Courtesy of Nuclear-Chicago, a subsidiary of G.D. Searle & Co., Des Plaines, Ill.)

entire available density range of the X-ray film. A similar contrast enhancement can be achieved with the mechanical dot printer by an attachment. This device moves a multicolored ink ribbon under the printer head in accordance with the output from a rate meter and thus reflects small changes in radioactivity by changes of the color of the dots. A basic problem with radioisotope scanners is that the detector must travel very slowly in order to give a high enough count rate for detecting small variations in activity. Therefore the scan of a larger organ can take a long time. One scanner reduces this time by employing ten detectors simultaneously. Another method is used in the *radioisotope camera* shown in Figure 14.12, which has one large, stationary scintillation crystal. The position of a light flash in this crystal is determined by means of a resistor matrix from the output signals of an array of several photomultiplier tubes mounted in contact with the rear surface of the crystal. The detection of a nuclear event at a certain point in the crystal causes a light flash at the corresponding location on the screen of a cathode-ray tube, which is photographed with

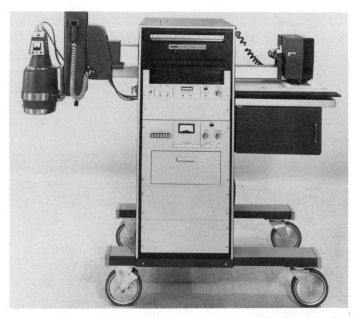

Figure 14.11. Radioisotope scanner (Magnascanner® 500/D). (Courtesy of Picker Corporation, North Haven, Conn.)

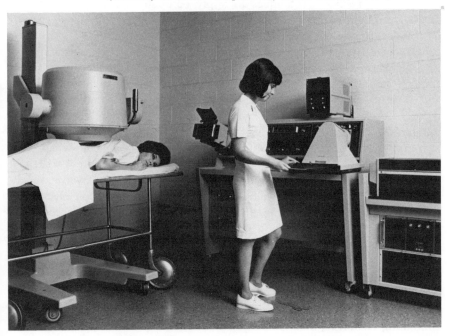

Figure 14.12. Radioisotope camera. (Courtesy of Nuclear-Chicago, a subsidiary of Searle & Co., Des Plaines, Ill.)

a Polaroid camera (left of technician) or with a special camera that uses X-ray film. Different types of collimators are used in this camera, depending on the geometry of the organ to be examined.

Scans of the thyroid gland can be obtained fairly easily with iodine 131. They show cysts as areas of reduced activity and possible malignant tumors as "hot nodules" with increased activity compared to the rest of the gland. Other organs are less-easily visualized and require the use of contrast enhancement in the scanner and the administration of large doses of short-lived radioisotopes. The logistics of obtaining such isotopes can be simplified by the use of technetium 99 m, which, although it has a half–life of only 6 hours, is the decay product of molybdenum 99, which has a half–life of 66 hours. The molybdenum 99 is contained in a special device aptly called a "cow" because the technetium 99m is "milked" from it by *eluting* it—letting a buffer solution trickle through the device. These short-lived radioisotopes do not occur naturally in the body, and, unlike iodine, the organs of the body do not have a natural selectivity to these elements. Physical effects, such as variations in blood flow, account for differences in the isotope distribution that outline the organs. The organs that can be visualized include the lungs, brain, and liver.

Despite the substantial technical effort involved in obtaining X-ray pictures or radioisotope scans, a very experienced physician is required to interpret the results. Techniques to apply computer image processing to this field are still in their infancy.

Hydrogen and carbon, the two elements that constitute the largest percentage of all organic substances, have useful radioisotopes that are only beta emitters. With these radioisotopes, many natural and synthetic substances, including chemicals, nutrients, and drugs, can be made radioactive and their pathways in the organism can be traced. The radioactivity of these isotopes, however, can be measured only in vitro, and special detectors have to be used. For older methods, the sample is placed in a *planchet,* a round, flat dish made of aluminum or stainless steel, in which the solvent is evaporated. In a *planchet counter* [shown in Figure 14.13(a)] the planchet becomes part of a Geiger-Muller tube. The thin layer in which the sample is spread and its close contact with the collection electrodes result in a fairly high counting efficiency for beta radiation. The counting cell is continuously purged by a flow of gas that removes ionization products.

For the soft beta radiation from tritium, a radioactive isotope of hydrogen, however, the sensitivity of the planchet counter is marginal and *liquid scintillation counters* are now widely used instead. In these devices the sample is placed in a small counting vial where it is mixed with a solvent containing chemicals that scintillate when struck by beta rays. The vial is

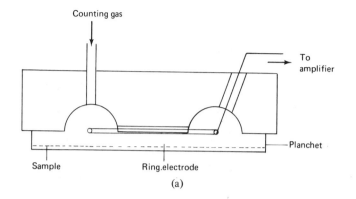

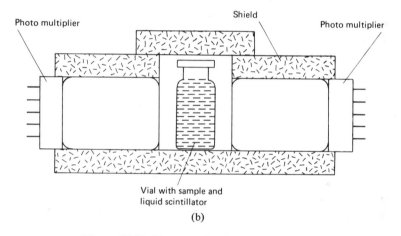

Figure 14.13. Detector for beta radiation: (a) Planchet or gas flow counter; (b) Liquid scintillation counter.

then placed in a detector [Figure 14.13(b)], in which it is positioned be-tween two photomultiplier tubes. The light signal picked up is very weak, and erroneous counts from tube noise must be reduced by a coincidence circuit, which passes only pulses that occur at the outputs of both tubes simultaneously. The remainder of the circuit is similar to the gamma measurement system shown in Figure 14.7. The low activities often en-countered in measurements of this type sometimes require very long count-ing times. This situation has lead to the development of systems that automatically change the samples and print out the results. Figure 14.14 shows one such system; it also subtracts background counts and corrects for variations in the counting efficiency.

Figure 14.14. Liquid scintillation spectrometer (Beckman LS 230). (Courtesy of Beckman Instruments, Inc., Fullerton, Calif.)

· 15 ·

THE COMPUTER
IN BIOMEDICAL
INSTRUMENTATION

The development of the digital computer and the application of computer techniques to medicine and allied fields have ushered in a new era of progress in medical research and health care. In fact, it may be shown someday that no other single event will have had the impact on the practice of medicine that the computer has created. Coupled with some of the more recent developments in biomedical instrumentation described in previous chapters, the advent of the computer has placed a powerful new tool in the hands of the physician and medical researcher, one by which they can greatly expand their capabilities. Although much of the software and some of the hardware are still in the development stage, the computer is already credited with a number of worthwhile accomplishments.

Although not usually considered a part of biomedical instrumentation per se, the digital computer plays an ever-increasing role in the overall man-instrument system. Thus an understanding of some of the basic concepts of digital computation is considered essential for anyone involved in the field of biomedical instrumentation. Furthermore, some familiarity with the more important applications of computers in clinical medicine and research should provide the biomedical engineer with an insight as to the ultimate utilization of certain of the measurements obtained from various biomedical instrumentation systems.

At the beginning of the computer era, analog computation methods were employed for many of the biomedical applications now performed by digital computers. Since most physiological data originate in the instrumentation system in analog form, the analog computer was well adapted to processing such data, particularly for certain types of applications. However, because of the availability of smaller, faster, and less-expensive digital computers for on-line use and the development of more-sophisticated software, the digital computer has been able to take over most of the functions of the analog computer, even those functions for which the latter was best suited intrinsically. In fact, except for certain simulation applications, general purpose analog computers are seldom used in the processing of biomedical data. Their components, however, enjoy widespread use in biomedical instrumentation systems. For this reason, some knowledge of analog computation and analog computer components is essential as part of the background of a biomedical instrumentation engineer or technician.

The major role of the biomedical instrumentation engineer or technician in connection with the application of digital computers to biomedical instrumentation systems does not generally involve actual design of the computer hardware or the programming of the computer. Rather, the engineer or technician must be primarily concerned with the interface of the computer with the instrumentation from which it is to receive data. Thus a knowledge of interfacing techniques and problems is necessary to round out this special part of the biomedical instrumentation field.

The objective of this chapter, then, is to bring to the reader a brief presentation of the basic concepts of digital computation, an overall view of some of the more important applications of computers in medicine, a description of analog computer components as part of the instrumentation system, and a discussion of interfacing techniques, including methods of analog-to-digital and digital-to-analog conversion.

15.1. THE DIGITAL COMPUTER

Although a wide variety of digital computers can be found in biomedical applications, ranging from the smallest of minicomputers to a large multimillion-dollar computer complex, they all contain four basic elements: an *arithmetic unit* to perform the mathematical and decision-making functions, a *memory* to store data and instructions, one or more *input-output* (I/O) devices to permit communication between the computer and the outside world, and a *control unit* to control the operation of the computer. A block diagram showing the relationship of these elements is presented in Figure 15.1.

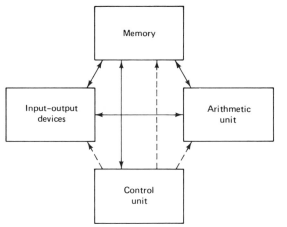

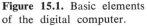

Figure 15.1. Basic elements of the digital computer.

Under the direction of the control unit, data from the instrumentation system or from some other source enter the computer via one of the input devices. The data may be transferred directly to memory, where it is stored until needed, or to the arithmetic unit for some initial processing. After processing, results are either stored in the memory for future recall, or they may be presented to the outside world via one or more of the output devices.

As its name implies, the digital computer accepts, manipulates, and presents data in digital form. Although various digital codes are used, each has as its base the binary system in which all values are represented by a set of 1s and 0s. Bistable elements are used in the computer to represent these values. Each binary digit is known as a *bit,* and the number of bits that are normally stored or manipulated together in the computer constitute the computer *word.* Thus a computer may be designated as a 12-bit, 16-bit, or 24-bit computer, depending on the word length. Some computers work with a variable-length word, dividing each value into eight-bit segments called *bytes.* In these machines, the word length may be any number of bytes up to some limit. Also, in some machines, both alphabetic characters and decimal numbers can be coded into six- or eight-bit groups called *characters.*

A set of bistable circuits sufficient to accommodate one computer word is called a *register.* In any computer, the arithmetic unit contains several registers. In most cases, a register has the capability of shifting data within itself on command. The arithmetic unit also contains circuitry to perform binary addition and substraction. In more sophisticated computers, hardware multiplication and division are also provided. Most arithmetic units are also capable of performing logic AND and OR functions, plus special functions that may be wired into a given type of computer. By utilizing a larger number of "hard-wired" functions and more sophisticated circuits to perform these functions, larger computers are able to carry out their mathematical and decision–making operations in less time than smaller computers.

The memory unit of the computer is used to store numbers or other information with which the computer will be required to operate at some later time. Such information must be readily accessible and retrievable. A computer memory can be visualized as a huge array of "mailboxes," each large enough to contain exactly one word of information and each identified by a number called an *address*. Each memory location (or "mailbox") has a unique address. If a word of data is to be entered into the computer memory, the instruction indicates a specific address to identify the location in memory into which the word is to be placed. Later, when that particular item of data is to be used, the instruction specifies the address from which it is to be obtained. It should be kept in mind that the address of a data word is independent of the value of the data word itself. When a word is read from memory, it is not removed but remains at that storage address until replaced by another word in that precise location.

In most computers, magnetic cores are used as the bistable elements that store data in the memory, although some of the newer computers are able to achieve greater speed by use of integrated circuit memories. The major disadvantage of the integrated circuit memories is that they are *volatile*. That is, if power is turned off, all information stored in the integrated circuit memory is lost. Such is usually not the case with core memories, for magnetic cores generally retain their magnetism even when the power is off.

Since each core or integrated circuit element contains one bit of information, each word of memory requires a number of elements equal to the word length of the computer. Even most of the small minicomputers contain at least 4000 words of this type of memory, which is known as *random access memory*. The large computers have random access memories approaching a million words.

Because random access memory is relatively expensive, it is usually limited in size to meet the immediate data-handling requirements of the applications for which the computer is to be used. To back up this primary memory, most computers have secondary memories, consisting of either a *magnetic disk* or *drum* or a *digital magnetic tape drive*.

Programming a digital computer consists of providing the control unit with a sequence of step-by-step instructions, called the *program*. Each instruction contains a numerical *operation code,* which designates the operation to be performed (like ADD, SUBTRACT, STORE), and an *address,* which gives the location in the memory from which a specific unit of data or information is to be obtained or into which a result is to be placed. Some computers have instructions that contain two or three addresses, so that a single instruction can indicate locations in the memory of two numbers to be processed as well as a location in which the results are to be stored.

The program itself is stored in a portion of the memory and is entered into the memory through one of the input devices. For each step of the computation, the control unit must *"fetch"* an instruction from the memory and *decode* that instruction (which is in numerical form) to determine which specific task is to be performed. It can then recall from memory the appropriate data and perform the specified operation.

The entire set of operations that a given computer is capable of performing is called its *repertoire* or *instruction set.* Even small digital computers have a large number of basic operations in their repertoires, including addition, subtraction, multiplication, and division, the logic functions of AND and OR, decision-making functions, reading from memory, storing in memory, and inputting or outputting information through the I/O (input-output) devices.

Regardless of the length of the computer word or the method used in representing data, most modern computers can be programmed by using symbolic terms, such as English words and algebraic formulas. A special program automatically translates instructions of this type into operation codes to which the computer is able to respond directly. In many cases, a single symbolic instruction generates a sequence of operation codes. The set of symbolic instructions and the rules for formatting these instructions constitute a *programming language.* One widely used language is FORTRAN, (abbreviated from FORmula TRANslation) which allows algebraic-like equations and symbols to be used in the program.

The input-output (I/O) devices of a computer (sometimes called *peripheral devices*) determine its versatility. Some of the more commonly used I/O systems include equipment to read and punch cards and/or paper tape, record and play back digital magnetic tapes, accept direct typewriter input, and provide a typewriter or line printer output. In its application with biomedical instrumentation, the I/O equipment might also include an analog-to-digital converter to convert data from analog form into the digital (usually binary) form required for computer input, a digital-to-analog converter to provide an analog representation of the output for display or control purposes, or a cathode-ray tube display. Analog-to-digital and digital-to-analog conversion, plus other aspects of interfacing the digital computer with biomedical instrumentation, are discussed in detail in later sections of this chapter.

In order to receive data from a data source, such as a biomedical instrumentation system, solve problems, and carry out calculations, the functions of the computer memory, arithmetic unit, and the various input and output devices must be coordinated. Moreover, instructions must be interpreted and circuitry throughout the computer must be appropriately interconnected for each step of the process. These and other control functions are performed by the control unit, which consists of several registers and counters for interpreting instructions, keeping track of the operation being performed, and generating control signals to be fed to the various parts of the computer.

Although the above-described components are common to all computers, their implementation can assume a wide variety of forms, ranging from the large-scale computer shown in Figure 15.2 to the minicomputer

Figure 15.2. Digital computer installation. (IBM System 360, Model 85.) (Courtesy of IBM Corporation.)

shown in Figure 15.3. The large-scale computer, often costing millions of dollars, is designed to process large amounts of data at high speeds, usually for a sizable number of users, either in a batch processing or time-sharing mode. *Batch processing* is a term used to define a method of operation in which all data for a given problem must be entered into the computer before processing begins. Once the data have been entered, the entire computational resources of the computer are devoted to that problem. When avail-

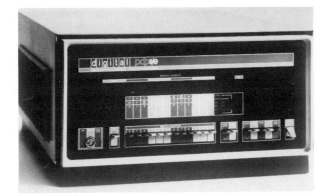

Figure 15.3. Minicomputer. (Courtesy of Digital Equipment Corporation, Maynard, Mass.)

able, the results are printed out or otherwise presented to the user, and the computer begins work on the next problem. In large computers, such as that shown in Figure 15.2, the results from the previous problem may be printed out while the current problem is being processed. At the same time, data for the next problem may be entered into the computer.

In contrast to batch processing, *time sharing* is a method whereby a number of users at various locations can be serviced simultaneously by the same computer. Each of the users submits data and receives results from the computer via a terminal connected to the computer. Although it may appear that the computer is able to solve a large number of problems and process many sets of data simultaneously, in practice such is not the case. In time sharing, the computer sequentially attends to each user, allotting a certain fraction of its time to each. The division of time depends on the nature of the problem being solved and a predetermined priority schedule. Provided the number of users is not excessive, responses to all users are essentially as rapid as if each user were alone on the computer.

The user's *terminal* is his interface with the computer. It can range from a simple typewriter unit to a very elaborate data acquisition system, perhaps including an analog-to-digital converter and possibly some form of graphics display. A terminal is said to be *local* if it is at the same location as the computer and directly wired to it. A *remote* terminal is a terminal situated at a location away from the computer, so that a direct cable connection cannot be made. Communication between the computer and a remote terminal is generally by telephone line. The modulator-demodulator device by which data are encoded and decoded for telephone-line communication between a computer and a terminal is called a *modem* (MODulator—DEModulator). The modem may either be a type that electrically connects to a leased telephone line or it may be equipped with an acoustical coupler in which the receiver of any telephone may be placed. Selection of a modem and method of transmission depend on the required rate of transmitting data and the percentage of time the communication system must be in operation.

In contrast to the large-scale computers described above, a *minicomputer* is a small, general-purpose computer designed to minimize cost but at the same time provide adequate computational capability to serve the needs of a given laboratory or hospital unit. When used for a particular type of problem, the minicomputer is *dedicated* to that task. In order for the minicomputer to work on a new problem, a new program normally has to be entered, and sometimes the computer must be physically disconnected from its previous source of data and connected to the new source. Because of its relatively low cost, the minicomputer often becomes a part of the instrumentation system itself.

Generally these small computers are used *on-line*. This means that the source of data is connected directly to the input of the computer and that the computer accepts datum values as they are generated. An *off-line* data source is one that is not directly connected to the computer. Data from an off-line source are usually recorded in some manner for later computer entry. Although minicomputers are normally operated on-line, on-line operation is also possible from a remote terminal of a large time-shared computer system, provided that the terminal is able to transmit data directly from the instrumentation to the computer.

A computer is said to operate in *real time* if the results of the analysis or processing can be made available in time to be used effectively as feedback to the data source. This implies that the natural time scale of the data source is matched by the computer and that computation can occur rapidly enough to permit the computer to "keep up" with the data rate. An on-line system does not necessarily imply real-time operation.

The capability and flexibility of a computer in any application depend on the size and configuration of the computer itself, the assortment of input and output devices with which it is equipped, and the programming support available. The actual cost of the computer installation must include the cost of all three. As indicated earlier, basic computers used in biomedical systems range from minicomputers to large-scale installations. In some applications, remote terminals tied in with a large commercial computer installation are used. The size and complexity of input and output equipment are equally varied. Programming or *software* support may be obtained from the computer manufacturer (often at an added cost), or the user may choose to hire the necessary programmers.

Although most general-purpose digital computers can be used for the acquisition and processing of physiological data, provided that the necessary input and output devices are attached, some manufacturers offer computer systems designed especially for this purpose. Such systems include many of the necessary peripheral devices and are often furnished complete with a package of "canned" programs so that the computer can be used directly for many applications. An example is shown in Figure 15.4. This

Figure 15.4. Digital computer system for processing physiological data (PDP-12). (Courtesy of Digital Equipment Corporation, Maynard, Mass.)

is a 12-bit computer with a standard 4096-word core memory. As an option, the core memory size can be increased to as many as 32,768 words. The core memory is supplemented by special addressable tape drives that can be used for either program or data storage or both. If more rapid secondary memory is required, a disk can be added as an optional feature. The system also includes a built-in analog-to-digital converter for handling physiological data in analog form, a cathode-ray oscilloscope display, and a conventional input-output typewriter with paper tape reader and punch.

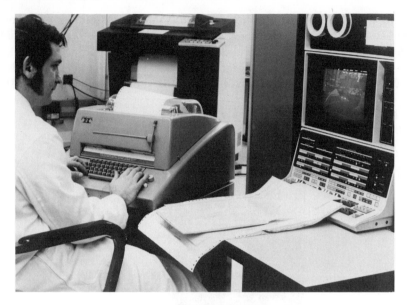

Figure 15.5. Computer system for the automated clinical labora-tory. (Courtesy of Digital Equipment Corporation, Maynard, Mass.)

Many other peripheral items can be added as optional equipment if needed for a given application. Special program packages, including several for biomedical appications, are available. Figure 15.5 shows the same com-puter in a configuration for processing data from automated clinical chemis-try equipment, such as that described in Chapter 13. When ordered for a special function such as this, the computer system comes as a complete package, including all necessary hardware and software, to permit the user to operate the system without any necessary programming capability. This type of system is often called a *turnkey* system, denoting the idea that all the user must do is turn on the key and the computer is set up to function as part of his instrumentation system.

15.2. DATA OUTPUT FROM BIOMEDICAL INSTRUMENTATION

Before discussing the problem of interfacing a digital com-puter with a biomedical instrumentation system, it is necessary to review the kinds of data that might be generated by an instrumentation system. By far the majority of biomedical instruments provide data in analog form —that is, a voltage proportional to the variable measured. There are, however, some exceptions in which digital data are generated directly. These exceptions include the outputs of digital counters, instruments in which the timing of switch closures constitute the output data (such as bar presses in an animal behavior experiment), or instruments with digital readouts. Data are also sometimes available as a digital code. Closely akin to these devices are analog instruments that also have a digital output, either in the form of a digital code or as a punched card, paper tape, or digital magnetic tape output. The major problem in accepting digital data from an instrumentation system is in the matching of codes and data formats.

Analog output data from biomedical instrumentation may originate in various forms. If, however, the analog signals are brought to a standard amplitude range and output impedance through proper amplification, the only pertinent difference among the signals is the frequency content of the data. The usual frequency ranges of several types of physiological data are discussed in earlier chapters of this textbook and summarized in Appendix B. Although there is sometimes considerable disagreement in the field as to what actually constitutes the frequency range required for a given measurement, the values in Appendix B represent ranges that should satisfy most requirements. It should be recognized that, for clinical purposes, a more restricted range is often suitable, whereas, for certain research requirements, the range must be extended to a wider bandwidth.

In addition to the physiological data described in Appendix B, analog output data from analytical laboratory equipment such as chemistry systems, gas chromatographs, spectrophotometers, and so on, often require computer analysis. The output of these instruments usually consists of very low frequency curves or values that can be read out at low data rates.

15.3. ANALOG COMPUTATION

As stated in the introduction to this chapter, analog computers at one time were used in preference to digital computers in many biomedical applications. The primary reasons were the relatively lower cost of the analog computer for certain types of problems and the fact that most physiological data originate in analog form. However, because of the availability of lower-cost, faster, and more-compact digital computers that can perform similar functions, the trend today is away from the analog computer in favor of the digital hardware. Furthermore, the greater availability of suitable software for processing and analysis of physiological data has helped reduce the programming problem, one of the major obstacles to the use of the digital computers in the laboratory or clinic. Nonetheless, general-purpose analog computers, such as the one shown in Figure 15.6,

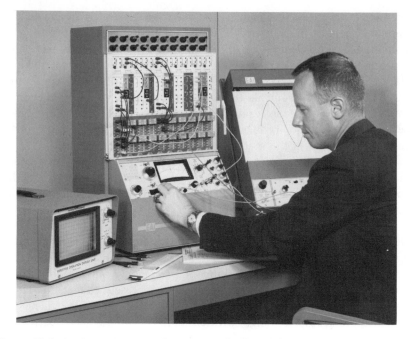

Figure 15.6. Analog computer. (Courtesy of Electronics Associates, Inc.)

are still used occasionally in some simulation applications and for other special purposes.

The primary difference between the analog computer and the digital computer is in the method by which data are represented. Where the digital computer represents variables in numerical form, the analog computer represents dependent variables by voltages that vary with respect to time (the independent variable). Programming of the analog computer consists of interconnecting circuit components that add, invert, integrate, multiply, or perform other mathematical operations on the voltages representing the variables. Programming is accomplished on a plugboard on the front panel of the computer, as shown in Figure 15.6. Outputs of the analog computer (also shown in the figure) can include a meter reading, an oscilloscope display, or a graphic plot. The analog variables can also be recorded on an analog magnetic tape recorder for future use.

The analog computer deals with voltages and time, both of which are continuous. Thus it is able to represent continuous values of all variables. Because all mathematical operations are performed by individual circuits interconnected for a given problem, all operations are performed simultaneously and in "real time." The number of analog computer components required for any given problem depends on the complexity of the problem. The accuracy of computation depends on the accuracy of the individual amplifiers and other components, which, in turn, depend on the quality of the components. The limit is usually about 0.1 percent. Unlike the digital computer, the analog computer cannot store large quantities of information, nor can it make decisions or alter its program in the event of certain predetermined results. Another limitation of the analog computer is its inability to deal with alphabetic information.

Even though the general-purpose analog computer is not as widely used as the digital computer, analog computation circuits are well adapted to a variety of biomedical instrumentation applications. Some of the more important of these components are described in the following paragraphs.

The active element of most analog computational devices is a stable dc amplifier which intrinsically has high gain. Most of this gain is sacrificed, however, to obtain greater stability through the use of negative feedback. This type of amplifier is called an *operational amplifier*. The negative feedback is provided by means of a feedback resistor, R_f, connected between the amplifier output and the inverting input as shown in Figure 15.7. Thus a portion of the output signal is subtracted from the input

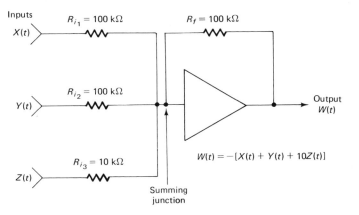

Figure 15.7. Operational amplifier as analog adder.

signal from which it is derived. The subtraction takes place at the input terminal of the amplifier, which is often called the *summing junction*.

The input signal reaches the summing junction through an input resistor R_i. With sufficient negative feedback, the gain of the circuit, A, is practically independent of the intrinsic gain of the amplifier, and is equal to the ratio of the feedback resistor to the input resistor. Thus:

$$A = \frac{R_f}{R_i}$$

When two or more inputs are simultaneously fed to the summing junction, each through its own input resistor, the various input signals are added and the output of the amplifier is the sum of the inputs, each input multiplied by the ratio of the feedback resistor to its respective input resistor. If all input resistors were equal in value to the feedback resistor, the output would become the direct mathematical sum of the inputs. Such a circuit, shown in Figure 15.7, is called a *summing amplifier* or analog *adder*. Because of the inverting characteristic of the operational amplifier when used in this fashion, the output is always a negative representation or inversion of the sum of the inputs.

By substituting a capacitor for the feedback resistor in an analog adder circuit as shown in Figure 15.8, the circuit can be changed into an *integra-*

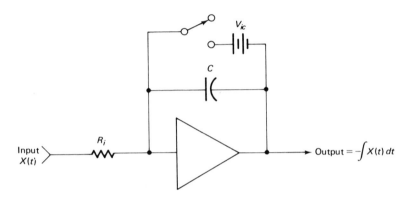

Figure 15.8. Operational amplifier as analog integrator.

tor which produces at its output the integral of the sum of the inputs. The value of the capacitor determines the integration rate. At the beginning of the integration interval the capacitor is charged to a voltage representing the initial conditions of the problem. If the initial condition is zero, the capacitor is discharged. The value of the integral is represented by the output voltage of the amplifier at the end of the interval of integration.

The product of two analog signals can be obtained by use of an analog *multiplier.* This is not an operational amplifier circuit, but is a special device that combines the effects of two currents by multiplication. Such a device may utilize the *Hall effect,* by which a potential is developed across a piece of semiconductor material proportional to the current through the material and the magnetic field to which it is exposed. If the current through the material is proportional to one input signal and the magnetic field can be made to vary with the second, the output voltage is proportional to the product of the two input currents. Other modern multipliers utilize the nonlinear characteristics of transistors to achieve multiplication.

Special analog computers have been developed for specific biomedical determinations. The analog cardiac output computers shown in Figures

6.30 and 6.31 and the pulmonary function indicator shown in Figures 8.6 and 8.7 of Chapter 8 are examples of such devices. Both types of instruments incorporate appropriate analog computational circuits of the types described above.

15.4. COMPUTER INPUT REQUIREMENTS

Except for computers with built-in conversion equipment, digital computers require data in digital form. Furthermore, the data must be entered in a format and at logic levels compatible with the particular computer system involved.

Some computers require data in a purely binary form. Others accept data as binary-coded-decimal digits. Still others accept alphabetic and numeric data combined in a specific coded form.

Most computers enter data sequentially, a word or a character at a time. The bits comprising the word or character are entered via parallel input lines. In some computer inputs, data are accepted at the input lines at a fixed rate, synchronized to the computer master clock (interval timing signals). In other situations, the data enter on an "interrupt" basis, such that the computer interrupts its other functions to accept a word of data whenever it is presented. Since computers usually work at extremely high rates of speed, data can usually be accepted at a very high rate, providing that the input device can operate that rapidly. It is generally more efficient to feed the computer a sequence of words at its maximum input rate rather than require that data be accepted at a slower rate.

In addition to matching the word or character format and the input rate requirements of the computer, it is also necessary that digital input data appear at the correct *logic levels* for the computer. The logic levels are simply the voltage levels that represent the two binary states, 0 and 1. Actually, any two arbitrary levels can be used, provided that the two are easily distinguishable. Generally the logic levels are dictated by the type of logic elements used in the particular computer. In some cases, the input signals must be pulses of precise duration with the rise times * specified by the computer manufacturer. In other situations, the input lines can remain at the levels for one word until the next word appears.

All necessary information for interfacing the computer to other equipment can normally be obtained from the computer manufacturer. In most cases, the analog-to-digital converter or other interface device is designed to match an analog input signal to a given computer. If, as discussed below, the data are recorded in digital form, the playback or reproduction system must be matched to the computer.

* Rise time is the time that a signal takes to rise from one value to another specified value, e.g., from 10 to 90% of full value. This is a common term in electric circuit theory.

15.5. ON-LINE COMPUTER INPUT

Figure 15.9 shows a block diagram of a typical on-line computer system. Both digital and analog input data are shown. The

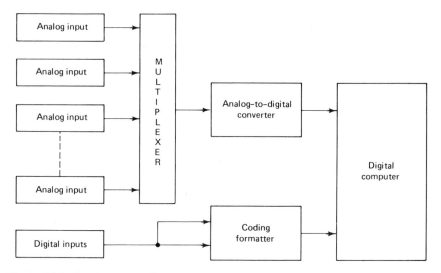

Figure 15.9. Inputs to an on-line computer system.

digital data simply require proper formatting and coding for computer entry, whereas the analog data must first be *digitized* by means of an analog-to-digital (A/D) converter. The subject of analog-to-digital conversion is treated in detail in Section 15.8. Where several analog signals are to be entered into the computer simultaneously through a common analog-to-digital converter, the signals are sampled sequentially by means of a *time multiplexer.* This device is simply a switching mechanism, synchronized to the analog-to-digital converter so that each input signal is converted in turn to digital form for computer entry.

The system shown in Figure 15.9 is characterized as on-line by the direct connection of all data sources to the computer. This connection may be accomplished by telephone lines or even by radio telemetry if the data sources are located some distance from the computer. The transmission can be performed either in analog form ahead of the analog-to-digital converter or in digital form between the A/D converter and the computer. Either way the system is still on-line, provided that data enter the computer as they are generated.

15.6. OFF-LINE COMPUTER INPUT

Where the flow of data from the biomedical instrumentation system to the computer is interrupted by recording or some other form of data storage, the system is said to be *off-line*. The possible forms of off-line systems are numerous. In one example the output of the instruments is manually read and punched into cards for computer entry. Although this system involves a high degree of manual intervention, it is a frequently used off-line system of computer entry with manual analog-to-digital conversion. More efficient off-line systems utilize either analog or digital magnetic tape or punched paper tape. Some typical off-line computer entry systems are shown in Figure 15.10.

The analog inputs to the off-line computer system (outputs from the various instruments) can be

1. recorded on magnetic tape in analog form for later playback into a multiplexer and analog-to-digital converter.
2. sampled by a multiplexer and digitized by an analog-to-digital converter as the data are generated and then recorded in digital form on a digital magnetic tape recorder.
3. when data rates are sufficiently slow, digitized by a low-speed analog-to-digital converter (digital voltmeter) and recorded on punched paper tape.

Data recorded on magnetic tape in analog form can be played back into a multiplexer and an analog-to-digital converter from another analog tape unit, possibly at the computer location. The digitized output from the A/D converter can either be entered directly into the computer or it can be recorded in digital form on a digital magnetic tape recorder. When data are recorded on digital magnetic tape, the digital tape can be played back on a digital tape transport at the computer location for computer entry.

The paper tape generated by a paper tape punch connected to the instrumentation system can be taken to a paper tape reader that is connected to the computer for data entry.

Obviously data generated in digital form can be recorded directly on either punched paper tape or magnetic tape and entered via the appropriate tape reader or transport at the computer location. Before the data can be stored on tape, however, they must be coded into characters of the proper length for the tape system employed. Punched paper tape uses an 8-bit character (seven bits plus a *parity* bit for error checking). Digital magnetic tape must be coded into 7- or 9-bit characters, depending on the type of tape recorder used. The 7-bit code consists of 6 data bits plus

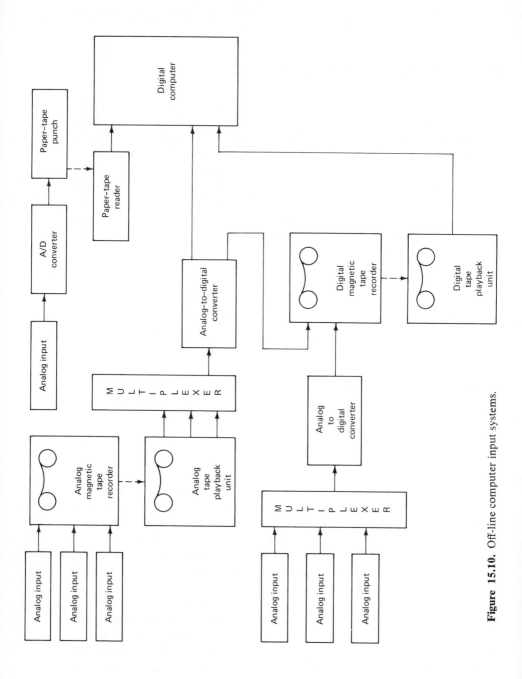

Figure 15.10. Off-line computer input systems.

parity, whereas the 9-bit code consists of 8 data bits plus parity. Both types are standard, but the type used must be compatible with the computer tape transport on which the tape is to be played back. Also, the *packing density,* the number of characters per linear inch of magnetic tape, must be compatible with the playback unit. Most current equipment has a packing density of 800 bits per inch (bpi), although the old standards of 556 and 200 bpi are still in use, and new higher-packing densities up to 1600 bpi are used on some machines.

Special data acquisition systems for off-line collection of data are available commercially. An example is shown in Figure 15.11. This system

Figure 15.11. Data acquisition system. (Courtesy of Datum, Inc., Anaheim, Calif.)

provides analog amplification, analog-to-digital conversion, formatting for digital magnetic tape recording, and an incremental digital tape recorder. These features are described in later sections of this chapter. Systems of this type are usually tailored to the specific application, although they utilize relatively standard components.

15.7. DIGITAL-TO-ANALOG CONVERTERS

In order to obtain a continuous analog signal from a sequence of values in digital form, two processes are required. The first

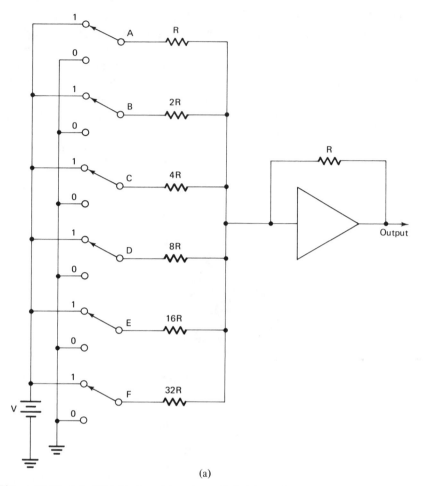

(a)

Figure 15.12. (a) Weighted resistor type digital-to-analog convertor. (All switches shown in Binary "1" position.) (b) Binary ladder type digital-to-analog converter. (All switches shown in Binary "1" position.)

is the generation of a voltage proportional to the value of each digital word as it appears in the sequence. The second is the estimation of values to fill in the intervals between the given digital points. Both of these processes are accomplished by a *digital-to-analog converter.*

Generation of a voltage proportional to a digital word can be accomplished in one of two ways. Both methods are illustrated in Figure 15.12. Figure 15.12(a) shows the weighted resistor (summing amplifier) digital-to-analog converter. This is the same circuit as the analog adder described in Section 15.3. The output of this circuit is the sum of the input voltages, but with each input voltage weighted or multiplied by the ratio of the feed-

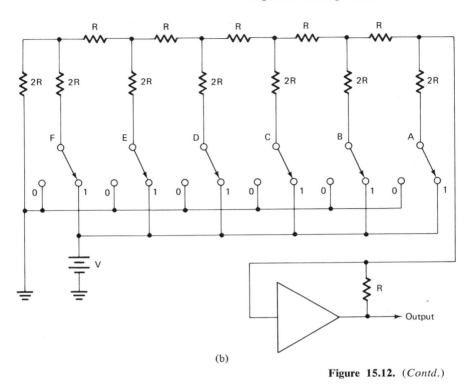

(b)

Figure 15.12. (*Contd.*)

back resistor to the input resistor through which that input voltage is applied. In the circuit shown in Figure 15.12(a), each bit of a 6-bit binary word controls the switch to one input. If a given bit has a value of 1, its corresponding switch places the appropriate input at a reference voltage V. If that bit has a value of 0, however, the input is set to ground (0 volts). The most significant bit (labelled A in the figure) then, contributes a voltage equal to V to the output of the circuit when that bit is a 1, but when that bit has a value of 0, it contributes nothing. Because the input resistor for bit B has twice the value of that for bit A, a 1 in bit B contributes exactly half the voltage of V to the output. Similarly, bit C contributes one fourth the voltage of V, and so on down to the least significant bit, F, which, when given a value of 1, contributes only $\frac{1}{32} V$. These contributions correspond exactly to the relative values of the bits in the binary word. Thus, the output of the operational amplifier is proportional to the sum of the value of all bits which have the value of 1, and consequently is proportional to the value represented by the digital word. For a binary word of greater length (a greater number of bits), an additional input resistor and switch are required for each additional bit. For an n-bit word, the input resistor for the least significant bit would have a value of $2^{n-1}R$.

Figure 15.12(b) shows a so-called *binary ladder circuit*. The output of the ladder circuit is connected to the input of an operational amplifier. As in the case of the analog adder, the ladder has an input corresponding to each bit of the binary word. Again each input has a switch controlled by the value of its corresponding bit. As before, when a bit has a value of 1, its input is switched to ground. The ladder network is so arranged that each input switched to voltage V contributes a voltage to the input of the amplifier proportional to the value of the corresponding binary bit, while the output voltage of the circuit is proportional to the sum of all bits with a value of 1. All resistors are either of value R or $2R$. The accuracy of this circuit is not dependent upon the absolute value of resistors, but rather upon their relative values. Also, the ladder is so arranged that, regardless of the combination of switch positions, the input impedance seen by the amplifier is constant and equal to R. In the circuit shown in Figure 15.12(b), switch A is controlled by the most significant bit and switch F is controlled by the least significant bit. To accommodate digital words of greater length, the network can be extended to provide an input for each additional bit which contributes the correct voltage for that bit.

In both types of digital-to-analog converters, the switching is usually done by solid-state switching circuits. Although many circuit configurations of this type are in use, they all essentially accomplish the same purpose of providing the reference voltage with a digital input of 1 and ground with an input of 0.

There are also several ways of estimating the value of the analog signal at the output of the converter between the occurrence of digital data points. The simplest, called *zero order hold,* assumes that the signal remains constant at the level of each digital value until the next one occurs. Then it jumps immediately to the level of the new value, where it again remains until another value is received. Unless abrupt changes in the data can be expected which could result in excessive error, this method is usually used. More complex (and more expensive) methods are also available, such as the *first order hold,* in which the signal at any time is caused to change at the same rate as it did between the two previous digital data points.

15.8. ANALOG-TO-DIGITAL CONVERSION

An analog-to-digital converter is a device that accepts a continuous analog voltage signal as input and from that signal generates a sequence of digital words that represent the analog voltage as it varies with time. There are actually two processes involved in the *digitizing* of analog data. The first is *sampling*—the process of measuring the analog voltage at discrete points in time. The sampled voltage must then be *quantized.* Quantizing is the selection of a digital word of specified length to represent the analog voltage.

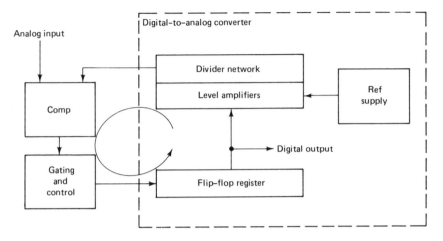

Figure 15.13. Analog-to-digital converter incorporating digital-to-analog converter. (Copyright 1964, Digital Equipment Corporation, Maynard, Mass. All rights reserved.)

The heart of an analog-to-digital converter is a digital-to-analog converter of the type described in Section 15.7. This fact is shown in the block diagram of an analog-to-digital converter in Figure 15.13. In this figure, the divider network (binary ladder) and reference supply constitute a circuit similar to that shown in Figure 15.12(b). The flip-flop register is a set of bistable (flip-flop) circuits, each of which can represent a value of binary 0 or 1 and can thus store one bit of a binary digital word. The entire set of flip-flops that constitutes the register represents each digital word to be generated by the converter. Through the level amplifiers, each flip-flop controls a corresponding input to the ladder network and together they produce an analog output with the same voltage as represented by the flip-flop register. At the time of sampling, this voltage is compared with the analog input voltage in an analog comparator circuit. When these two voltages differ, the bits in the flip-flop register are adjusted through appropriate gating and control circuitry until agreement is reached. At that time, the value represented by the flip-flop register is the nearest digital equivalent to the analog input voltage and is caused to appear at the output of the converter.

Although nearly all analog-to-digital converters use this comparison method of matching the value of the register with the input voltage, the methods by which the digital value of the register is adjusted to match the input signal can differ widely. The simplest method is to reset the register to zero at the beginning of each conversion and simply have it count up until agreement is achieved. A converter of this type is called a *counter converter*. Its conversion steps are illustrated in Figure 15.14. Even with a rapid counting rate, the average time for a conversion is quite long. For

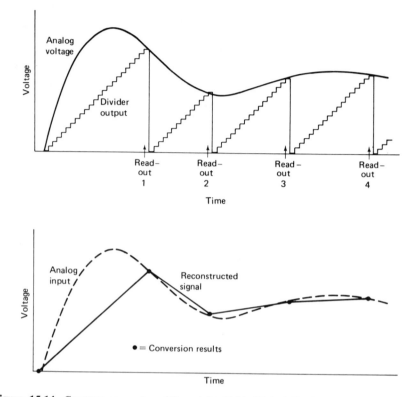

Figure 15.14. Counter converter. (Copyright 1964, Digital Equipment Corporation, Maynard, Mass. All rights reserved.)

example, in order to count up the full value of an 11-bit-plus-sign word, 2047 counts would be required for each conversion. The average number of counts would be about half of that. Thus, with a 1-MHz counting rate, the average conversion time would be greater than 1 millisecond. This assumes that the sign is detected independently. Counter converters are often used in digital voltmeters where rapid digitizing rates are not required.

A much faster converter for signals that do not change their levels abruptly is the *continuous converter* which has its conversion steps shown in Figure 15.15. Again a counter is used, but this counter can count either up or down and is not reset to zero for each reading. The initial reading after a rapid change in the signal takes as long as the counter converter, but afterward the converter simply follows the variations in the signal by counting up or down as the level changes. Obviously this type of converter could not be used with a multiplexer that sequentially samples values from two or more analog inputs.

Conversion steps of the most common type of analog-to-digital

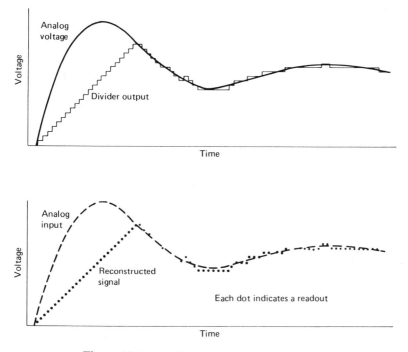

Figure 15.15. Continuous converter. (Copyright 1964, Digital Equipment Corporation, Maynard, Mass. All rights reserved.)

converter are shown in Figure 15.16. This device is called a *successive-approximation* converter. In this conversion method, each bit is successively tested to determine whether its addition to the value of the register would cause the input signal to be exceeded. If not, that particular bit is set to a 1. If the bit would have caused the value of the register to be greater than that of the input signal, then the bit is left at 0. The process begins at the bit representing the largest value (most significant bit) and continues from "left to right" down the register. The advantage of this type of system is that the conversion time is fixed and does not depend on the input signal. Furthermore, this type of converter gives a good response to large, rapid changes in input, such as might be expected with a multiplexer. In order to avoid changes in the input signal during the time the converter is in the process of checking each bit, a sample-and-hold circuit is often used to read the voltage at the beginning of each conversion period and to maintain that voltage during conversion period. The result is a closer approximation of the analog signal, as shown in Figure 15.17.

Important factors in selecting an analog-to-digital converter are the resolution of the quantizing process, the conversion rate, and the conver-

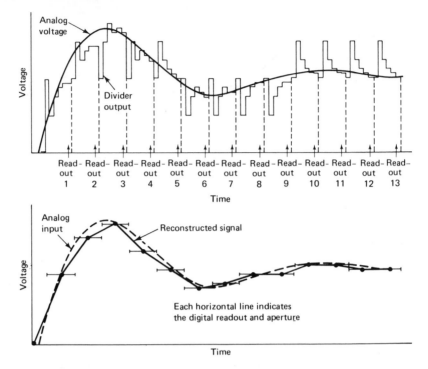

Figure 15.16. Successive approximation converter. (Copyright 1964, Digital Equipment Corporation, Maynard, Mass. All rights reserved.)

sion aperture time. Also to be considered are the logic levels at the output and the method of coding provided.

The quantizing *resolution* of the converter is determined by the number of bits in the output word. An 11-bit-plus-sign word, for example, is capable of dividing the full range of the input signal into 4095 increments of level. This number includes 2047 positive increments and a similar number of negative increments, plus zero. The accuracy of any voltage reading, then, cannot exceed about 1 in 2000, or about 0.05 percent of full scale. Most physiological data do not require that degree of accuracy, however, for many transducers cannot provide accuracies much better than 1.0 percent. But since the cost of one or two additional bits of resolution is relatively low, it usually pays to provide for somewhat greater accuracy than that actually needed.

The *conversion rate* of an analog-to-digital converter depends on the conversion method used and the speed of the control circuitry. Extremely high rates of conversion are available. Shannon's sampling theorem requires that, in order to reproduce a periodic signal without severe distortion, the sampling rate be at least twice the highest frequency component

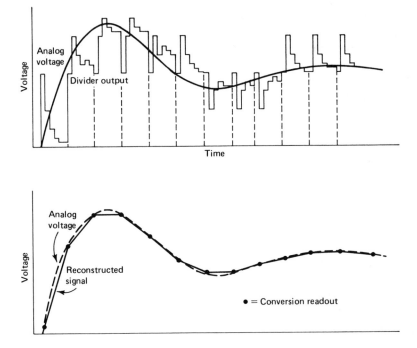

Figure 15.17. Successive approximation converter with sample-and-hold. (Copyright 1964, Digital Equipment Corporation, Maynard, Mass. All rights reserved.)

that the system is able to pass. For nonperiodic waveforms, it is generally good practice to use a sampling rate of at least four or five times the highest frequency component. Obviously the higher the sampling rate, the more accurate will be the representation of the analog signal; but higher digitizing rates mean that more data must be stored and handled by the computer. Usually this results in a greater computation cost.

The *aperture time* is the period of time during which the analog signal is actually being sampled for conversion. A long aperture time might result in a change of data during the sampling interval. Most modern analog-to-digital converters have sufficiently short aperture times for the conversion rates at which they operate.

The digital output of most analog-to-digital converters is a binary word of a length determined by the size of the flip-flop register of the converter. This word length may not be compatible with the length of word required by the computer into which the digital data are to be fed, or the number of bits that can be accommodated at a time by a digital magnetic or punched paper tape recorder. In addition, the *logic levels* (the voltage levels used to represent 0 and 1) and perhaps the data rate of the converter may not be compatible with those of the devices required to receive the data. To handle such problems as these, a *data formatter* is often included with the analog-to-digital converter. The formatter consists of circuitry capable of rearranging binary data and sometimes temporarily storing values so that the data produced by the analog-to-digital converter can be changed to meet the input requirements of the receiving device.

In many situations it has been found that the cost of suitable analog-to-digital conversion equipment, formatting equipment, and other required control hardware approaches the cost of a minicomputer system which incorporates this equipment. There are currently several computer-centered data acquisitions systems on the market. Such systems can often be used directly for a limited amount of data processing as well as data acquisition, or they can be used as input devices for larger computers.

15.9. MULTIPLEXING

The process of sequentially taking readings from two or more analog data channels with a single analog-to-digital converter is called *time multiplexing*. If N data channels are multiplexed into a converter which operates at R conversions per second, and all channels are converted at the same rate, the conversion rate for any given channel is R/N conversions per second. This means that with multiplexing, the conversion rate of the converter must be the required conversion rate for each channel multiplied by the number of channels. If it is important that all of the channels be converted in exact time-correspondence, sample-and-hold circuitry must be incorporated into the multiplexer to "hold" values until they can be digitized.

15.10. SELECTIVE DIGITIZING

One of the major problems in constant-rate conversion of data from analog to digital form is the large amount of data produced. For example, a two-hour measurement of the arterial blood pressure waveform digitized at a rate of 100 conversions per second (a reasonable rate for such data) would produce about 720,000 readings to be entered into the computer. If, however, the measurement were taken as part of an experiment to determine the effects of some drug on the systolic and diastolic blood pressure, only the readings of the systolic and diastolic peaks would be required for the analysis (see Figure 15.18). At a heart rate of 75 beats per minute, about 9,000 systolic and 9,000 diastolic values would occur in the 2-hour period. This means that of the original 720,000 values converted for computer entry, only about 18,000 are actually used in the computer and the other 702,000 readings must be thrown away after the computer has recognized the systolic and diastolic points.

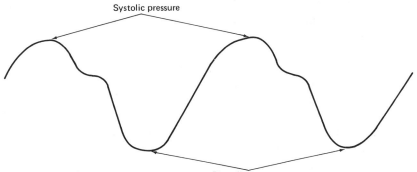

Figure 15.18. Arterial blood pressure waveform.

Where excessive data may prove to be a problem, a method of *selective digitizing* can be used in which analog circuitry is devised to recognize those features of the analog waveform important to the computer analysis and to control the analog-to-digital converter so that only the values of important features are digitized. In order to utilize this concept, the user must be able to specify in advance those features of the analog data that are required for analysis. Further, he must be able to specify certain characteristics of the signal which the instrumentation can recognize so it can identify those features. These characteristics must be of such a nature that some type of electronic circuit can recognize them without ambiguity.

Where these conditions are met, the principle of selective digitizing can be applied to greatly reduce the amount of data entered into the computer thereby improving the efficiency of the computation. This concept is not limited to blood pressure data, but can in some circumstances be applied to EEG, ECG, EMG, plethysmograph data, respiratory signals, or any type of physiological data in which certain features of the signal are actually used in the analysis. In some cases, analog preprocessing may be desirable to partially reduce the data in analog form before it is digitized.

For off-line selective digitizing, an *incremental digital magnetic tape recorder* can be used. This type of tape recorder advances the tape one character at a time, and records on command so that digital characters are recorded as they are received.

15.11. BIOMEDICAL COMPUTER APPLICATIONS

Applications of the digital computer in medicine and related fields are so numerous that even listing all of them is beyond the scope of this textbook. Most of these applications, however, utilize a few basic capabilities of the computer which provide an insight to ways in which computers can be used in conjunction with biomedical instrumentation. These basic capabilities include:

1. *Data acquisition.* The reading of instruments and transcribing of data can be done automatically under control of the computer. This not only results in a substantial saving of time and effort, but also reduces the number of errors in the data. When data are expected at irregular intervals, the computer can continuously scan all input sources and accept data whenever they are actually produced. If the data originate in analog form, the computer usually controls the sampling and digitizing process as well as identification and formatting of the data. In some cases, the computer can be programmed to reject unacceptable readings and provide an indication of possible trouble in the associated instrumentation. Sometimes the computer provides automatic calibration of each input source.

2. *Storage and retrieval.* The ability of the digital computer to store and retrieve large quantities of data is well known. The biomedical field provides ample opportunities to make use of this capability. In a modern hospital, large amounts of data are accumulated from many sources. These include admission and discharge information, physicians' reports, laboratory test results and several other kinds of information associated with each patient. In addition, the hospital also generates a considerable amount of non-patient-oriented data, such as pharmacy records, inventories of all types, and accounting records. Without a computer, the storage of this vast amount of information is both space- and time-consuming. Manual retrieval of the data is tedious, and for some types of information, almost impossible. The digital computer, however, can serve as an automated filing system in which information can be automatically entered as it is generated. These files can be stored as long as necessary and updated whenever appropriate. Any or all of the information can be retrieved on command whenever desired.

3. *Assimilation and organization of information.* In addition to simply storing and retrieving data, the digital computer can be programmed to organize the stored information in various ways. For example, data related to a given patient can originate at various times and from many different sources within the hospital. To generate a patient data file, the computer must assimilate all information from each patient as it arrives and file it with data already in the computer for that patient. The computer can then organize the data and prepare a complete report including all tests for the day. Further, if desired, any given test result can be reported along with similar data from previous days

to permit the physician to observe day-to-day changes. The data can also be manipulated in other ways for special purposes as appropriate. While these types of data manipulation are relatively simple with a computer, they can be extremely cumbersome if performed manually, particularly where large amounts of data are involved.

4. *Data reduction and transformation.* The sequence of numbers resulting from digitizing an analog physiological signal such as the ECG or EEG would be quite useless if retrieved from the computer in raw form. To obtain meaningful information from such data, some form of data reduction or transformation is necessary to represent the data as a set of specific parameters. These parameters can then be analyzed, compared with other parameters, or otherwise manipulated. For example, the electroencephalogram (EEG) signal can be subjected to *Fourier transformation* to obtain a frequency spectrum of the signal. Further analysis can then be performed using the frequency-related parameters rather than the raw EEG data. Various computer programs are available to perform the transformations, which otherwise would be too time-consuming to be practical. The electrocardiogram (ECG) signal can also be subjected to data reduction methods. In one such method, the ECG is represented by the means and standard deviations of a number of important amplitudes and durations. The subject of computer analysis of the ECG is covered in more detail in Section 15.11.1.

5. *Mathematical operations.* Many important physiological variables cannot be measured directly, but rather must be calculated from other variables which are accessible. For example, many of the respiratory parameters described in Chapter 8 can be calculated from the results of a few simple breathing tests and gas concentration measurements. If a digital computer is connected on-line with the measuring instruments, the calculated results can be obtained while the patient is still connected to the instruments. This not only enables the physician to conduct further tests if the results so indicate, but can also inform him immediately if any measurements were not properly made and require repetition.

6. *Pattern recognition.* In order to reduce certain types of physiological data into useful parameters, it is often necessary that important features of the physiological waveform be identified. For example, analysis of the ECG waveform requires that the important amplitudes and intervals of the electrocardiogram be recognized and identified. Digital computer programs are available to search the data representing the ECG signal for certain predetermined characteristics that identify each of the important peaks. In Section 15.11.1, the technique by which this is accomplished is described. Similar methods are used in

other pattern recognition problems, but since each type of pattern has unique features that must be identified, programming for pattern recognition is a highly specialized process.

7. *Limit detection.* In applications involving monitoring and screening, it is often necessary to determine when a measured variable exceeds certain limits. For example, in the analysis of the electrocardiogram, each important parameter of the ECG can be checked to determine whether it falls within a preestablished "normal" range. By comparison of the measured parameter with each limit of the range, the computer can indicate which parameters exceed the limit and the amount by which they deviate from normal. Using this technique, patients can be screened to select those with ECG irregularities that should receive further attention. In most cases, the "normal" range is defined in advance, but sometimes, the computer is programmed to establish normal ranges for each patient based upon the averages of repeated measures taken under specified conditions.

8. *Statistical analysis of data.* In the diagnosis of disease, it is often necessary to select one most likely cause out of a set of possible causes associated with a given set of observed symptoms, measurements, and test results. Similarly, medical research investigators must decide at times whether an observed change or condition in a person or animal is due to some treatment imposed by the researcher, or whether the result could be attributed to some other cause or just to chance alone. Both of these situations require the use of inferential statistical procedures, some of which are quite complex. Fortunately, most statistical methods lend themselves well to computer techniques, especially when large numbers of variables must be analyzed together or where data from a large number of patients are used. Even simple descriptive statistics, such as means, standard deviations, and frequency distributions can be computerized, resulting in significant savings of time and effort.

9. *Data presentation.* An important characteristic of any instrumentation and data processing system is its ability to present the results of measurements and analyses to its users. By virtue of appropriate output devices, a digital computer can provide information in a number of useful forms. Table printouts, graphs, and charts can be produced automatically, with features clearly labelled using both alphabetic and numeric symbols. If the necessary computer peripherals are available, plots and cathode-ray-tube displays can also be generated. In addition to controlling the output devices, the computer can be programmed to organize the data for presentation in the most meaningful form possible, thus providing the user with a clear and accurate report of his results.

10. *Control functions.* Digital computers can be equipped to provide analog or digital output signals that can be used to control other devices. In this case, the computer can be programmed to influence or control the physiological, chemical, or other measurement process from which its input data are being generated. In essence, the computer can be used to provide feedback to the source of its data. For example, while reading and analyzing the results of a chemical process, the computer can be made to control the rate, quantity, or concentration of reagents added to the process, or it could control the heating element of a temperature bath. By controlling these and other possible inputs, the process can be regulated to achieve desired results. In addition, the computer can be programmed to recognize certain characteristics of the measured results that would indicate possible sources of error. Sometimes other parameters are monitored in addition to the actual results to increase the sensitivity of the computer to conditions which may result in erroneous measurements. The computer can be made to automatically compensate for some sources of error, such as a gradual drift in the baseline, by either altering the process itself or by mathematically adjusting the results before printing them out. When more serious types of error occur, the computer can alert the operator to the condition or, if necessary, can automatically stop the process.

The extent to which each of the above described capabilities can actually be utilized in a given situation depends on the available *hardware* (equipment and circuitry in the computer and associated devices) and *software* (computer programs) in the system. Obviously, some of these capabilities require greater resources than others.

Following are some specific examples of computer applications in clinical medicine and research. Although they represent only a few of the many possible ways in which computers can be used in medicine and biology, they serve to illustrate the role of each of the above-described capabilities. In each example, the computer techniques are described in conjunction with their associated biomedical instrumentation.

15.11.1. COMPUTER ANALYSIS OF THE ELECTROCARDIO-GRAM. The greatest progress to date in the use of computers for the clinical analysis of physiological data has occurred in the field of cardiology. There are several reasons for this. First of all, ECG potentials are relatively easy to measure. Secondly, the ECG is an extremely useful indicator for both screening and diagnosis. In addition, certain abnormalities of the ECG are quite well defined and can be readily identified.

Measurement of the electrocardiogram for computer analysis is essentially the same as is used for manual ECG interpretation, except that some presently used computer methods require special lead configurations and sometimes simultaneous measurement of three or more leads. See Chapters 3 and 6 for details of the electrocardiogram and its measurement. Most computerized systems use the 12 standard leads described in Chapter 6. There are some more elaborate systems, however, that simultaneously measure three orthogonal components of the ECG vector. For some of these systems, a special orthogonal lead configuration is used.

Entry of the ECG into a digital computer requires that the analog ECG signals be converted into digital form. Although some attempts have been made to partially reduce the ECG data in analog form by selective digitizing, nearly all presently used systems incorporate an analog-to-digital converter operating at a constant rate. The actual sampling rate depends upon the desired bandwidth of the signal to be analyzed. Sampling rates ranging from 100 readings per second up to 1000 readings per second are in current use. Analog filtering is often used ahead of the converter to eliminate noise and interference above the upper limit of the desired frequency band.

To increase the general usefulness and practicality of computerized ECG analysis, equipment is available for telephone or radiotelemetry transmission of the ECG signal from a remote location, such as a physician's office or even an ambulance, to the computer. The ECG data can be transmitted either in analog or digital form, depending upon the configuration of the equipment. A system for transmitting ECG data in digital form is shown in Figure 15.19. This equipment consists of an electrocardiograph, an analog-to-digital converter, and a telephone interface. The telephone interface of the sending unit is shown in detail in Figure 15.20. At the left is an *acoustical coupler* designed to accommodate the telephone receiver. The digitized ECG modulates an audible tone which is sent to the computer via ordinary telephone equipment. Identification of the patient and other pertinent information is transmitted by means of a digital code generated within the system and transmitted along with the ECG data.

The receiving unit at the computer location, shown in Figure 15.21, receives and demodulates the audible tone to retrieve the digital data for computer entry. In some systems of this type, results can be sent back to a Teletype® unit at the remote location.

Once inside the computer, the ECG signal can be subjected to some additional smoothing by means of digital filtering methods. This smoothing process eliminates high frequency variations in the signal that might otherwise be mistaken for features of the ECG.

Pattern recognition techniques are next employed to identify the various features of the ECG. These features are defined in Chapter 3. The most stable reference point of the ECG pattern, and one of the most reliably identified, is the downward slope between the R and S waves of the

Beckman Telesender System for Computerized ECG

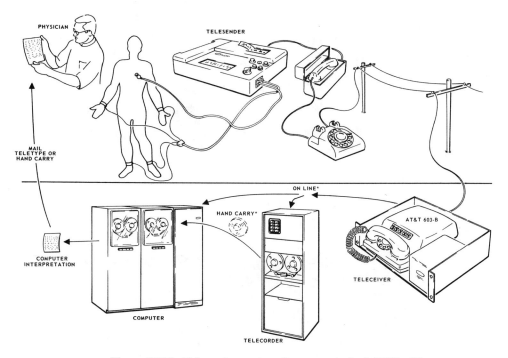

Figure 15.19. Telesender system for computerized ECG. (Courtesy of Beckman Instruments, Inc., Fullerton, Calif.)

QRS complex. This slope can be characterized as the most negative peak that occurs in the first derivative of the ECG waveform. To recognize this point, the ECG signal must be differentiated to obtain a signal representing the first derivative, and the first derivative signal must be scanned to locate its most negative peaks. Other tests are then applied to both the ECG and its derivative to verify that a true RS slope has been located.

From this reference point, the computer scans the ECG data in a backward direction with respect to time to locate the positive peak just preceding the reference. This peak is identified as the R wave. The negative peak of the ECG just subsequent to the reference is the S wave, and the negative peak just ahead of the R wave is the Q wave.

A predetermined interval of the ECG signal prior to the QRS complex is scanned for a positive peak to locate the P wave. Actually, the P wave is often identified on the basis of both the ECG waveform and its first derivative. The T wave is identified as a peak within a predetermined interval of the ECG signal following the QRS complex. In most computer programs, identification of the various waves is based on at least two leads.

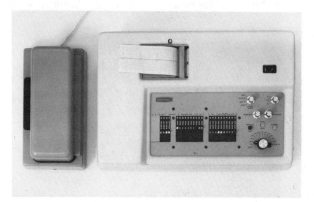

Figure 15.20. Telesender. (Courtesy of Beckman Instruments, Inc., Fullerton, Calif.)

Figure 15.21. Teleceiver. (Courtesy of Beckman Instruments, Inc., Fullerton, Calif.)

The baseline of the ECG waveform is usually defined as a straight line from the onset of the P wave in one ECG cycle to the onset of the P wave in the next cycle. The amplitude of each of the waves (P, Q, R, S, and T) is measured with respect to that baseline. Also, a few points along the S-T segment are measured to determine their deviation from the baseline. Deviations from the baseline of the ECG signal as well as characteristics of the first derivative waveform are used to locate the onset and ending times of all waves. From this information the duration of each wave and the intervals between waves are measured. The duration of the QRS complex, the P-R interval, and the S-T interval are especially significant.

Each of the measured amplitudes, durations, and intervals is a characteristic parameter of the ECG signal. Another important parameter is the heart rate (determined by measuring the time intervals between successive R waves). Each of these parameters can be averaged over several cycles with the means and standard deviations being printed out for each of the leads measured.

For screening purposes, each of the parameters can also be checked to see if it falls within a normal range for that parameter. Any parameters which lie outside the normal range are indicated on the computer-generated report. A report of this type is shown in Table 15.1. This is the result of a test run on a 36-year-old male who was presumably normal, but was found by this screening analysis to have bradycardia (slow heart rate).

Identification and other patient information is printed at the top. The mean values for the various parameters are then presented in a matrix form. The columns represent the 12 standard leads while the rows indicate the parameters. Data from lead V3 were purposely omitted to show the response of the system to missing data. Below this matrix, values for the P-R, QRS, and Q-T intervals and the heart rate for each of the leads are printed out. The heart rate varies from lead to lead because in this system each lead is measured at a different time. Calibration information for each lead and the calculated angle of the axis of the heart (see Chapter 6) for each portion of the ECG cycle are also given. At the bottom of the printout are indications of any noted abnormalities. In the example, the condition of bradycardia (heart rate below 60 beats per minute) is noted as well as the absence of data from one lead.

In more sophisticated systems for computer analysis of the ECG, additional ways of representing the ECG are derived to further aid in distinguishing an abnormal ECG from a normal one. One such representation is a three-dimensional time-variant vector derived from the simultaneous measurement of three orthogonal leads. The behavior of this vector tells much more about the electrical activity of the heart than does the instantaneous calculation of the axis angle for a given portion of the ECG cycle.

Another parameter is the *time integral* of the ECG waveform. To obtain this integral, the areas of each wave above and below the baseline are determined and the sum of the areas below the baseline (negative) is subtracted from the sum of the areas above the baseline (positive). This integral can be determined for any portion of the ECG cycle. The sum of the time integral of the QRS complex and that of the T wave is sometimes called the *ventricular gradient,* and is believed to indicate the difference in the time course of depolarization and repolarization of the ventricles. The time integrals of the three orthogonal leads can be added vectorially to obtain three-dimensional time integrals.

Some systems for computer analysis of the ECG use statistical methods in an attempt to classify ECG patterns as various types of abnormalities or as being normal. Obviously, the more information available about the ECG, the better will be the discriminating ability of the computer programs. Multivariate statistical analysis techniques are sometimes employed,

RUN

TABLE 15.1. ECG COMPUTER ANALYSIS DATA

H456789A   13:54   11/5/70

U.S.P.H.S. CERTIFIED E.C.G. PROGRAM PROCESSED BY THE BECKMAN HEARTLINE
FOR   BECKMAN INSTRUMENTS, INCORPORATED   LOC 10
STAT
PAT 123456789   DATE 11- 5-70   SERIAL 126   OPERATOR 5
36 YR   MALE   5 FT 11 IN   190 LBS
BP   NORMAL   MEDS NONE

	I	II	III	AVR	AVL	AVF	V1	V2	V3	V4	V5	V6	
PA	.08	.13	.00	-.07	.05	.08	.05	.12		.12	.08	.07	PA
PD	.13	.12	.00	.08	.09	.10	.05	.10		.10	.11	.08	PD
Q/SA	-.07	.00	.00	-.91	-.11	.00	.00	.00		.00	.00	.00	Q/SA
Q/SD	.02	.00	.00	.06	.02	.00	.00	.00		.00	.00	.00	Q/SD
RA	.86	.97	.13	.00	.61	.58	.16	.41		1.67	1.72	1.33	RA
RD	.05	.09	.05	.00	.05	.09	.02	.03		.08	.10	.09	RD
SA	-.10	.00	-.21	.00	-.08	.00	-.95	-2.64		.00	.00	.00	SA
SD	.01	.00	.02	.00	.02	.00	.05	.07		.00	.00	.00	SD
RPA	.00	.00	.07	.00	.00	.00	.00	.00		.00	.00	.00	RPA
RPD	.00	.00	.02	.00	.00	.00	.00	.00		.00	.00	.00	RPD
STO	.03	.00	.00	-.03	.03	.02	-.03	.09		.08	.01	.00	STO
STM	.03	-.01	-.02	-.01	.04	.02	.04	.29		.04	.01	.00	STM
STE	.04	.00	-.04	-.02	.06	.00	.03	.38		.08	.04	.02	STE
TA	.28	.27	.07	-.30	.24	.19	-.15	1.15		.61	.43	.34	TA
PR	.16	.18	.00	.21	.15	.19	.17	.19		.19	.18	.18	PR
QRS	.08	.09	.09	.06	.09	.09	.07	.10		.08	.10	.09	QRS
QT	.38	.39	.43	.37	.39	.39	.38	.39		.39	.40	.41	QT
RATE	60	71	61	55	59	58	56	54		62	53	56	RATE
CODE	3	2	2	3	2	2	3	3	A	3	2	2	CODE
CAL	99	99	99	99	99	99	99	99		99	99	99	CAL

	P	QRS	T	Q	R	S	STO		ST-T	QRS-T
AXIS IN										
DEGREES	53	47	28		37	253	23		05	19

MSDL APPROVED VERSION   •
 D 41-42-25-11   •
1131 RATE UNDER 60   • BRADYCARDIA
     1 LEAD NOT MEASURED   •
                               • ATYPICAL ECG
                               • ------------ M.D.

TIME 1 SECS.

both for one-dimensional and three-dimensional data. Because of the wide interpersonal variation even among normals, accurate computer classification is difficult.

15.11.2. THE DIGITAL COMPUTER IN THE CLINICAL CHEMISTRY LABORATORY. The modern clinical laboratory includes various types of automated instruments for the routine analysis of blood, urine and other body fluids and tissues. Some of these devices are described in Chapter 13. While automated equipment can be used for most laboratory tests, there are still many determinations which are performed manually, either because of insufficient volume for certain tests or because satisfactory automated tests have not yet been devised. As a result, data from the clinical laboratory are generated in many forms, many of which require manual transcription of the test results.

In the chemistry laboratory, Autoanalyzers® and other types of automated clinical chemistry equipment produce charts on which the test results are recorded. To produce laboratory reports which eventually become a part of the patients' records, data must be transcribed from these charts and combined with results from manually performed tests. Care must be taken to assure that data are accurately transcribed and that each test result is associated with the correct patient information.

To accommodate the large output of test results from the automated clinical chemistry equipment and to assimilate those data with patient information and the results of manually performed tests, a number of clinical chemistry laboratories have installed computer systems for data acquisition and processing. Computers of various sizes and in various configurations can be used in such systems, depending upon the extent to which the computer participates in the operation of the laboratory. In a highly automated system, the computer accepts test requisitions, prepares lists for blood drawing, schedules the loading of sample trays, reads test results, provides on-line quality control of the process, assimilates data, performs calculations, prepares reports and stores data for possible comparison with future test results.

In a typical computerized system, the medical staff may order tests directly via a remote terminal on the hospital ward or by use of machine-readable requisition forms which are automatically read by computer input equipment in the laboratory. From this requisition information, the computer schedules the drawing of blood by printing out blood drawing lists and preprinted specimen labels. These labels contain identification information to be used for all tests, automated and manual, from a given patient during that day. As the specimens arrive in the laboratory, the computer prepares a loading list which assigns a specific sample position in the analyzer loading tray for each test.

Patient information is entered into the computer either at the time the patient is admitted to the hospital or when the medical staff orders tests. This information is usually entered by keyboard, either from a remote terminal or in the laboratory.

Once a test run is begun, the outputs of all automated instruments are automatically read by the computer. This is usually accomplished by means of retransmitting slide wires attached to the recorder pens which produce an analog voltage proportional to the output of each instrument. These analog voltages are sampled and converted to digital form by means of a time-multiplexer and an analog-to-digital converter. The computer is programmed to recognize legitimate peaks as they arrive and to reject questionable or improperly shaped peaks. The computer also performs the necessary calculations to convert the value of each measured peak into medically useful units. By virtue of its position in the sequence of measured peaks, each test result is identified and associated with the correct patient. Control samples, placed randomly (by computer assignment) throughout the run are used to periodically check the calibration of the system. By monitoring these control samples and the measured values from patient samples, the computer is able to perform "on-line quality control." In some cases, the computer automatically corrects the output values for drift and certain other types of error. In case of severe error the computer provides a warning to the operator, who may then choose to stop the test because of equipment malfunction.

The computer, after assimilating data from all automatically performed tests, may also receive results from manually performed tests. These manual test results may be entered by keyboard or via machine readable data sheets especially prepared for each type of test. Once all test results have been received, credibility checks can be run to search for any impossible or unlikely combinations of results or any impossible changes in a given patient's test results from one day to the next. Once all results have been checked and verified, the computer provides a physician's report, either in printed form or on a cathode-ray-tube terminal. This terminal can either be located in the laboratory, on the patient's ward, or in the physician's office. In addition, the computer incorporates the test results into its patient filing system, so that whenever desired, the physician can request a profile of test results for a given patient over a specified number of days. Such a profile allows the physician to note changes in a patient's condition over time.

Another feature of most clinical laboratory computer systems is a capability for handling emergency requests. Such emergencies often require that a specimen of blood or urine be entered into the system ahead of routine samples. Since patient identification is usually controlled by the position of each sample in the sample tray of the automated instrument, changes in sample positions to accommodate emergency needs must also be made known to the computer, either by keyboard notification of each change or by some automatic means of reading sample cup labels.

Provision must also be made for a physician to obtain results of a specific test before other tests on that patient have been completed and prior to the normal reporting of results. The inquiry is usually made by keyboard, either at the computer or from a remote terminal. Results of that specific test, if available, are given at the same terminal. If the test has not been completed at the time of the inquiry, the physician is so notified.

15.11.3. THE DIGITAL COMPUTER IN THE CARDIAC CATHE-
TERIZATION LABORATORY. In Chapter 6, catheterization methods for direct
measurement of blood pressure were described. In these procedures, the
physician places one or more catheters within the chambers of the heart, the
aorta, or some other part of the cardiovascular system. To assure that each
catheter is properly placed, the physician is guided by fluoroscopy as he
follows the movement of the catheter on the image intensifier of an X-ray
machine. Through these catheters, blood pressure measurements can be
obtained from which pressure gradients across heart valves, vascular resistance, and other factors that assist the physician in locating and defining
abnormalities of the heart and circulatory system can be determined.
Catheterization may also be involved in the dye dilution method of measuring cardiac output.

Because of the nature of the procedure and a certain risk which is involved, cardiac catheterization is performed only when other diagnostic
procedures are insufficient. When a patient is brought into the catheterization laboratory, every effort is made to reduce the time he is actually catheterized to a minimum. This means that any assistance the physician can
receive in the placement of the catheter and in being assured that his measurements are correct is extremely valuable, both in terms of the physician's time and the safety and comfort of the patient.

Without computer assistance, the physician (or an assistant) must
laboriously calculate the necessary gradients, resistance values and other
parameters. This process usually requires at least an hour or two. If the
results show that some part of the system did not produce suitable input
data (for example, a catheter tip in the wrong location), the patient must
be brought back into the catheterization laboratory and subjected to the
risk a second time.

With an on-line computer, calculated results are available almost immediately. Thus, the physician can use these values to help him guide
the catheter to the correct location and greatly reduce the possibility of
erroneous placement. Also, he receives immediate feedback if something
should go wrong with the procedure. This permits him to correct the situation and avoids the necessity of bringing the patient back a second time.
An additional advantage of having immediate results lies in the physician's
ability to perform additional tests, if the results so indicate, while the patient
is still catheterized.

Data entered into the computer from the catheterization laboratory
generally include pressure values from one or more catheters, an ECG
signal (usually used as a timing reference), and possibly a measure of dye
concentration. These analog signals are sampled and converted into digital
form by means of a computer-controlled multiplexer and analog-to-digital
converter. Conversion rates are usually at least 100 samples per second

for the pressure and ECG signals. The dye concentration values are taken at a much lower rate.

In some computer systems, automatic calibration is provided by having the computer adjust the gain and baseline levels of the analog measurement equipment in response to known calibration levels. In a typical program, the ECG and blood pressure signals are sampled for a total of eleven heart cycles. The first and last are removed to assure that only complete cycles are used. The remaining nine are then smoothed by digital filtering, and subjected to pattern recognition techniques similar to those described for the ECG in Section 15.11.1. From this analysis, a number of vital parameters are generated for each cycle. To eliminate the effect of abnormal beats, such as those caused by premature ventricular contractions, selected parameters of the nine cycles are rank ordered and the middle three are chosen to be averaged to produce the reported values. A report may be generated periodically, or the data may be stored until called for by the physician.

· 16 ·

ELECTRICAL
SAFETY OF
MEDICAL EQUIPMENT

In the United States deaths from electricity account for about 1 percent of all accidental deaths. Accidental electrocutions can occur in the household or as a result of industrial accidents. At first glance the hospital does not seem to present any more danger from electrical accidents than other locations. In fact, not until the last few years was it discovered that under certain conditions hospital patients may be much more susceptible to danger from electrical currents than under normal circumstances and that special precautions must be taken in some parts of the hospital. The conditions under which such dangers can occur are often difficult to apprehend, and controversies have developed around some of the proposed safety measures to protect against these dangers. Nevertheless, it can be expected that eventually the issues in question will be settled and that the use of certain design features to increase electrical safety will become commonplace in the construction of new hospitals and in new electromedical equipment.

371

16.1. PHYSIOLOGICAL EFFECTS OF ELECTRICAL CURRENTS

In order for electricity to have any effect on the body, the latter must become part of an electric circuit. At least two connections must exist between the body and an external source of voltage to effect a current flow. The magnitude of the current depends on the voltage between the connections and on the electrical resistance of the body. Most body tissue contains a high percentage of water; therefore, it is a fairly good electrical conductor. The part of the body that is located between the two points of electrical contact forms an *inhomogeneous volume conductor,* in which the distribution of the current flow is determined by the local tissue conductivity.

Basically, the electrical current can affect the tissue in two different ways.* First, the electrical energy dissipated in the tissue resistance can cause a temperature increase. If a high enough temperature is reached, tissue damage (burns) can occur. With household current, electrical burns are usually limited to localized damage at or near the contact points, where the density of the current is the greatest. In industrial accidents with high voltage, as well as in lightning accidents, the dissipated electrical energy can be sufficient to cause burns involving larger parts of the body. In *electrosurgery,* the concentrated current from a radio-frequency generator with a frequency of 2.5 or 4 MHz is used to cut tissue or coagulate small blood vessels.

Second, as shown in Chapter 10, the transmission of impulses through sensory and motor nerves involves electrochemical action potentials. An extraneous electric current of sufficient magnitude can cause local voltages that can trigger action potentials and stimulate nerves. When sensory nerves are stimulated in this way, the electric current causes a "tingling" or "prickling" sensation, which at sufficient intensity becomes unpleasant and even painful. The stimulation of motor nerves or muscles causes the contraction of muscle fibers in the muscles or muscle groups affected. A high enough intensity of the stimulation can cause tetanus of the muscle, in which all possible fibers are contracted, and the maximal possible muscle force is exerted.

The extent of the stimulation of a certain nerve or muscle depends on the potential difference across its cells and the local density of the cur-

---

* A third type of injury can sometimes be observed under skin electrodes through which a small dc current has been flowing for an extended time interval. These injuries are due to electrolytic decomposition of perspiration into corrosive substances and are, therefore, actual chemical burns.

rent flowing through the tissue. An electric current flowing through the body can be hazardous or fatal if it causes local current densities in vital organs that are sufficient to interfere with the functioning of the organs. The degree to which any given organ is affected depends on the magnitude of the current and the location of the electrical contact points on the body with respect to the organ.

The organ most susceptible to electric current is the heart. A tetanizing stimulation of the heart results in complete myocardial contraction, which stops the pumping action of the heart and interrupts the blood circulation. If the circulation is not resumed within a few minutes, first, brain damage, then death results, because of the lack of oxygen supplied to the brain's tissues. However, if the tetanizing current is removed within a short time, the heartbeat resumes spontaneously.

A current of lower intensity that excites only a portion of the heart's muscle fibers can be more dangerous than a current sufficient to tetanize the entire heart. This partial excitation can change the electrical propagation patterns in the myocardium, desynchronizing the activity of the heart and causing random, ineffectual muscle activity. This condition is called *fibrillation*. When fibrillation occurs in the ventricles, the heart ceases to pump blood. *Ventricular fibrillation* is normally not reversible, and once induced, regular heart rhythm is not resumed when the current is removed. In order to restore the pumping action of the heart, the heart muscle must be tetanized by a sufficiently strong current pulse from an external *defibrillator* (see Section 7.4). The complete contraction of all heart muscle fibers resynchronizes the myocardium, after which, hopefully, the heart resumes its normal contractions. Ventricular fibrillation is the most frequent cause of death in fatal electric accidents.

Respiratory paralysis can also occur if the muscles of the thorax are tetanized by an electric current flowing through the chest or through the respiratory control center of the brain. Such a current is likely to affect the heart also, because of its location.

The magnitude of electric current required to produce a certain physiological effect in a person is influenced by many factors. Figure 16.1 shows the approximate current range and the resulting effects for 1-second exposures to various levels of 60-Hz alternating current applied externally to the body. For those physiological effects that involve the heart or respiration, it is assumed that the current is introduced into the body by electrical contact with the extremities in such a way that the current path includes the chest region (arm-to-arm or arm-to-diagonal leg).

For most people, the perception threshold of the skin for light finger contact is approximately 500 μA, although much lower current intensities can be detected with the tongue. With a firm grasp of the hand, the threshold is about 1 mA. A current with an intensity not exceeding 5 mA

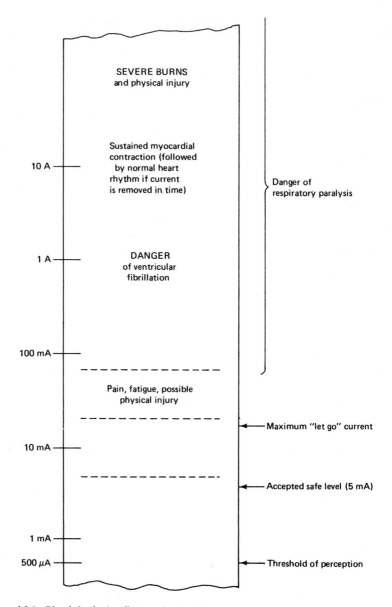

Figure 16.1. Physiological effects of electrical current from 1-second external contact with the body (60 Hz ac).

is generally not considered harmful, although the sensation at this level can be rather unpleasant and painful. When at least one of the contacts with the source of electricity is made by grasping an electrical conductor with the hand, currents in excess of about 10 or 20 mA can tetanize the arm muscles and make it impossible to "let go" of the conductor. The minimum current level at which this condition takes place is therefore called the *let-go current*. Ventricular fibrillation can occur at currents above about 100 mA, while currents in excess of about 1 or 2 amperes can tetanize the heart muscle and cause sustained contraction of the heart, often accompanied by respiratory paralysis.

Data on these effects are rare for obvious reasons and are generally limited to accidents in which the magnitude of the current could be reconstructed, or to experimentation with animals. From the data available it appears that the current required to cause ventricular fibrillation increases with the body weight and that a higher current is required if the current is applied for a very short duration. From experiments in the current range of the perception threshold and let-go current, it is known that the effects of the current are almost independent of frequency up to about 1000 Hz. Above that limit, the current must be increased proportionally with the frequency in order to have the same effect. It can be assumed that, at higher current levels, a similar relationship exists between current effects and frequency.

In the foregoing considerations, the electrical intensity is always described in terms of electrical current. The voltage required to cause the current flow depends solely on the electrical resistance that the body offers to the current. This resistance is affected by numerous factors and can vary from a few ohms to several megohms. The largest part of the body resistance is normally represented by the resistance of the skin. The inverse of this resistance, the skin conductance, is proportional to the contact area and also depends on the condition of the skin. Intact, dry skin has a conductivity of as low as 2.5 microohm per cm^2. This low conductivity is mainly caused by the horny, outermost layer of the skin, the epithelium, which provides a natural protection against electrical danger. When this layer is permeated by a conductive fluid, however, the skin conductivity can increase by two orders of magnitude. If the skin is cut, or if conductive objects like hypodermic needles are introduced through the skin, the skin resistance is effectively bypassed. When this situation occurs, the resistance measured between the contacts is determined only by the tissue in the current path, which can be as low as 500 ohms. Electrode paste used in the measurement of bioelectric potentials (see Chapters 4, 6, and 10) also has the purpose of reducing the skin resistivity by electrolyte action and mechanical abrasion. Many medical procedures require the introduction of conductive objects into the body, either through natural openings or through incisions in the skin. In many instances, therefore, the hospital patient is deprived of the natural protection against electrical dangers that the skin normally provides. Because of the resulting low resistance, dangerously high currents can be caused by voltages of a magnitude that normally would be rendered safe by the high skin resistance.

To make matters worse, in certain modern medical procedures electrically conductive catheters are introduced, not only into the body but also directly into the heart. Such catheters are used for diagnostic purposes (*cardiac catheterization*) to withdraw blood samples from various parts of the heart, to inject radiopaque dye for angiographic procedures, or to measure the blood pressure (see Chapter 6). In these cases, the catheter provides an electrical connection to the heart through the fluid column that is contained within the insulating catheter, although this column does have a fairly high electrical resistance (between 150 kohm and 2 megohm, depending on the size of the catheter). For the treatment of heart block with an electrical pacemaker (see Section 7.2), a pacing catheter is inserted into the heart to provide a low-resistance connection. With either type of catheter, the electrically-conductive catheter can form one of the two contacts that connect the body to an external electric circuit, as shown in Figure 16.2. When this is the case, all the current flowing through the catheter

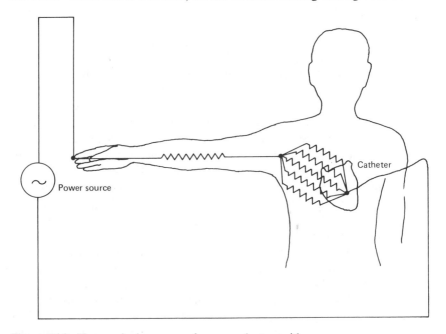

Figure 16.2. Human body as a volume conductor with one contact being a catheter in the heart.

passes directly through the heart. The current density in the cardiac muscle can, therefore, be several orders of magnitude higher than when the same current is applied to a contact more remote from the heart. As a result, a patient with any type of cardiac catheter is much more sensitive to electric current than he would otherwise be. Thus such a patient is said to be *electrically susceptible*. The effect of an electric current applied directly to the heart, in such a patient, is called *microshock*. In this context, the effect of current when no direct connection to the heart exists is referred to as *macroshock*.

Information on the current necessary to cause ventricular fibrillation when applied directly to the heart is limited. In dog experiments, fibrillation could be achieved with as little as 20 μA. In a few measurements on humans whose hearts had to be stopped during open-heart surgery, currents of at least 180 μA were required. To provide a margin of safety, contemporary standards and specifications for medical equipment set much lower limits. For equipment to be used with electrically susceptible patients, the current that accidentally can flow into the patient (*risk current*) should not exceed 10 μA. For equipment used only with patients who are not electrically susceptible, this limit is 500 μA. Under older standards, currents as high as 5 mA were allowed.

16.2. SHOCK HAZARDS FROM ELECTRICAL EQUIPMENT

An example of an electric-power-distribution system, as it may be found in a hospital, is shown in somewhat simplified form in Figure 16.3. From the main hospital substation, the power is distributed to individual buildings at 4800 volts, usually through underground cables. A stepdown transformer in each building has a secondary winding for 230 volts that is center tapped and thus can provide two circuits of 115 volts each. This center tap is grounded to the earth by a connection to a ground rod or water pipe near the building's substation. Heavy electrical devices, such as large air conditioners, ovens, and X-ray machines, operate on 230 volts from the two ungrounded terminals of the transformer secondary. Lights and normal wall receptacles receive 115 volts through a black "hot" wire from one of the ungrounded terminals of the transformer secondary and a white "neutral" wire that is connected to the grounded center tap, as shown in Figure 16.3.

In all modern installations, each wall receptacle has a third contact, called the *equipment ground*. This contact is connected separately to the ground (earth) at the building substation, either through the galvanized steel conduit that protects the other conductors or through a separate ground conductor. The use of the conduit as a ground connector, although in agreement with the National Electrical Code (NEC) of 1971, has its dangers, because corrosion and loose connectors can increase the resistance

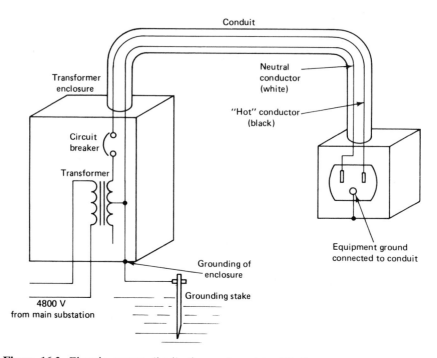

Figure 16.3. Electric power distribution system (simplified).

of the conduit to a dangerous level. A separate ground connector (which the 1971 NEC requires only for electrically susceptible patient locations) is therefore preferable for the entire power-distribution system in hospitals.

In order to be exposed to an electrical macroshock hazard, a person must come in contact with both the hot and the neutral conductors simultaneously, or with both hot conductors of a 230-volt circuit. However, because the neutral wire is connected to ground, the same shock hazard exists between the hot wire and any conductive object that is in any way connected to ground. Included would be such items as a room radiator, water pipes, or metallic building structures. In the design of electrical equipment, great care is taken to prevent personnel from accidentally contacting the hot wire by the use of suitable insulating materials and the observation of safe distances between conductors and equipment cases. Through insulation breakdown, wear, and mechanical damage, however, contact between a hot wire and an equipment case can accidentally occur. If the equipment case were not grounded, any person touching the equipment while being in contact with a grounded object would then be exposed to a severe shock hazard, as shown in Figure 16.4(a).

The purpose of the equipment ground contact on the wall receptacle

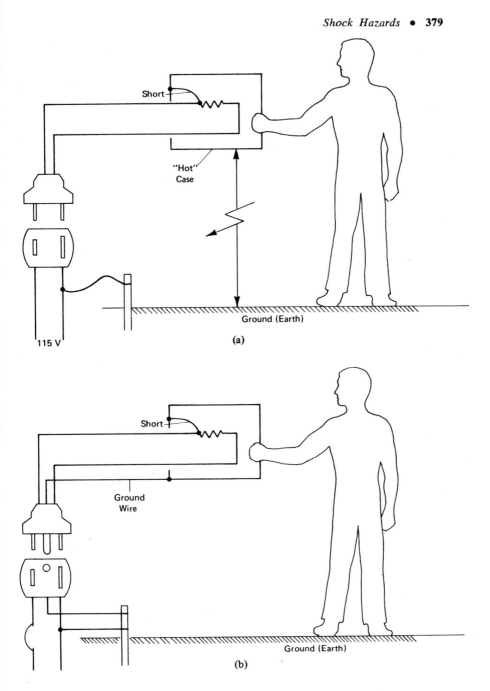

Figure 16.4. Ground shock hazards.

is to reduce this danger. Through the third (green) conductor in the power cord and a round or U-shaped pin on the power plug, a connection to ground is provided for the equipment case [Figure 16.4(b)]. When an accidental contact between the hot connector and the equipment case occurs, the current can return through this equipment-ground connection to ground without creating an electrical hazard. If the accidental contact forms a low enough resistance, the current flow will be large enough to trip the circuit breaker for the hot conductor.

The integrity of this equipment-ground connection, therefore, is of utmost importance. An interruption of its continuity, because of a broken wire or ground pin or because of the use of a three-to-two prong adapter (cheater plug), destroys its protective value completely. Even if the ground connection is not completely interrupted but merely has a resistance higher than about 1 ohm, the voltage developed across this resistance due to excessive fault current can elevate the equipment case to a potential that can pose a macroshock hazard.

Although a macroshock hazard usually occurs only as the result of insulation breakdown, microshock hazards can be created by equipment with perfectly intact insulation. Currents of a magnitude large enough to present a microshock hazard can result merely from the capacitive coupling between the hot wire and the case in electrical equipment. At a line voltage of 115 volts, a capacitance of only 220 pF causes a current of 10 μA to flow. Capacitances of many times this amount can occur when conductors with large surface areas are in the vicinity of conductive parts connected to the case, as in transformers and motors, or in heating elements. Many household appliances, lamps, and electrically operated therapeutic and diagnostic devices found in the hospital, therefore, have capacitive leakage currents in excess of 10 μA. Although perfectly safe for normal operation, these devices can create a microshock hazard for electrically susceptible patients. One example of how such a microshock accident could happen is shown in Figure 16.5. A patient has a catheter inserted intravenously to monitor his right atrial pressure. The catheter is part of a strain-gage-type pressure transducer system that is connected to a line-operated pressure monitor. This arrangement establishes a ground connection to the patient's heart through the fluid column in the catheter, the pressure transducer, and the equipment ground lead in the line cord of the pressure monitor. This might be the only ground connection to the body of the patient.

Under these conditions a microshock hazard is created by any conductive connection between the patient and an *ungrounded* electrical device having a leakage current of more than 10 μA. If this happens, the path of the leakage current includes the heart of the patient, because it is at this point that the body of the patient is connected to ground. Figure 16.5 shows the patient touching the device directly, but a conductive connection

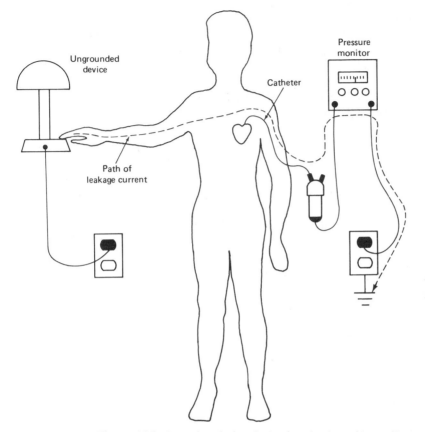

Figure 16.5. Scenario of electrical microshock accident. (See explanation in text.)

can also be established by another person who touches the device and the patient simultaneously.

Figure 16.5 shows only one of the many possible ways in which microshock accidents may happen. What is required in every case, however, is a direct electrical connection to the heart from outside the body. Cardiac catheters for diagnostic or pacing purposes are used mainly with patients who have some heart disease in the first place. Thus in such patients, ventricular fibrillation often occurs as a natural result of the disease. This explains why the danger of microshock-induced fibrillation has gone unnoticed for a long time and why any estimates of the number of such accidents that might have occurred are merely guesswork.

Although only normal safety precautions for the prevention of macroshock hazards are required in most parts of the hospital, special measures must be taken in those areas in which electrically susceptible patients are present. In addition to the cardiac-care and intensive-care units, these areas include the cardiac catheterization room and the surgical suites in which thoracic surgery is performed.

16.3. SPECIAL SAFETY MEASURES FOR ELECTRICALLY SUSCEPTIBLE PATIENTS

Numerous measures have been proposed to reduce the danger of microshock electrocutions. Some that seem to be of obvious value and that can be incorporated into new equipment at fairly low expense have found wide usage. Others that are more expensive and of disputable value have been subject to controversy. In any case, it should be appreciated that design features of instruments alone cannot guarantee electrical safety if the instruments are not properly used and maintained. Many accidents are known to be caused by human error, and proper training of all personnel who work around electrically susceptible patients is extremely important.

The safety measures commonly used basically have the purpose of ensuring that no voltage differences can exist between objects that may come in contact with the patient, and that patients are prevented from coming in contact with any grounded or conductive objects in the first place. As a second line of defense, steps may be taken to reduce leakage currents to below 10 μA in order to maintain microshock safety, even if the ground integrity of the equipment should be lost.

To ensure that all conductive objects in the vicinity of the patient are at the same potential, an *equipotential grounding system* should be used, as required by the National Electrical Code, 1971, for all electrically susceptible patient areas. In such a system (Figure 16.6), all electrical outlets in a patient's room or cubicle are clustered together on one panel. Equipment-ground contacts from all the outlets are tied together by means of a *ground bus*. All metal objects in the room are also connected to this bus (the reference ground) by separate leads, particularly metal door and window frames; outlets for oxygen, air, and vacuum; faucets and metal sinks; metal racks and stands for patient-monitoring equipment; light fixtures; and overhead racks for the mounting of bottles of intravenous infusions. The reference ground point of each patient room or cubicle is then connected to a common tie point for the ward, which is connected to the equipment ground (earth) for the building. This system ensures that all objects which might come in contact with the patient are at the same potential, provided that the individual equipment-ground connections of all electrical devices are intact.

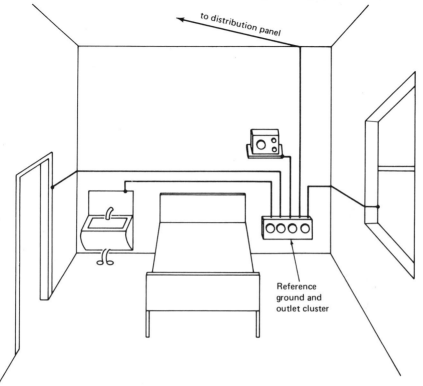

to distribution panel

Reference
ground and
outlet cluster

Figure 16.6. Principle of an equipotential grounding system in one room or cubicle of an intensive care or cardiac care unit.

As an additional safeguard in case any of these ground connections should be accidentally interrupted, the outer cases of all pieces of equipment could be insulated. This *dual insulation* can sometimes be easily achieved by using nonconductive materials for the equipment case. The same results can be obtained with a metallic case if the case is completely insulated from the chassis of the equipment.

Instruments for the monitoring of the life signs of the patient, by necessity, require connections to the patient. In older ECG machines and ECG monitors, one of the patient leads was directly connected to the neutral line terminal, or to equipment ground with only a 5-mA fuse for patient protection, as shown in Figure 16.7(a). The use of this type of device with electrically susceptible patients can be hazardous. Even with such equipment, however, a certain degree of protection can be provided by the use of *current limiters* inserted in all patient leads as shown in Figure 16.7(b). Current limiters are special semiconductor diodes that, for small-signal

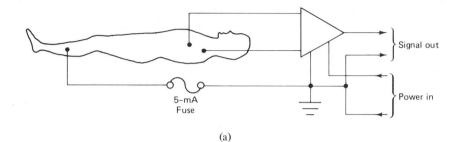

(a)

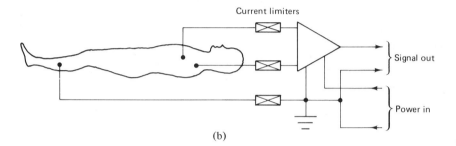

(b)

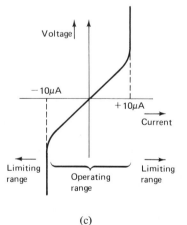

(c)

Figure 16.7. Current limiters. (a) Input circuit of older ECG machine or ECG monitor; (b) The same circuit modernized by the addition of current limiters; (c) Electrical characteristics of current limiter.

levels, behave like resistors of several kilohms ["operating range" in Figure 16.7(c)], but that, for larger signals, are driven into saturation and thus limit the current flow to a safe level.

Modern ECG monitors and ECG machines avoid these problems by using input amplifiers that are electrically isolated from the rest of the instrument. One of the methods by which this condition can be achieved is shown in Figure 16.8. Here the power to the input amplifier and the

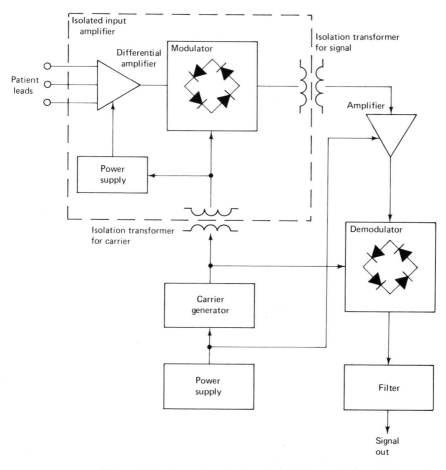

Figure 16.8. Input circuit of modern ECG machine or ECG monitor with isolated patient leads achieved by the use of a carrier amplifier.

signal from the amplifier are each coupled magnetically by means of transformers, so that the entire input module is electrically isolated. Because of the frequency range of ECG signals, direct transformer coupling is impractical, and an amplitude modulation scheme is used to convert the signal frequency to a range more favorable for the use of transformers. The carrier signal at a frequency of several kilohertz is also used to power the floating dc power supply for the differential amplifier in the isolated input module. Other manufacturers of such equipment employ optical coupling with light-emitting diodes and photodiodes for signal isolation, while the floating power supply consists of solar cells that are powered from a small light bulb.

Older strain-gage-type pressure transducers for blood pressure measurements provide an electrical connection between the fluid column in the catheter and the case of the unit to which the transducer was connected. Modern transducers are electrically isolated and often have a plastic case. The catheters used with these transducers have the purpose of providing a hydraulic connection, while the electrical connection also achieved is only incidental. It has therefore been proposed that catheter walls be made of electrically conductive material so that any current flowing through the fluid column is dissipated before it can reach the heart. Pacing catheters, on the other hand, are inserted to provide an electrical connection. These catheters and the pacemakers connected to them constitute the most dangerous possible points for accidental contact with the patient. For this reason, the trend is toward battery-operated pacemakers that can be fastened to or worn by the patient. Special care must be taken when handling any catheter or pacemaker.

An obvious way to reduce the microshock hazard would seem to be to limit the leakage current in all devices to a level below 10 μA, so that the devices would be safe even if the equipment-ground connections were lost. This situation is not easily achieved, however, because a standard 6-ft line cord, alone, contributes about 6 μA of leakage current, due to the capacity between the hot conductor and the equipment-ground conductor. Thus it is advantageous to keep line cords as short as possible. By so doing, and if care is taken in the design of the power supply, the leakage current in patient-monitoring equipment can be kept at a safe level. In other equipment, especially items containing motors or heaters, a reduction of the leakage current is more difficult. For such equipment, leakage current can be reduced to a safe level by the use of an isolation transformer. Such a transformer is shown in Figure 16.9. The secondary winding provides a line voltage supply that is not grounded and that should result in a negligible current through the leakage capacity to ground. Although the isolation transformer can be very effective in actually reducing the leakage current, it has limitations that should not be overlooked. The capacitive coupling between primary and secondary windings, which results in transformer leakage, can be kept very low by careful design and shielding. The ground capacity of the secondary winding is more limiting, for it can have

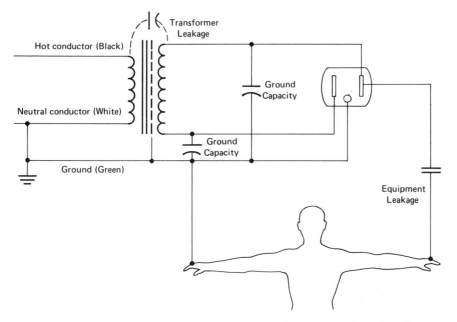

Figure 16.9. Isolation transformer for the reduction of equipment leakage current.

the effect of providing an undesirable ground connection for the supposedly isolated transformer output. Thus isolation transformers are most effective if this ground capacity is kept small. This condition can be achieved by installing the transformer close to the equipment and by providing a separate transformer for each piece of equipment. The use of large isolation transformers to power all the equipment in a patient room or cubicle seems to be of limited value in reducing the microshock hazard because of the large ground capacity that is unavoidable in such an arrangement.

BIBLIOGRAPHY

GENERAL

The following works may be considered as general references throughout the book. Other references are cited for particular chapters.

Best, C. M. and N. B. Taylor, *The Living Body, a Text in Human Physiology*. New York: Holt, Rinehart & Winston, Inc., 1971.

Dorland's *Illustrated Medical Dictionary*, 24th edition. Philadelphia: W. B. Saunders Co., 1965.

Evans, W. F., *Anatomy and Physiology, the Basic Principles*. Englewood Cliffs, N.J.: Prentice-Hall, Inc., 1971.

Geddes, L. A. and L. E. Baker, *Principles of Applied Biomedical Instrumentation*. New York: John Wiley & Sons, Inc., 1968.

Guyton, A. C., *Textbook of Medical Physiology*. Philadelphia: W. B. Saunders Co., 1971.

389

Mountcastle, V. B. (ed.), *Medical Physiology,* 12th ed., Vols. I and II. St. Louis, Mo.: The C. V. Mosby Co., 1968.

Schmidt-Nielsen, K., *Animal Physiology.* Englewood Cliffs, N.J.: Prentice-Hall, Inc., 1971.

Strong, P., *Biophysical Measurements.* Tektronix, Inc., Beaverton, Ore., 1970.

Yanof, H. M., *Biomedical Electronics.* Philadelphia: F. A. Davis Company, 1965.

CHAPTER 3

Brazier, Mary A. B., *The Electrical Activity of the Nervous System.* Baltimore: Williams & Wilkins, 1968.

Burch, G. E. and T. Winsor, *A Primer of Electrocardiography.* Philadelphia: Lea & Febiger, 1960.

Keele, C. A. and E. Neil, *Samson Wright's Applied Physiology.* New York: Oxford University Press, Inc., 1961.

Scher, A. M., "The Electrocardiogram," *Scientific American,* 205, No. 5 (November, 1961), 132–141.

Thompson, R. F., *Foundations of Physiological Psychology.* New York: Harper & Row, Publishers, 1967.

CHAPTER 4

Bair, E. J., *Introduction to Chemical Instrumentation.* New York: McGraw-Hill Book Company, 1962.

Bates, R. G., *Determination of pH-Theory and Practice.* New York: John Wiley & Sons, Inc., 1954.

Brown, J. H. V., J. E. Jacobs, and L. Stark, *Biomedical Engineering.* Philadelphia: F. A. Davis Company, 1971.

Durst, R. A. (ed.), *Ion-Selective Electrodes.* National Bureau of Standards Publication No. 314, 1969.

Hill, O. W. and R. S. Khandpur, "The Performance of Transistor ECG Amplifier," *World Medical Instruments,* 7, No. 5 (May, 1969), 12–22.

Pfeiffer, E. A., "Electrical Stimulation of Sensory Nerves with Skin Electrodes for Research, Diagnosis, Communication and Behavioral Conditioning: A Survey," *Medical and Biological Engineering,* 6 (1968), 637–651.

CHAPTERS 5 AND 6

Burch, G. E. and T. Winsor, *A Primer of Electrocardiography,* 4th edition. Philadelphia: Lea & Febiger, 1960.

Burch, G. E. and N. P. De Pasquale, *Primer of Clinical Measurement of Blood Pressure.* St. Louis. Mo.: The C. V. Mosby Co., 1962.

Cappelen, C. (ed.), "New Findings in Blood Flowmetry," *Universitetsforlaget.* Olso, Norway, 1968.

Cardiac Output Computer. Promotional Material available from Columbus Instruments, Columbus, Ohio, 1970.

ECG Measurements, Application Note AN711. Hewlett Packard Medical Electronics Division, Waltham, Mass., 1970.

Franklin, D. L., "Techniques for Measurement of Blood Flow Through Intact Vessels," *Med. Electronics and Biol. Engineering,* 3 (1965), 27–37.

The Measurement of Cardiac Output by the Dye Dilution Method. Monograph available from Lexington Instruments Corporation, 241 Crescent Street, Waltham, Mass., 1963.

Pressure Measurements, Application Note AN710. Hewlett Packard Medical Electronics Division, Waltham, Mass., 1969.

"Recommendations for Standardization of Instruments in Electrocardiography and Vectorcardiography," AHA Subcommittee Report, *IEEE Transactions on Biomed. Eng.,* January, 1967, 60–68.

Sphygmomanometers—Principles and Precepts. Copiague, N.Y.: W. A. Baum Co., 1965.

Warbasse, R. J. et al., "Physiologic Evaluation of a Catheter Tip Electromagnetic Velocity Probe," *The American Journal of Cardiology,* 23 (March, 1969), 424–433.

Zierler, K. L., "Circulation Times and the Theory of Indicator Dilution Methods for Determining Blood Flow and Volume," in *Handbook of Physiology,* Sec. 2, Vol. 1, J. Field (ed.), American Physiological Society, 1959, 585–615.

CHAPTER 7

Hendrie, W. A., "Patient Monitoring," *World Medical Instrumentation,* Vol. 7, No. 7 (July, 1969).

Planning a Patient Monitoring System. Hewlett Packard Medical Electronics Division, Waltham, Mass., 1967.

Note: Many instrument manufacturers have published description booklets on patient monitoring systems.

CHAPTER 8

Bass, B. H., *Pulmonary Function in Clinical Medicine.* Springfield, Ill.: Charles C. Thomas, Publisher, 1964.

Cherniak, R. M. and L. Cherniak, *Respiration in Health and Disease.* Philadelphia and London: W. B. Saunders Co., 1965.

Comroe, Julius H. et al., *The Lung Clinical Physiology and Pulmonary Function Tests.* Chicago, Ill.: Year Book Medical Publishers, 1963.

Filley, Giles F., *Pulmonary Insufficiency and Respiratory Failure.* Philadelphia: Lea & Febiger, 1968.

Gaensler, Edward A. and George W. Wright, "Evaluation of Respiratory Impairment," *Archives of Environmental Health,* Vol. 12 (February, 1966).

Gambino, S. R., *"Workshop Manual on Blood pH, P_{CO_2}, Oxygen Saturaiton, and P_{O_2},"* American Society of Clinical Pathologists Council on Clinical Chemistry, 1963.

The Riker Pulmonitor, Vol. 1 in a series of monographs on spirometry, published by the Riker Laboratories, Northridge, California.

Woolmer, Ronald F., *A Symposium of pH and Blood Gas Measurements.* Boston: Little, Brown and Company, 1959.

CHAPTER 9

Barnes, R. B., "Thermography of the Human Body," *Science,* 140, No. 3569 (May 24, 1963), 870–877.

Bartholomew, D., *Electrical Measurements and Instruments.* Boston: Allyn & Bacon, Inc., 1963.

Keele, C. A. and E. Neil, *Samson Wright's Applied Physiology.* New York: Oxford University Press, Inc., 1961.

Lawson, R. N. and L. L. Alt, "Skin Temperature Recording with Phosphors: A New Technique," *Canadian Medical Assn. Journal,* 92 (February 6, 1965), 255–260.

Lion, K. S., *Instrumentation in Scientific Research.* New York: McGraw-Hill Book Company, 1959.

Segal, B. L., "Echocardiography, Clinical Application in Mitral Stenosis," *JAMA,* 1951 (January 17, 1966), 161–166.

Thompson, R. F., *Foundations of Physiological Psychology.* New York: Harper & Row, Publishers, 1967.

CHAPTER 10

Brazier, Mary A. B., *The Electrical Activity of the Nervous System,* 3rd ed. Baltimore: The Williams & Wilkins Co., 1968.

Thompson, R. F., *Foundations of Physiological Psychology.* New York: Harper & Row, Publishers, 1967.

Venables, P. H. and Martin, I. (eds.), *A Manual of Psycho-Physiological Methods.* New York: John Wiley & Sons, Inc., 1967.

CHAPTER 11

Békésy, G. V., "A New Audiometer," *Acta Oto-Laryngologica,* 34 (1947), 411–422.

Grings, W. W., *Laboratory Instrumentation in Psychology.* Palo Alto, Calif.: The National Press, 1954.

Schwitzgebel, R. L., "Survey of Electro-mechanical Devices for Behavior Modification," *Psychological Bulletin,* 70, No. 6 (1968), 444–459.

Sidowsky, J. R., "Buyers guide for the Behavioral Scientist," *American Psychologist,* 24 (1969), 309–384.

Stebbins, W. C., "Behavioral Techniques," in *Methods in Medical Research,* B. F. Rushmer (ed.), Vol. 11, Year Book Medical Publishers, Inc., Chicago, Ill., 1966.

Venables, P. H. and I. Martin, *A Manual of Psychophysiological Methods,* New York: John Wiley & Sons, Inc., 1967.

Note: See also the following company manuals on the programming of behavioral experiments:

Bits of Digi—An Introductory Manual to Digital Logic Packages. BRS— Foringer Div. of Tech. Serv., Inc., Beltsville, Md., 4th Edition, 1970.

Lafayette Data Systems Operation and Program Manual, Lafayette Instrument Company, Lafayette, Ind., 1970.

Solid State Control—A Handbook for the Behavioral Laboratory. Lehigh Valley Electronics, Fogelsville, Penna., 1970.

CHAPTER 12

Cromwell, Leslie, "Use of Telemetry in Cardiovascular Research," in *Proceedings of the 1968 National Telemetry Conference IEEE,* Houston, Texas, April, 1968.

————, "Some Advances in Techniques for Remote Monitoring of Blood Pressure," in *IEEE Conference Record, Fifth Annual Rocky Mountain Bioengineering Symposium,* Denver, Colo., May, 1968.

————, "Investigation of Cardiovascular Phenomena by the Use of Biotelemetry," in *Journal of Physiology* (London), 198, No. 2 (September, 1968), 114.

————, "Biotelemetry Applied to the Measurement of Blood Pressure." Doctoral Dissertation, U. C. L. A., December, 1967.

Franklin, D. L., R. L. Van Citters, and N. W. Watson, "Applications of Telemetry to Measurement of Blood Flow and the Pressure in Unrestrained Animals," in L. Winner (ed.), *Proceedings of the National Telemetry Conference,* 233–234. New York, 1965.

Hanish, H. M., *Biolink Telemetry Systems Application Notes.* Biocom., Inc., Culver City, Calif., 1971.

Konigsberg, E., "A Pressure Transducer for Chronic Intravascular Implantation," *Fourth National Biomedical Sciences Instrumentation Symposium,* Anaheim, Calif., May, 1966.

Mackay, R. S., *Bio-Medical Telemetry*. New York: John Wiley & Sons, Inc., 1968.

CHAPTER 13

Annino, J. S., *Clinical Chemistry*. Boston: Little, Brown and Company, 1964.

Lee, L. W., *Elementary Principles of Laboratory Instruments*. St. Louis, Mo.: The C. V. Mosby Co., 1970.

White, W. L., M. M. Erickson, and S. C. Stevens, *Practical Automation for the Clinical Laboratory*. St. Louis, Mo.: The C. V. Mosby Co., 1965.

Willard, H. H., L. L. Merritt, and J. A. Dean, *Instrumental Methods of Analysis*. Princeton, N.J.: D. Van Nostrand Co., Inc., 1965.

CHAPTER 14

Blahd, W. H., *Nuclear Medicine*. New York: McGraw-Hill Book Company, 1965.

Chase, G. D. and J. L. Rabinowitz, *Principles of Radioisotope Methodology*, 3rd edition. Minneapolis, Minn.: Burgess Publishing Co., 1967.

Jaundrell-Thompson, F. and W. J. Ashworth, *X-Ray Physics and Equipment*. Philadelphia: F. A. Davis Company, 1970.

Quimby, E. H., Feitelberg, S., and Gross, W., *Radioactive Nuclides in Medicine and Biology: Basic Physics and Instrumentation*. Philadelphia: Lea & Febiger, 1970.

Selman, J., *The Fundamentals of X-Ray and Radium Physics*. Springfield, Ill.: Charles C. Thomas, Publishers, 1965.

Silver, S., *Radioactive Nuclides in Medicine and Biology: Medicine*. Philadelphia: Lea & Febiger, 1968.

CHAPTER 15

Huskey, H. D. and G. A. Korn (eds.), *Computer Handbook*. New York: McGraw-Hill Book Company, 1962.

Milhorn, H. T., Jr., *The Application of Control Theory to Physiological Systems*. Philadelphia: W. B. Saunders Co., 1966.

Stacy, R. W. and B. D. Waxman (eds.), *Computers in Biomedical Research,* Vols. I, II, and III. New York: Academic Press, Inc., 1965.

Stephenson, B. W., *Analog-Digital Conversion Handbook.* Digital Equipment Company, Maynard, Mass., 1964.

Ware, W. H., *Digital Computer Technology and Design,* 2 Vols. New York: John Wiley & Sons, Inc., 1963.

Weibell, F. J., "An Efficient Method of Digitizing and Analyzing Transient Physiological Data. Masters Thesis, U.C.L.A., 1969.

CHAPTER 16

Bruner, J. M. R., "Hazards of Electrical Apparatus," *Anesthesiology,* 28 (1967), 396–424.

Keesey, J. C. and F. S. Letcher, "Minimum Threshold for Physiological Responses to Flow of Alternating Electric Current Through the Human Body at Power-Transmission Frequencies," *Research Report MR 005.08–0030B #1,* Naval Medical Research Institute, Bethesda, Md., September, 1969.

Manual for the Safe Use of Electricity in Hospitals, 76BM, National Fire Protection Association (NFPA), 60 Batterymarch Street, Boston, Mass., 1971.

National Electrical Code 1971. National Fire Protection Association, 60 Batterymarch Street, Boston, Mass., 1971.

"Patient Safety," Application Note AN 718. Hewlett-Packard, Inc., Medical Electronic Division, Waltham, Mass., 1970.

Pfeiffer, E. A., "Electrical Stimulation of Sensory Nerves with Skin Electrodes for Research, Diagnosis, Communication and Behavioral Conditioning: A Survey," *Medical and Biological Engineering,* 6 (1968), 637–651.

Pfeiffer, E. A. and F. J. Weibell, "Safe Current Limits: Are They Too Low?" *Biomedical Safety and Standards,* Vol. 2, No. 8 (August 10, 1972), pp. 92–94.

APPENDICES

• A •

MEDICAL TERMINOLOGY
AND GLOSSARY

A.1. MEDICAL TERMINOLOGY

One of the problems of an interdisciplinary field is that of communication between the disciplinary components that make up the field. Engineers and technicians have to learn enough physiology, anatomy, and medical terminology to be able to discuss problems intelligently with members of the medical profession.

The typical technical person faces enough difficulty with language, but when confronted with medical terminology, his problems are compounded. However, with a few simple rules, medical terminology can be understood more easily. Most medical words have either a Latin or Greek origin, or, as in engineering, chemistry, and physics, the surnames of prominent researchers are used.

Most words consist of a root or base which is modified by a prefix or suffix or both. The root is often abbreviated when the prefix or suffix is added.

The list below gives some of the more common roots, prefixes, and suffixes.

399

PREFIXES

a	without or not	*mal*	bad
ab	away from	*medio*	middle
ad	to	*mes*	middle
an	absence of	*meta*	more than
ante	before	*micro*	small
antero	in front	*ortho*	straight, correct
anti	against	*para*	beside
bi	two	*patho*	disease
brady	slow	*peri*	outside
dia	through	*poly*	several
dys	difficult, painful	*pseudo*	false
endo	within	*quadri*	fourfold
epi	upon	*retro*	backward
eu	well, good	*sub*	beneath
ex	away from	*supra*	above
exo	outside	*tachy*	extreme rapidity
hyper	over	*trans*	across
hypo	under or less	*tri*	three
infra	below	*ultra*	beyond
intra	within	*uni*	single or one

ROOTS

aden	gland	*gaster*	stomach
arteria	artery	*haemo* or *hemo*	blood
arthros	joint	*hepar*	liver
auris	ear	*hydro*	water
brachion	arm	*hystera*	womb
bronchus	windpipe	*kystis*	bladder
cardium	heart	*larynx*	throat
cephalos	brain	*myelos*	marrow
cholecyst	gall bladder	*nasus*	nose
colon	intestine	*nephros*	kidney
costa	rib	*neuron*	neuron
cranium	head	*odons*	tooth
derma	skin	*odynia*	pain
enteron	intestine	*optikas*	eye
epithelium	skin	*os*	bone
esophagus	gullet	*osteon*	bone

ostium	mouth	*pyretos*	fever
otis	ear	*ren*	kidney
pes	foot	*rhin*	nose
pharynx	throat	*rhythmos*	rhythm
phlebos	vein	*spondylos*	vertebra
pleura	chest	*stoma*	mouth
pneumones	lungs	*thorax*	chest
psyche	mind	*trachea*	windpipe
pulmones	lungs	*trophe*	nutrition
pyelos	kidney	*vene*	vein
pyon	pus	*vesica*	bladder

SUFFIXES

algia	pain	*emia*	blood
centeses	puncture	*iasis*	a process
clasia	remedy	*itis*	inflammation
ectasis	dilatation	*oma*	swelling, tumor
ectomy	cut	*sclerosis*	hardening
edema	swelling		

ia is also used as a suffix in many combinations and it indicates a state or condition.

Examples of how the words are formed are easily illustrated. Arteriosclerosis means hardening of the arteries. The heart is the *cardium* and the loose sac in which it is contained is called the *pericardium* (outside of the heart). If the pericardium is diseased, it is called *pericarditis*. Note that some letters are dropped or changed for the new word, but the construction is easily recognizable.

Another example would be the root *trophe,* literally meaning nutrition. The prefix *a* means absence of, and *hyper* means over. Therefore, *atrophy* is to waste away, and *hypertrophy* is to enlarge.

English usage is sometimes peculiar and utilizes the Greek and Latin words together. An example is the kidney—*ren* in Latin and *nephros* in Greek. We talk about kidney function as *renal* function, but inflammation of the kidney is *nephritis*.

Descriptions for relative position are frequently used in medical usage. These are:

anterior	situated in front of; forward part of.
distal	away from the center of the body.
dorsal	a position more toward the back of some object of reference.

frontal	situated at the front.
inferior	situated or directed below.
lateral	a position more towards the side or flank.
proximal	toward the center of the body.
sagital	relating to the median plane of the body or any plane parallel to it.
superior	situated or directed above

A.2. MEDICAL GLOSSARY

Throughout this book many biomedical words have been used which are possibly unfamiliar to the reader. To help achieve a better understanding, the following glossary of medical terms is presented in alphabetical order for easy reference. There are many sources for these definitions, including the authors' own interpretation, but among well-known reference books used are various Webster's Dictionaries (G. & C. Merriam Co., Springfield, Mass.) and Dorland's Illustrated Medical Dictionary, 24th ed. (W. B. Saunders Co., Philadelphia, 1965).

A

acetylcholine: a reversible acetic acid ester of choline having important physiological functions such as the transmission of a nerve impulse across a synapse.

acidosis: a condition of lowered blood bicarbonate (decreased pH).

afferent: conveying towards center or towards the brain.

alkalosis: a condition of increased blood bicarbonate (increased pH).

alveoli: air sacs in the lungs formed at the terminals of a bronchiole. It is through a thin membrane (0.001 mm thick) in the alveoli that the oxygen enters the blood stream.

anaerobic: growing only in the absence of molecular oxygen.

anoxic: oxygen insufficient to support life.

aorta: the great trunk artery that carries blood from the heart to be distributed by branch arteries through the body.

aortic valve: outlet valve from left ventricle to the aorta.

apnea: absence of breathing.

arrhythmia: an alteration in rhythm of the heartbeat either in time or force.

arteriole: one of the small terminal twigs of an artery that ends in capillaries.

artery: a vessel through which the blood is pumped away from the heart.

atrioventricular: located between an atrium and ventricle of the heart.

atrium: an anatomical cavity or passage; especially a main chamber of the heart into which blood returns from circulation.

auscultation: the act of listening for sounds in the body.

autonomic: acting independently of volition; relating to, affecting, or controlled by, the autonomic nervous system.

axon: a usually long and single nerve-cell process that, as a rule, conducts impulses away from the cell body of a neuron.

B

baroreceptors: nerve receptors in the blood vessels, especially the carotid sinus, sensitive to blood pressure.

bifurcation: branching—as in blood vessels.

bioelectricity: the electrical phenomena which appear in living tissues.

brachial: relating to the arm or a comparable process.

bradycardia: a slow heart rate.

bronchus: either of two primary divisions of the trachea that lead respectively into the right and the left lung; broadly—bronchial tube.

bundle of His: a small band of cardiac muscle fibers transmitting the waves of depolarization from the atria to the ventricles during cardiac contraction.

C

cannula: a small tube for insertion into a body cavity or blood vessel.

Capacity, Functional Residual: the volume of gas remaining in the lungs at the resting expiratory level. The resting end-expiratory level is used as the base line because it varies less than the end-inspiratory position.

Capacity, Inspiratory: the maximal volume of gas that can be inspired from the resting expiratory level.

capillaries: any of the smallest vessels of the blood-vascular system connecting arterioles with venules and forming networks throughout.

cardiac: pertaining to the heart.

cardiac arrest: standstill of normal heartbeat.

cardiac output: the product of the heart rate and stroke volume.

cardiology: the study of the heart, its action and diseases.

cardiovascular: relating to the heart and blood vessels.

catheter: a tubular medical device inserted into canals, vessels, passageways, or body cavities, usually to permit injection or withdrawal of fluids or to keep a passage open.

cell: a small, usually microscopic, mass of protoplasm bounded externally by a semipermeable membrane, usually including one or more nuclei and various nonliving products, capable (alone or interacting with other cells) of performing all the fundamental functions of life, and of forming the least structural aggregate of living matter capable of functioning as an independent unit.

cerebellum: a large, dorsally projecting part of the brain especially concerned with the coordination of muscles and the maintenance of bodily equilibrium.

cerebrum: the enlarged anterior or upper part of the brain.

coronary artery and sinus: vessels carrying blood to and from the walls of the heart itself.

cortex: the outer or superficial part of an organ or body structure; especially the outer layer of gray matter of the cerebrum and cerebellum.

cortical: of, relating to, or consisting of, the cortex.

cranium: the part of the head that encloses the brain.

cytoplasm: the protoplasm of a cell exclusive of that of the nucleus.

D

defibrillation: the correction of fibrillation of the heart.

defibrillator: an apparatus used to counteract fibrillation (very rapid irregular contractions of the muscle fibers of the heart) by application of electric impulses to the heart.

dendrite: any of the usual branching protoplasmic processes that conduct impulses toward the body of a nerve cell.

depolarize: to cause to become partially or wholly unpolarized.

diastole: a rhythmically recurrent expansion, especially the dilatation of the cavities of the heart as they fill with blood.

diastolic: of or pertaining to the diastole, e.g., diastolic blood pressure.

dicrotic: having a double beat; being or relating to the second expansion of the artery that occurs during the diastole of the heart (hence dicrotic notch in the blood pressure wave).

dyspnea: difficulty in breathing.

E

ECG: abbreviation for electrocardiogram.

ectopic: located away from the normal position.

EEG: abbreviation for electroencephalogram.

efferent: conveying away from a center.

electrocardiogram: a record of the electrical activity of the heart.

electrocardiograph: an instrument used for the measurement of the electrical activity of the heart.

electrode: a device used to interface ionic potentials and currents.

electroencephalogram: the tracing of brain waves made by an electroencephalograph.

electroencephalograph: an instrument for measuring and recording electrical activity from the brain (brain waves).

electrolyte: a nonmetallic electric conductor in which current is carried by the movement of ions.

electromyogram: the tracing of muscular action potentials by an electromyograph.

electromyograph: an instrument for measurement of muscle potentials.

electromyography: the recording of the changes in electric potential of muscle.

electrophysiology: the science of physiology in its relations to electricity; the study of the electric reactions of the body in health.

embolus: an abnormal particle (air, clot or fat) circulating in the blood.

embryo: a human or animal offspring prior to emergence from the womb or egg; hence, a beginning or undeveloped stage of anything.

EMG: abbreviation for electromyography.

epilepsy: any of a variety of disorders marked by disturbed electrical rhythms of the central nervous system, typically manifested by convulsive attacks, usually with clouding of consciousness.

Expiratory Reserve Volume: that volume capable of being expired at the end-expiratory level of a quiet expiration.

external respiration: movement of gases in and out of lungs.

extracellular: situated or occurring outside a cell or the cells of the body.

extracorporeal: situated or occurring outside the body.

extrasystole: premature contraction of the heart independent of normal rhythm.

F

fibrillation: spontaneous contraction of individual muscle fibers; specifically, nonsynchronized activity of the heart.

fluoroscopy: process of using an instrument to observe the internal structure of an opaque object (as the living body) by means of X rays.

Forced Expiratory Flow ($FEF_{200-1200}$): the average rate of flow for a specified portion of the forced expiratory volume, usually between 200 and 1200 mL. (formerly called Maximum Expiratory Flow rate)

Forced Expiratory Volume (qualified by the subscript indicating time interval in seconds, FEV_T—e.g., $FEV_{1.0}$): the volume of gas exhaled over a given time interval during the performance of a forced vital capacity. FEV can be expressed as a percentage of the forced vital capacity ($FEV_{T\%}$).

Forced Midexpiratory Flow (FEF$_{25-75\%}$): the average rate of flow during the middle half of the forced expiratory volume.

Forced Vital Capacity (FVC): the maximum volume of gas that can be expelled as forcefully and rapidly as possible after maximum inspiration.

Functional Residual Capacity: see Capacity, Functional Residual.

G

galvanic: uninterrupted current derived from a chemical battery.

ganglion: any collection or mass of nerve cells outside the central nervous system that serves as a center of nervous influence.

H

heart block: a delay or interference of the conduction mechanism whereby impulses do not go through all or a major part of the myocardium.

heparin: an acid occurring in tissues, mostly in the liver. It can be produced chemically and can make the blood incoaguable if injected into the blood stream intravenously.

hyperventilation: abnormally prolonged, rapid deep breathing or overbreathing.

hypoventilation: decrease of air in the lungs below the normal amount.

hypoxia: lack of oxygen.

I

infarct: an area of necrosis in a tissue or organ resulting from obstruction of the local circulation by a thrombus or embolus.

inferior vena cava: main vein feeding back to the heart from systemic circulation below the heart.

Inspiratory Capacity: see Capacity, Inspiratory.

Inspiratory Reserve Volume: maximal volume of gas that can be inspired from the end-inspiratory position.

ion: an atom or group of atoms that carries a positive or negative electric charge as a result of having lost or gained one or more electrons.

ischemic: a localized anemia due to an obstructed circulation.

isoelectric: uniformly electric throughout; having the same electric potential, and hence giving off no current.

isometric: having the same length: a muscle acts isometrically when it applies a force without changing its length.

isotonic: having the same tone: a muscle acts isotonically when it changes length without appreciably changing the force it exerts.

isotropic: exhibiting properties with the same values when measured along axes in all directions.

K

Korotkoff sounds: sounds produced by sudden pulsation of blood being forced through a partially occluded artery and heard during auscultatory blood pressure determination.

L

latency: time delay between stimulus and response.

lobe: a somewhat rounded projection or division of a body organ or part.

lumen: the cavity of a tubular organ or instrument.

Lung Capacity, Total: the amount of gas contained in the lung at the end of maximal inspiration.

M

Maximal Breathing Capacity: same as Maximal Voluntary Ventilation.

Maximal Voluntary Ventilation: the volume of air which a subject can breathe with maximal effort over a given time interval.

membrane: a thin layer of tissue which covers a surface or divides a space or organ.

metabolism: the sum of all the physical and chemical processes by which the living organized substance is produced and maintained.

mitral stenosis: a narrowing of the left atrioventricular orifice.

mitral valve: valve between the left atrium and ventricle of the heart.

motor: a muscle, nerve or center that effects or produces movement.

myelin: the fat-like substance forming a sheath around certain nerve fibers.

myocardium: the walls of the chamber of the heart which contain the musculature that acts during the pumping of blood.

myograph: an apparatus for recording the effects of a muscular contraction.

N

necrosis: death of tissue, usually as individual cells, groups of cells, or in small localized areas.

nerve: a cordlike structure which conveys impulses from one part of the body to another. A nerve consists of a bundle of nerve fibers either efferent or afferent or both.

neuron: a nerve cell with its processes, collaterals, and terminations—regarded as a structural unit of the nervous system.

nodes of Ranvier: nodes produced by constrictions of the myelin sheath of a nerve fiber at intervals at about 1 mm.

O

oxyhemoglobin: a compound of oxygen and hemoglobin formed in the lungs —the means whereby oxygen is carried through the arteries to the body tissues.

P

partial pressure of oxygen in air: the pressure of the oxygen contained in air. Since air is about 21 percent oxygen, partial pressure is 21% of 760 mm of mercury, or 159 mm Hg. That is, oxygen needs can be supplied by pure oxygen at 159 mm Hg which is equivalent to breathing air at 760 mm Hg P_{O_2} (at sea level).

perfuse: to pour over or through.

permeate: to pass through the pores or interstices.

plethysmography: the recording of the changes in the volume of a body part as modified by the circulation of the blood in it.

pneumograph: an instrument for recording the thoracic movements or volume change during respiration.

prosthesis: an artificial substitute for a missing or diseased part.

pulmonary: relating to, functioning like, or associated with the lungs.

pulmonary atelectasis: lung collapse.

Pulmonary Minute Volume (pulmonary ventilation): volume of air respired per minute = tidal volume $\times$ breaths/min.

pulse pressure: the difference between systolic and diastolic blood pressure (usually about 40 mm Hg.)

R

radioisotope: an isotope which is radioactive, produced artificially from the element by the action of neutrons, protons, deuterons, or alpha particles in the chain-reacting pile or in the cyclotron. Radioisotopes are used as tracers or indicators by being added to the stable compound under observation, so that the course of the latter in the body (human or animal) can be detected and followed by the radioactivity thus added to it. The stable element so treated is said to be "labeled" or "tagged."

Residual Capacity: see Capacity, Residual Functional.

Residual Volume: air left in the lungs after deep exhale (about 1.2 liters).

respiratory center: the center in the medulla oblongata that controls breathing.

respiratory quotient: ratio of volume of exhaled CO_2 to the volume of consumed O_2 (0.85).

S

semilunar pulmonary valve: outlet valve from the right ventricle into the pulmonary artery.

sinoatrial node: the pacemaker of the heart—a microscopic collection of atypical cardiac muscle fibers which is responsible for initiating each cycle or cardiac contraction.

sphygmomanometer: an instrument for measuring blood pressure, especially arterial blood pressure.

spirometer: an instrument for measuring the air entering and leaving the lungs.

stenosis: narrowing of a duct or canal.

stroke volume: amount of blood pumped during each heartbeat (diastolic volume of the ventricle minus the volume of blood in the ventricle at the end of systole).

superior vena cava: main vein feeding back to the heart from systemic circulation above the heart.

synapse: the point at which a nervous impulse passes from one neuron to another.

systemic: pertaining to or affecting the body as a whole.

systole: the contraction, or period of contraction, of the heart, especially that of the ventricles. It coincides with the interval between the first and second heart sound, during which blood is forced into the aorta and the pulmonary trunk.

systolic: of or pertaining to systole, e.g., systolic blood pressure.

T

tachycardia: relatively rapid heart action.

thorax: the part of the body of man and other mammals between the neck and the abdomen.

thrombus: a clot of blood formed within a blood vessel and remaining attached to its place of origin.

Tidal Volume: volume of gas inspired or expired during each quiet respiration cycle.

tissue: an aggregation of similarly specialized cells united in the performance of a particular function.

trachea: the main trunk of the system of tubes by which air passes to and from the lungs.

tricuspid valve: the valve connecting the right atrium to the right ventricle.

V

vasoconstriction: narrowing of the lumen of blood vessels especially as a result of vasomotor action.

vasodilation: dilation or opening of blood vessel by vasomotor action.

vasomotor: having to do with the musculature that affects the caliber of a blood vessel.

ventricle: a chamber of the heart which receives blood from a corresponding atrium and from which blood is forced into the arteries.

ventricular fibrillation: convulsive non-synchronized activity of the ventricles of the heart.

venule: a small vein; especially one of the minute veins connecting the capillary bed with the larger systemic veins.

Vital Capacity: volume of air which can be exhaled after the deepest possible inhalation.

PHYSIOLOGICAL
MEASUREMENTS
SUMMARY

B.1. BIOELECTRIC POTENTIALS

ELECTROCARDIOGRAM (ECG or EKG). A record of the electrical activity of the heart. Electrical potentials: 0.1 to 4 mV peak amplitude. Frequency response requirement: dc to 100 Hz. Used to measure heart rate, arrhythmia, and abnormalities in the heart. Also serves as timing references for many cardiovascular measurements. Measured with electrodes at the surface of the body.

ELECTROENCEPHALOGRAM (EEG). A record of the electrical activity of the brain. Electrical potentials: 10 to 100 microvolts peak amplitude. Frequency requirement: dc to 100 Hz. Used for recognition of certain patterns, frequency analysis, evoked potentials, etc. Measured with surface electrodes on the scalp and with needle electrodes just beneath the surface or driven into specific locations within the brain.

ELECTROMYOGRAM (EMG). A record of muscle potentials, usually from skeletal muscle. Electrical potentials: 50 microvolts to 1

411

millivolt peak amplitude. Required frequency response: 10 to 3000 Hz. Used as indicator of muscle action, for measuring fatigue, and so on. Measured with surface electrodes or needle electrodes penetrating the muscle fibers.

OTHER BIOELECTRIC POTENTIALS

1. Electroretinogram—a record of potentials from the retina.
2. Electro-oculogram—a record of corneal-retinal potentials associated with eye movements.
3. Electrogastrogram—a record of muscle potentials associated with motility of the GI tract.
4. Individual nerve action potentials—potentials generated by information being transmitted by the nervous system.

B.2. SKIN RESISTANCE MEASUREMENTS

GALVANIC SKIN RESPONSE (GSR). Measurement of the electrical resistance of the skin and tissue path between two electrodes. A variation of resistance from 1000 to over 500,000 ohms. Variations are associated with activity of the autonomic nervous system. Used to measure autonomic responses. Principle behind "lie detection" equipment. Variations occur with bandwidth from 0.1 to 5 Hz. Measured with surface electrodes.

BASAL SKIN RESISTANCE (BSR). Same as GSR, except that the BSR is a measure of the slow baseline changes instead of the variations caused by the autonomic system. Frequency response requirements: dc to 0.5 Hz.

B.3. CARDIOVASCULAR MEASUREMENTS

BLOOD PRESSURE MEASUREMENTS

1. *Arterial:* Pressure variations from 30 to 400 mm Hg. Pulsating pressure with each heart beat. Frequency response requirements: dc to 30 Hz. Measured at various points in the arterial circulatory system. Measured directly by implanted pressure transducer; transducer connected to catheter in blood stream, or manometer; indirectly by sphygmomanometer, etc.

2. *Venous:* Pressure variations from 0 to 15 mm. Hg. An almost static pressure with some variations with each heart beat. Frequency response requirement: 0 to 30 Hz. Measured at various points in the venous circulatory system. Measured by manometer, implanted pressure transducer, or external transducer connected to catheter.

BLOOD VOLUME MEASUREMENTS

1. *Systemic volume:* Measure of total blood volume in the system. Measured by injection of an indicator such as a dye and subsequent measurement of indicator concentration.
2. *Plethysmograph measurement:* A measure of local blood volume changes in limbs or digits. This is an actual change in volume measured as a displacement change in a closed cup or tube. Volume pulsations occur at rate of heart beat. Required frequency response: dc to 40 Hz. Can also be measured indirectly with photoelectric device or tissue impedance measurement. Used to measure effectiveness of circulation, and in pulse-wave velocity measurements.

BLOOD FLOW MEASUREMENTS. A measure of the velocity of blood in a major vessel. In a vessel of a known diameter, this can be calibrated as flow and is most successfully accomplished in arterial vessels. Range is from -0.5 to $+1650$ ml/sec. Required frequency response: dc to 50 Hz. Used to estimate heart output and circulation. Requires exposure of the vessel. Flow transducer surrounds vessel. Methods of measurement include electromagnetic and ultrasonic principles.

BALLISTOCARDIOGRAM. Slight movement of body due to forces exerted by beating of the heart and pumping of blood. Patient placed on special platform. Movement measured by accelerometer. Required frequency response: dc to 40 Hz. Used to detect certain heart abnormalities.

PULSE AND CARDIOVASCULAR SOUND MEASUREMENTS

1. *Pulse pressure measurements:* Pressure variations at surface of the body due to arterial blood pulsations. Used for timing of pulse waves, pulse-wave velocity measurements, and as an indirect indicator of arterial blood pressure variations. Required frequency response: 0.1 to 40 Hz. Measured by low frequency microphone or crystal pressure pickup.
2. *Heart sounds:* An electrically amplified version of the sounds normally picked up by the conventional stethoscope. Frequency response: 30 to 150 Hz. Picked up by microphone.

3. *Phonocardiogram:* A graphic display of the sounds generated by the heart and picked up by a microphone at the surface of the body. Frequency response required is 5 to 2000 Hz. Measured by special crystal transducer or microphone.
4. *Vibrocardiogram:* A measure of the movement of the chest due to the heart beat. Frequency response required: 0.1 to 50 Hz. Special pressure or displacement transducer placed on the appropriate point on the chest.
5. *Apex cardiogram:* A measurement of the pressure variations at the point where the apex of the heart beats against the rib cage. Frequency response required: 0.1 to 50 Hz. Measured with special pressure sensitive microphone or crystal transducer.

B.4. RESPIRATION MEASUREMENTS

RESPIRATION FLOW MEASUREMENTS. A measurement of the rate at which air is inspired or expired. Range: 250 to 3000 ml/sec, peak. Frequency response required: 0 to 20 Hz. Used to determine breathing rate, minute volume, depth of respiration. Measured by pneumotachometer or as the derivative of volume measurement.

RESPIRATION VOLUME. Measurement of quantity of air breathed in or out during a single breathing cycle or over a given period of time. Frequency response required: 0 to 10 Hz. Used for determination of various respiration functions. Measured by integration of respiration flow rate measurements or by collection of expired air over a given period. Indirect measurement by belt transducer, impedance pneumograph, or whole-body plethysmograph. (See Appendix C for values of pulmonary capacities, volumes, and lung function tests.)

B.5. TEMPERATURE MEASUREMENTS

SYSTEMIC TEMPERATURE. A measure of the basic temperature of the complete organism. Measured by thermometer, rectal or oral, or by rectal or oral thermistor probe.

LOCAL SKIN TEMPERATURE. Measurement of the skin temperature at a specific part of the body surface. Measured by thermistors placed at the surface of the skin, infrared thermometer or thermograph, heat-quenched phosphors, or heat-sensitive liquid crystals.

B.6. PHYSICAL MOVEMENTS

Various measurements of *displacement, velocity, force,* or *acceleration.* Measured by transducers sensitive to the parameter desired or derived indirectly from related parameters. Special measurement of movement by ultrasound techniques.

B.7. BEHAVIORAL CHARACTERISTICS

Measurement of response of organism to various stimuli. Responses measured may be any of the above, or may be subjective. Includes such measures as speech, visual and sound perception, tactile perception, smell, and taste. Measuring devices include generation of the appropriate stimulus as well as transducers for the various responses.

· C ·

AVERAGE NORMAL VALUES FOR LUNG FUNCTION TESTS

(20–30 year-old male)

1. VOLUMES AND CAPACITIES

Tidal Volume TV	600 ml
Inspiratory Reserve Volume IRV	3000 ml
Expiratory Reserve Volume ERV	1200 ml
Residual Volume RV	1200 ml
Total Lung Capacity TLC	6000 ml
Vital Capacity VC	4800 ml
Inspiratory Capacity IC	3600 ml
Functional Residual Capacity FRC	2400 ml

2. VENTILATORY FUNCTION TESTS

Forced Vital Capacity FVC	4800 ml
Forced Expiratory Volume $FEV_{1.0}$	3840 ml
$FEV_{3.0}$	4650 ml
Percentage Expired $FEV_{1.0\%}$	80%
$FEV_{3.0\%}$	97%

Forced Expiratory Flow $FEF_{200-1200}$	9.1 ± 2.0 Liters/sec
Forced Midexpiratory Flow $FEF_{25-75\%}$	4.7 ± 1.1 Liters/sec
Maximum Voluntary Ventilation MVV	160 Liters/min
Respiratory Dead Space V_D	150 ml
Alveolar Ventilation V_A	4700 ml/min

3. DISTRIBUTION OF INSPIRED GAS

Single Breath Nitrogen Washout	$< 1.5\%$ N_2
Pulmonary Mixing Index Thin O_2 Washout	$< 1.5\%$ N_2

4. ALVEOLAR GAS

Oxygen Partial Pressure P_{AO_2}	105 mm Hg
Carbon Dioxide Partial Pressure P_{ACO_2}	40 mm Hg

5. GAS EXCHANGE AND DIFFUSION

Oxygen Consumption V_{O_2}	250 mL
Carbon Dioxide Output V_{CO_2}	200 mL
Respiratory Exchange Ratio R	0.82–0.86
Diffusing Capacity of Lung O_2 D_{LO_2}	> 15 ml/min/mm Hg
Diffusing Capacity CO Steady State D_{LCO_2}	14–18 ml/min/mm Hg
Diffusing Capacity CO_1 Single Breath D_{LCO_2}	25–35 ml/min/mm Hg
Fraction CO Removed F_{CO}	$> 50\%$

6. ARTERIAL BLOOD

Oxygen Partial Pressure P_{aO_2}	90 mm Hg
Oxygen Content C_{aO_2}	19.5 Vols %
Oxygen Capacity	201 Vols %
Oxygen Percent Saturation of Hemoglobin S_{aO_2}	97%
Carbon Dioxide Partial Pressure	40 mm Hg
pH	7.38–7.42

7. ALVEOLAR-ARTERIAL PRESSURE DIFFERENCE

Oxygen Gradient (Alveolar — arterial)	5–15 mm Hg
Carbon Dioxide Gradient (arterial — Alveolar)	< 2 mm Hg

8. VENTILATION/BLOOD FLOW

Alveolar Ventilation/Pulmonary Blood Flow 0.8

9. MECHANICS OF BREATHING

Compliance of Lung and Thorax C_{L+T}	0.1 Liter/cm H_2O
Compliance of Lungs C_L	0.2 Liter/cm H_2O
Airway Resistance	1.6 cm H_2O/Liter/sec
Work of Breathing (Rest)	0.5 kg meters/min
Work of Breathing (Maximal)	10 kg meters/breath

Notes: (a) 1, 2, 5 and 8 increase with size and exercise. (b) Arterial blood assumes normal hemoglobin content of blood of 15 grams per 100 mL.

· D ·

PROBLEMS AND
EXERCISES

A.1. INTRODUCTION

This book is both a reference and a textbook. In the latter function, problems and exercises are needed to aid the student. In a book of this nature, which is primarily descriptive, quantitative problems are not as necessary as in the usual technical book. A few have been provided, but most of these exercises are designed to test the student's knowledge of the key portions of the text, and provide an opportunity to expand on it. The problems are relatively short and do not include long essay-type questions. Such questions are left to the instructor to pose.

CHAPTER 1

1.1. There are many factors to consider in the design and application of a medical instrumentation system. Discuss what you think are the ten most important and state why.

1.2. The book lists a number of qualities important to a medical instrumentation system. Suggest one additional quality not listed and state your reasons.

1.3. How would you state the sensitivity characteristics of the following instruments:
  (a) an electrocardiograph to give a 2-inch deflection on a recorder for a 2-mV peak reading?
  (b) an electroencephalograph to give a 1.5-cm deflection for a 50-μV peak reading?
  (c) a thermistor-temperature measuring system to record body temperature at a normal value plus or minus 5 percent on a 3-inch scale?

1.4. Check elsewhere in this book or in other references for the required frequency response of:
  (a) an electromyogram
  (b) blood flow measurements
  (c) phonocardiogram
  (d) plethysmogram

1.5. Discuss the possibility of other errors not listed in Section 1.3.

1.6. By using the table of roots, prefixes, and suffixes in Appendix A, determine what the following medical names mean:
  (a) periodontitis
  (b) bradyrhythmia
  (c) tachycardia
  (d) endo-esophagus
  (e) exostosectomy
  (f) hepatitis
  (g) dysentery
  (h) epidermitis

1.7. Name six body functions and relate them to a field or topic normally studied as an engineering-type subject—for example, cardiovascular system and fluid mechanics.

1.8. The text lists three basic differences that contribute to communication problems between the physician and the engineer. What are they? How can they be overcome? Can you think of any others?

CHAPTER 2

2.1. Discuss the major differences encountered between measurements in a physiological system as distinct from a physical system.

2.2. What are the objectives of a biomedical instrumentation system?

2.3. Explain the difference between in vivo and in vitro measurements.

2.4. Name the major physiological systems of the body.

2.5. What specific features might be incorporated into an instrument designed for clinical use as opposed to one designed for research purposes?

2.6. In designing an instrumentation system for measurement of physiological variables, which of the components shown in Figure 2.1 should be determined first? Why? Which would you next determine?

2.7. Draw a diagram showing the hydraulic (cardiovascular) system of the body, using the terminology common to the engineering analogy given in this chapter.

CHAPTER 3

3.1. Draw an action potential wave form and label the amplitude and time values.

3.2. Explain polarization, depolarization, and repolarization.

3.3. What is a biopotential? Name six types of biopotential sources.

3.4. Explain the electrical action of the sinoatrial node.

3.5. Do you think the electroencephalogram is subject to frequency discrimination? Explain.

3.6. How are the potentials in muscle fibers measured, and what is the record called that is obtained?

3.7. How does an evoked EEG response differ from a conventional electroencephalogram?

CHAPTER 4

4.1. Name the three basic types of electrodes for measurement of bioelectric potentials.

4.2. For a patient, which type of electrode would be the least traumatic?

4.3. Why are microelectrodes sometimes needed?

4.4. What are the problems involved in using flat electrodes in terms of interference or high impedance between electrode and skin? How could you help eliminate this problem?

4.5. What do you understand by the term "reference electrode"?

4.6. What is a glass electrode used for?

4.7. What is an ear clip electrode used for?

4.8. Why are the partial pressure of oxygen and the partial pressure of carbon dioxide useful physical parameters? Explain briefly how each can be measured.

4.9. Calculate the potential difference across a membrane separating two very dilute solutions of a monovalent ion, one concentration being 100 times as great as the other. Assume a body temperature of 37°C.

4.10. What is the major advantage of floating-type skin surface electrodes?

4.11. What is the hydrogen ion concentration of blood with a pH of 7.4?

CHAPTER 5

5.1. A patient has a cardiac output of 4 liters per minute, a heart rate of 86 beats per minute, and a blood volume of 5 liters. Calculate the stroke volume and the mean circulation time. What is the mean blood velocity in the aorta (in feet per second) when the vessel has a diameter of 30 mm?

5.2. Explain the operation of the heart and the cardiovascular system briefly. Draw an analogous electric circuit and show how Ohm's law and Kirchoff's laws could apply in the analog.

5.3. Develop a time-phase diagram showing the correlation of the mechanical pumping of the heart, including the opening of the valves, with the electrical-excitation events.

5.4. Draw the waveshape of blood pressure on a time base and explain it. What is the dicrotic notch?

5.5. What is the difference in the information contained in a phonocardiogram and an electrocardiogram?

5.6. In a harmonic analysis of the following waveforms, what range of frequencies could be expected in the human being:
(a) the ECG
(b) the phonocardiogram
(c) the blood pressure wave
(d) the blood flow wave.

5.7. Would you expect blood flow to obey Bernoulli's equation even with reservations? Explain why.

5.8. If a person stands up, does his blood pressure increase? Why?

5.9. If a person eats a large meal, does his heart rate increase? Why?

5.10. What part of the cardiovascular system normally contains the greatest volume of blood?

5.11. Define systole and diastole.

CHAPTER 6

6.1. Draw an electrocardiogram (in lead II), labeling the critical features. Include typical amplitudes and time intervals for a normal person.

6.2. A differential amplifier has a positive input terminal, a negative input terminal, and a ground connection. ECG electrodes from a patient are connected to the positive and negative terminals, and a reference electrode is connected to ground. A disturbance signal develops on the patient's body. This will appear as a voltage from the positive terminal to ground and a similar voltage from the negative terminal to ground. How does the differential amplifier amplify the ECG signal while not essentially amplifying the disturbance signals? Draw a sketch showing the patient connected to the amplifier.

6.3. Why are the vector sums of the projections on the frontal-plane cardiac vector at any instant onto the three axes of the Einthoven triangle zero?

6.4. Explain the difference between indirect and direct measurement of blood pressure.

6.5. The "thermostromuhr" and the indicator-dilution method with cool saline as an indicator both use thermistors for detectors. What is the difference between the two methods?

6.6. For a cardiac-output determination, 5 mg of cardio-green was injected into a patient and a calibration mixture with a concentration of 5 mg per liter was prepared from a previously withdrawn blood sample. The calibration mixture gave a deflection of $\frac{5}{4}$ cm on the recorder used, which had a paper speed of 1 cm per second. The area under the extrapolated curve (obtained by the Hamilton method) was 86 cm^2. What is the cardiac output in liters per minute? (*Answer:* 0.872 liter/min)

6.7. Explain the basic operation of the following blood pressure transducers:
(a) a resistance-bridge type
(b) a linear variable differential transformer type

6.8. Explain what is meant by plethysmography; discuss one way to make measurements and their clinical implications.

6.9. You are to measure the blood pressure of a dog during heavy exercise on a treadmill by using a catheter-type resistance strain gage transducer.

What is the desirable frequency response for your whole system? Explain.

6.10. Laplace's law can be used for cylindrical blood vessels. Simply stated in this context, the tension in the wall of a vessel is the product of the radius and internal pressure. Given that 1 mm Hg is equivalent to approximately 1300 dynes per square centimeter, find the tension in the wall of

(a) an aorta with a mean pressure of 100 mm Hg and a diameter of 2.4 cm. (*Answer:* 1.56×10^5 dynes/cm²)

(b) a capillary with a mean pressure of 25 mm Hg and a diameter of 8 microns. (*Answer:* 13 dynes/cm²)

(c) the superior vena cava with a mean pressure of 10 mm Hg and 3 cm diameter. (*Answer:* 19.5×10^3 dynes/cm²)

6.11. Assume that blood flow obeys Bernoulli's equation:

$$\frac{p}{w} + \frac{v^2}{2g} + z = \text{constant}$$

where p = the pressure
  w = the specific weight
  v = the velocity
  g = the gravitational constant
  z = the elevation head

The three terms are often referred to as the pressure head, the velocity head, and the potential or elevation head, respectively. In measurements on a patient, the elevation is a constant, and so the equation can be expressed as

$$p + \frac{wv^2}{2g} = \text{constant}$$

A certain type of blood pressure transducer positioned in the aorta will measure this value, but since the lateral blood pressure is simply the p term, the $wv^2/2g$ represents an error. If the density of the blood w/g is estimated to be 1.03 grams per cubic centimeter, and the blood is flowing at a velocity of 100 cm per second, calculate the error in blood pressure measurement. Given 1 mm Hg is equivalent to 1330 dynes/cm². (*Answer:* 3.88 mm) Do you consider this to be a significant error in the aorta?

6.12. You are employed by a hospital research unit on a certain project to measure the blood pressure and blood flow in the femoral artery of an anesthetized dog lying on an operating table.

(a) Design a system to do this by (i) describing the transducers, if any, you would use. (ii) specifying all necessary instrumentation. (iii) discussing surgical or medical methods used to ensure that your physiological measurements are taken correctly—for example, catheterization, implantation, and so on. Draw block diagrams to illustrate.

(b) How would you zero and calibrate your blood pressure measurements?

6.15. Blood shows certain conductive properties. Discuss an instrument that uses this property.

6.14. Discuss the advantages and disadvantages of four types of blood pressure transducer that can either be implanted or placed in the bloodstream through a catheter.

6.15. Draw a simplified model using block diagrams to show how the brain, pressoreceptors, and hormonal secretion could control the heart rate. Use the brain and the heart as your feed-forward loop and other parameters as feedback loops.

6.16. Why is the impedance plethysmograph sometimes called a pseudo-plethysmograph?

6.17. Explain the difference between a phonocardiogram and a vibrocardiogram. How do transducer requirements differ for these two measurements?

CHAPTER 7

7.1. Design a coronary-care hospital suite. Show all rooms in a layout plan. Illustrate all your instrumentation systems by block diagrams.

7.2. Discuss warning devices to be used in intensive-care units.

7.3. Explain the operation of a pacemaker and why it is needed.

7.4. What do you understand by fibrillation? How do you compensate for it? Draw a circuit of a direct current defibrillator.

7.5. What part of the electrocardiogram is the most useful for determining heart rate? Explain.

7.6. A certain patient-monitoring unit has an input amplifier with a common-mode rejection ratio of 100,000 to 1 at 60 Hz. At other frequencies, the common-mode rejection ratio is 1000:1. Do you consider these ratios adequate for monitoring the ECG? Explain.

7.7. Discuss possible causes of a patient-monitoring system falsely indicating an excessive high heart rate.

7.8. What is a "demand" pacemaker and when is it used?

CHAPTER 8

8.1. Using the correct anatomical and physical terms, explain the process of respiration, tracing the taking of a breath of air through the mouth to the using of the oxygenated blood in the muscle of an athlete's leg.

8.2. How many lobes are there in the lungs? Explain.

8.3. Boyle's law is an important law in physics. How does it relate to the breathing process? (*Hint: PV* is a constant at a constant temperature.)

8.4. What is the difference between death by carbon monoxide poisoning and death by strangulation? Explain.

8.5. Define the important lung capacities and explain them.

8.6. A person has a total lung capacity of 5.95 liters. If the volume of air left in the lungs at the end of maximal expiration is 1.19 liters, what is his vital capacity? (*Answer:* 4.76 liters)

8.7. If the volume of air expired and inspired during each respiratory cycle varies from 0.5 to 3.9. liters during exercise, what is this value called and what does it mean?

8.8. During a typical day, a person works for 8 hours, rests for 4 hours, walks for 1 hour, eats for 2 hours, and sleeps for 9 hours. How many pounds of oxygen would he consume during the whole day? (During sleep and rest he can be assumed to consume 0.05 pound per hour; during eating, this figure will double; during walking, consumption will treble; and during work it will quadruple.) (*Answer:* 2.6 pounds)

8.9. Explain the operation of a pulmonary function indicator.

8.10. Since the lungs contain no musculature, what causes them to expand and contract in breathing?

8.11. For what measurements can a spirometer be used? What basic lung volumes and capacities cannot be measured with a spirometer? Why?

CHAPTER 9

9.1. Discuss four different types of transducers, explaining what they measure and the principles involved.

9.2. Explain the difference between a thermistor and a thermocouple in temperature measurement.

9.3. You have invented a device that changes its resistance linearly as a function of ozone in a sample of air (smog). Draw a transducer-type circuit excited by an audio oscillator and explain how it would operate and how you would use it.

9.4. What do you understand by the term gage factor?

9.5. Why is skin surface temperature lower than systemic temperature measured orally?

9.6. Discuss the relationship between displacement, velocity, acceleration, and force.

9.7. Explain the difference between isometric and isotonic transducers.

9.8. What is a mercury strain gage? Describe its operation and list as many biomedical applications as you can.

9.9. Which of the following types of physical transducers, in basic form, are capable of a direct measurement of displacement and which are primarily velocity transducers?
(a) potentiometer transducer
(b) piezoelectric crystal
(c) differential transformer
(d) bonded strain gage
(e) unbonded strain gage
(f) capacitance transducer
(g) induction-type transducer
(h) Doppler-type ultrasound transducer

CHAPTER 10

10.1. What is the difference between afferent and efferent nerves?

10.2. Explain the difference between a motor nerve and a sensory nerve.

10.3. How does the action of the sympathetic nervous system differ from that of the parasympathetic system? Quote an example from a body system.

10.4. What is a neuronal spike? Draw a typical spike showing amplitude and duration.

10.5. What is a 10–20-electrode placement system and with what bioelectric instrument is it used?

10.6. Discuss some possible uses of electromyography.

10.7. Draw a sketch of a neuron and label the cell body, dendrite, axon, and axon hillock.

10.8. What are the nodes of Ranvier and what useful purpose do they serve?

10.9. Explain the way in which a neuronal spike is transmitted from one neuron to another.

10.10. What are graded potentials?

10.11. Explain the function of the
   (a) cerebral cortex
   (b) cerebellum
   (c) reticular activation system
   (d) hypothalamus

10.12. What is a spinal reflex and how is it related to the functions of the brain?

10.13. If the same neuronal spike were measured intracellularly and extracellullary, what would be the difference between the two measurements?

10.14. What are the differences in amplification and bandwidth requirement of amplifiers for ECG, EMG, and EEG?

CHAPTER 11

11.1. You want to determine what concentration of salt in water can be detected by the human taste sense. How would you set up the experiment?

11.2. For a "differential response" experiment, an animal box contains two lamps and one bar. The positive reinforcement is food, dispensed by a magnetic feeder, the negative reinforcement is electric shock. Devise a simple programming circuit, using relays, that causes positive reinforcement when the animal presses the bar while either light is on and negative reinforcement if the animal presses the bar while both lights are on. Bar pressing while no light is on shall have no effect.

11.3. List some of the possible difficulties that might be encountered in using GSR measurements as a lie detector test.

11.4. Explain the principle of the Békésy audiometer.

11.5. What is a cumulative recorder and how is it used?

11.6. You have been assigned the task of measuring all possible responses of the autonomic nervous system. Design a system for providing various forms of stimuli that would be expected to actuate the autonomic system for measurement of each response. Describe the type of instrumentation you would use.

CHAPTER 12

12.1. List some advantages and disadvantages of biotelemetry.

12.2. Draw a block diagram of a system to send an electrocardiogram from an ambulance to a hospital by telemetry.

12.3. Why do you think measurements of physiological parameters on an unanesthetized animal may be more useful than those on an anesthetized one?

12.4. What do you see as some of the problems of telemeterized systems in the future?

12.5. It is desirable to monitor the temperature of a man very accurately while he is climbing a mountain and then record the data on tape for later computer analysis. You are to remain in a cabin at the foot of the mountain. Explain how you would do this accurately, and draw a block diagram of any equipment stages used in your system.

12.6. Explain how four physiological parameters can be monitored and telemetered simultaneously.

12.7. If subdermal needles connected to a telemetry transmitter are implanted into a muscle, explain how a trained physician might recognize different effects from another room by using a sense other than vision to monitor.

12.8. Design a hospital with a telemetry system, explaining why you would telemeterize the functions you have selected.

CHAPTER 13

13.1. List the most important components of the blood.

13.2. List the main types of blood tests and explain each briefly.

13.3. What do you understand by the term blood count?

13.4. Describe the operation of a blood counter.

13.5. Describe the colorimetric method of determining chemical concentration.

13.6. When counting red blood cells with one of the automatic counting methods described in Section 13.1, you will, by necessity, also count the white blood cells in the process. Why is the error introduced by this negligible? Why must the blood be diluted for all the automatic blood cell counters? How do the automatic cell counters avoid counting the platelets?

13.7. For a glucose determination, a standard with a known glucose concentration of 80 mg per 100 ml is used. After the color reaction has taken place, this standard shows a transmittance of 38 percent. A patient sample shows a transmittance of 46 percent. What is the glucose concentration in this patient sample? Another patient sample shows a transmittance below 10 percent, which is hard to read accurately. What can be done to this sample to bring the transmittance into a more suitable range, and what correction has to be applied in the calculation?

13.8. Explain the difference between the continuous-flow method and the discrete sample method of automated clinical chemistry equipment. What are some of the shortcomings of each?

CHAPTER 14

14.1. In both X-ray and radioisotope procedures, potentially harmful ionizing radiation is used for diagnostic purposes. Why is the safe radiation intensity for X rays much higher than that for isotope methods?

14.2. X-ray and radioisotope methods for diagnostic purposes both make use of the tissue-penetrating properties of radiation. What is the principal difference between the two methods?

14.3. Why is the use of radioisotopes for in vivo methods limited to those isotopes that emit gamma radiation?

14.4. Describe the principle of visualizing body organs by radioisotope methods.

CHAPTER 15

15.1. Define each of the following terms as related to digital computation:
(a) word length
(b) register
(c) memory
(d) character
(e) address
(f) byte
(g) time sharing
(h) modem
(i) real time
(j) on line
(k) software

15.2. Describe the processes required to enter each of the following types of data into a digital computer:

(a) numerical data written in tabular form on sheets of paper
(b) the output of a pneumotachograph transducer
(c) the output of a digital electronic counter
(d) an electrocardiogram signal

15.3. Assuming two given functions of time, $f_1(t)$ and $f_2(t)$, draw an analog circuit to provide

$$y(t) = \int f_1(t) \, dt + f_2(t)$$

15.4. What role does a digital-to-analog converter play in an analog-to-digital converter?

15.5. An analog-to-digital converter is required for conversion of three channels of electrocardiogram signal. Assuming a possible amplitude error of ± 1 percent in the analog signal and a bandwidth of dc to 100 Hz, what is the minimum conversion rate and word length that could be used in this converter? Could the continuous converter of Figure 15.15 be used for this purpose? Explain.

15.6. Several applications of digital computers to medicine are given in Chapter 15. Can you suggest other possible applications?

CHAPTER 16

16.1. Name two different ways in which electricity can harm the body.

16.2. List the various effects of electrical current that occur with increasing current intensity.

16.3. What is the difference between electrical macroshock and microshock? In what parts of the hospital are microshock hazards likely to exist?

16.4. What is the basic purpose of the safety measures used with electrically susceptible patients?

16.5. Why is it so important to maintain the integrity of the grounding system for protection against microshock?

16.6. A fluid-filled catheter is used to measure blood pressure in the right atrium of the heart. Resistance of the fluid path is 1 Megohm. The external end of the catheter is grounded to the equipment ground of a receptacle at the left side of the patient's bed. The patient's right leg is grounded via a patient monitor to another receptacle at the right side of the patient's bed. Due to a malfunction in a vacuum cleaner, a fault current of 10 amperes flows through the ground wire connecting the two receptacles. What is the maximum allowable resistance for the ground wire connecting the receptacles to prevent exceeding the safe current limit for microshock in the patient?

INDEX*

* Medical terms are only indexed for reference in the body of the text. For further definition, please see the glossary commencing on page 402.

433